THE McGRAW-HILL CIVIL ENGINEERING PE EXAM DEPTH GUIDE

Structural Engineering

THE McGRAW-HILL CIVIL ENGINEERING PE EXAM DEPTH GUIDE
Structural Engineering

M. Myint Lwin, PE, SE
Chyuan-Shen Lee, Ph.D., PE, SE
J.J. Lee, Ph.D., PE, SE

McGRAW-HILL
New York   Chicago   San Francisco   Lisbon   London   Madrid
Mexico City   Milan   New Delhi   San Juan   Seoul
Singapore   Sydney   Toronto

Cataloging-in-Publication Data is on file with the Library of Congress

McGraw-Hill
A Division of The McGraw-Hill Companies

Copyright © 2001 by The McGraw-Hill Companies, Inc. All rights reserved. Printed in the United States of America. Except as permitted under the United States Copyright Act of 1976, no part of this publication may be reproduced or distributed in any form or by any means, or stored in a data base or retrieval system, without the prior written permission of the publisher.

1 2 3 4 5 6 7 8 9 0   AGM/AGM   0 7 6 5 4 3 2 1

ISBN 0-07-136181-2

The sponsoring editor for this book was Larry S. Hager and the production supervisor was Sherri Souffrance. It was set in Times Roman by Lone Wolf Enterprises, Ltd.

Printed and bound by Quebecor/Martinsburg.

This book is printed on recycled, acid-free paper containing a minimum of 50% recycled, de-inked fiber.

McGraw-Hill books are available at special quantity discounts to use as premiums and sales promotions, or for use in corporate training programs. For more information, please write to the Director of Special Sales, McGraw-Hill, Professional Publishing, Two Penn Plaza, New York, NY 10121-2298. Or contact your local bookstore.

Information contained in this work has been obtained by The McGraw-Hill Companies, Inc. ("McGraw-Hill") from sources believed to be reliable. However, neither McGraw-Hill nor its authors guarantee the accuracy or completeness of any information published herein, and neither McGraw-Hill nor its authors shall be responsible for any errors, omissions, or damages arising out of use of this information. This work is published with the understanding that McGraw-Hill and its authors are supplying information but are not attempting to render engineering or other professional services. If such services are required, the assistance of an appropriate professional should be sought.

CONTENTS

Preface xiii

About the Authors xiv

CHAPTER 1: PROPERTIES OF MATERIALS 1.1

    1.1 Introduction 1.1
    1.2 Stress and Strain 1.1
    1.3 Test Specimens 1.2
    1.4 Normal Stress 1.2
    1.5 Normal Strain 1.3
    1.6 Stress-Strain Diagrams 1.3
    1.7 Hooke's Law 1.4
    1.8 Modulus of Elasticity 1.4
    1.9 Proportional Limit 1.4
    1.10 Yield Point 1.5
    1.11 Strain Hardening 1.5
    1.12 Ultimate Strength and Breaking Strength 1.6
    1.13 Percentage Elongation 1.6
    1.14 Percentage Reduction in Area 1.7
    1.15 Working Stress 1.7
    1.16 Secant Modulus 1.7
    1.17 Tangent Modulus 1.7
    1.18 Poissons's Ratio 1.7
    1.19 Fatigue Life 1.8
    1.20 Ductility 1.8
    1.21 Modulus of Resilience 1.9
    1.22 Hardness 1.10
    1.23 Fracture Toughness 1.11
    1.24 Brittle Fracture 1.12

1.25 Creep and Shrinkage 1.12
1.26 Relaxation 1.13
1.27 Generalized Form of Hooke's Law 1.13
1.28 Material Testing 1.13

CHAPTER 2: PROPERTIES OF SECTIONS 2.1

2.1 Section Property 2.1
2.2 Centroid of an Area 2.1
2.3 Centroid of a Line 2.4
2.4 Centroid of a Volume 2.4
2.5 Theorems of Pappus-Guldinus 2.4
2.6 Moment of Inertia 2.6
2.7 Transfer of Axes 2.7
2.8 Methods for Determining Moment of Inertia 2.10
    2.8.1 Method 1: Moment of Inertia for Typical Sections 2.10
    2.8.2 Method 2: Moment of Inertia by Elements 2.11
    2.8.3 Method 3: Moment of Inertia by Areas 2.13
2.9 Product of Inertia 2.16
2.10 Transfer of Axes for Product of Inertia 2.17
2.11 Inclined Axes 2.17
2.12 Mohr's Circle 2.20
2.13 Radius of Gyration 2.20

CHAPTER 3: STRENGTH OF MATERIALS 3.1

3.1 Introduction 3.1
3.2 Tension and Compressibility 3.1
3.3 Determinate Force System 3.3
3.4 Indeterminate Force System 3.3
3.5 Elastic Analysis 3.5
3.6 Plastic Analysis 3.5
3.7 Strain Energy in Tension and Compression Members 3.7
3.8 Shear Stress 3.7
3.9 Shear Strain 3.8
3.10 Shear Modulus 3.9
3.11 Shear Deformation and Strain Energy in Pure Shear 3.9
3.12 Torsion 3.9
3.13 Torsion Shearing Stress and Strain 3.10
3.14 Torsion Resistance 3.12

3.15 Shearing Force and Bending Moment 3.13
3.16 Load, Shear, and Moment Relationships 3.15
3.17 Shear and Bending Moment Diagram 3.16
3.18 Stresses in Beams 3.18
3.19 Loads Acting on Beams 3.19
3.20 Neutral Axis 3.20
3.21 Section Modulus 3.20
3.22 Plastic Moment 3.22
3.23 Plastic Section Modulus 3.25
3.24 Deflection of Beams 3.25
3.25 Methods for Determining Beam Deflection 3.26
    3.25.1 The Double Integration Method 3.26
    3.25.2 The Moment-Area Method 3.29
    3.25.3 The Elastic Weight Method 3.33
    3.25.4 The Method of Superposition 3.34
3.26 Statically Determinate Beams 3.36
3.27 Statically Indeterminate Beams 3.36
3.28 Shear Center 3.36
3.29 Unsymmetric Bending 3.37
3.30 Curved Beams 3.38
3.31 Plastic Deformations of Beams 3.41
3.32 Plastic Hinge 3.41
3.33 Collapse Mechanism 3.42
3.34 Columns 3.42
3.35 Critical Buckling Load of a Column 3.42
3.36 Slenderness Ratio 3.43
3.37 Effective Length of a Column 3.43
3.38 Beam Columns 3.44
3.39 Combined Stresses 3.46
3.40 Principal Stresses and Planes 3.47
3.41 Determining Principal Stresses Using Mohr's Circle 3.55

CHAPTER 4: PRINCIPLES OF STATICS　　　　　　　　　　　　　　　　4.1

    4.1 Introduction 4.1
    4.2 Basic Concepts 4.2
    4.3 Scalar and Vector Quantities 4.2
    4.4 Newton's Laws 4.3
    4.5 System of Forces 4.3

4.6 Composition and Resolution of Forces 4.4
4.7 Moment and Couple 4.7
4.8 Varignon's Theorem 4.8
4.9 Static Friction 4.10
4.10 Equilibrium 4.10
    4.10.1 Equilibrium in Two Dimensions 4.10
    4.10.2 Equilibrium in Three Dimensions 4.11
4.11 Free-Body Diagram 4.11
4.12 Structures 4.14
4.13 Trusses 4.14
4.14 Determinacy 4.15
4.15 Influence Lines for Trusses 4.17
4.16 Method of Joints 4.18
4.17 Method of Sections 4.21
4.18 Method of Superposition 4.23
4.19 Flexible Cables 4.23
4.20 Parabolic Cables 4.24
4.21 Catenary Cables 4.27

CHAPTER 5: INTRODUCTION TO DESIGN AND ANALYSIS 5.1

5.1 Introduction 5.1
5.2 Responsibilities of Structural Designer 5.1
    5.2.1 Safety 5.2
    5.2.2 Aesthetics 5.2
    5.2.3 Constructibility 5.2
    5.2.4 Economy 5.3
5.3 Design Methods 5.3
    5.3.2 Strength Design Method 5.4
    5.3.3 Load and Resistance Factor Design (LRFD) 5.5
5.4 Design Loads 5.6
5.5 Design Specifications and Codes 5.8

CHAPTER 6: CONCRETE DESIGN 6.1

6.1 Introduction 6.1
6.2 Mechanical Properties of Concrete 6.2
    6.2.1 Compressive Strength 6.2
    6.2.2 Tensile Strength 6.2
    6.2.3 Stress-Strain Relationship 6.3

6.2.4 Modulus of Elasticity 6.3
6.2.5 Creep 6.4
6.2.6 Shrinkage 6.4
6.2.7 Thermal Coefficient 6.4
6.2.8 Unit Weight 6.4
6.3 Reinforcement
    6.3.1 Grades 6.4
    6.3.2 Sizes 6.5
    6.3.3 Development Length 6.5
    6.3.4 Splice 6.5
    6.3.5 Lengths 6.5
    6.3.6 Concrete Protection (Cover) of Reinforcement 6.6
6.4 Concrete Quality, Proportioning, Placing, and Curing 6.6
    6.4.1 Types of Concrete 6.6
    6.4.2 Aggregates, Water, Admixture 6.7
    6.4.3 Proportioning 6.7
    6.4.4 Placing and Curing 6.8
6.5 Design for Flexural (Pure Bending) Loading 6.8
    6.5.1 Assumptions 6.8
    6.5.2 Rectangular Singly Reinforced Beam 6.9
    6.5.3 Rectangular Doubly Reinforced Beam 6.12
    6.5.4 Check Crack Width Limitation 6.15
    6.5.5 Detailing 6.16
6.6 Design for Axial and Flexural Loading 6.16
    6.6.1 (Pure) Axial Loading 6.16
    6.6.2 Combined Axial and Flexural Loading 6.17
    6.6.3 Detailing 6.20
    6.6.4 Long Columns 6.21
6.7 Design for Shear 6.21
    6.7.1 Shear Strength (Contribution of Concrete and of Reinforcement 6.21
    6.7.2 Shear Friction 6.24
6.8 Design of Walls 6.25
6.9 Design of Footings 6.30
    6.9.1 Sizing of the Footing 6.30
    6.9.2 Flexure Check 6.30
    6.9.3 Shear Check (Beam Shear and Punching Shear) 6.30
    6.9.4 Bearing/Dowels 6.31
6.10 Introduction to Prestressed Concrete 6.36
6.11 Summary 6.40

CHAPTER 7: STEEL DESIGN 7.1

- 7.1 Introduction 7.1
- 7.2 Attributes of Structural Steels 7.2
- 7.3 Tension Members 7.2
  - 7.3.1 Design Tensile Strength 7.2
  - 7.3.2 Gross Area, A_g 7.4
  - 7.3.3 Net Area, A_n 7.4
  - 7.3.4 Effective Net Area, A_e 7.6
  - 7.3.5 Design of Tension Members 7.10
- 7.4 Compression Members 7.12
  - 7.4.1 Classification of Steel Sections 7.12
  - 7.4.2 Column Formulas 7.14
- 7.5 Beams 7.17
  - 7.5.1 Design for Flexure 7.17
  - 7.5.2 Beam Design Charts 7.22
  - 7.5.3 Design Shear Strength 7.24
  - 7.5.4 Deflections of Beams 7.25
- 7.6 Bending and Axial Force 7.26
- 7.7 Bolted Connections 7.33
  - 7.7.1 General Provisions 7.34
  - 7.7.2 Snug-Tight and Full-Tensioned Bolts 7.34
  - 7.7.3 Types of Connections 7.35
  - 7.7.4 Minimum Spacing and Edge Distance 7.36
  - 7.7.5 Maximum Spacing and Edge Distances 7.36
  - 7.7.6 Minimum Strength of Connections 7.36
  - 7.7.7 Design Tension of Shear Strength 7.37
  - 7.7.8 Combined Tension and Shear in Bearing-Type Connections 7.37
  - 7.7.9 Bearing Strength at Bolt Holes 7.38
  - 7.7.10 Slip-Critical Connections Designed at Service Loads 7.41
  - 7.7.11 Design Rupture Strength 7.42
- 7.8 Welded Connections 7.45
  - 7.8.1 Welding Code 7.46
  - 7.8.2 Types of Welding 7.46
  - 7.8.3 Types of Welds 7.46
  - 7.8.4 Fillet Weld 7.46
  - 7.8.5 Complete Penetration Groove Weld 7.47
  - 7.8.6 Nominal Strength of Weld 7.47
- 7.9 Composite Beams 7.51
  - 7.9.1 Effective Width 7.52
  - 7.9.2 Strength of Beams with Shear Connectors 7.52

7.9.3 Strength During Construction 7.53
7.9.4 Design Shear Strength 7.53
7.9.5 Shear Connectors 7.53
7.9.6 Required Number of Shear Connectors 7.53
7.9.7 Shear Connector Placement and Spacing 7.54
7.9.8 Neutral Axis in Concrete Slab 7.54
7.9.9 Deflection of Composite Section 7.56
7.10 Bearing Plates 7.56

CHAPTER 8: MASONRY DESIGN 8.1

8.1 Introduction 8.1
8.2 Materials 8.2
    8.2.1 Masonry Units 8.2
    8.2.2 Mortar 8.2
    8.2.3 Grout 8.4
    8.2.4 Reinforcing Accessories 8.4
    8.2.5 Modulus of Elasticity of Materials (UBC 2106.2.12 8.5
    8.2.6 Design Data and Section Properties 8.5
8.3 General Design Requirements 8.6
    8.3.1 Working Stress Design Method 8.7
    8.3.2 Strength-Design Method 8.20
    8.3.2.1 Strength Requirements 8.21
    8.3.3 Empirical Design Method 8.28
8.4 Design Examples—Working Stress-Design 8.31
    8.4.2 Reinforced Masonry Column and Pilaster Design 8.44
    8.4.3 Reinforced Masonry Wall Design for Out-of-Plane Loads 8.49
    8.4.4 Reinforced Masonry Wall Design for In-Plane Loads
        (Shear Wall Design) 8.57
8.5 Design-Examples—Strength-Design Method 8.71
    8.5.1 Load Factors Using Strength Design 8.71
    8.5.2 Lintel Design 8.72
    8.5.3 Reinforced Masonry column and Pilaster Design 8.74

CHAPTER 9: INTRODUCTION TO SEISMIC DESIGN 9.1

9.1 General 9.1
9.2 Seismic Hazard 9.2
9.3 Seismic Zones 9.3
9.4 Site Characteristics 9.3

9.5 Earthquake Risk Mitigation 9.5
9.6 1997 Uniform Building Code (UBC) Earthquake Provisions 9.6
    9.6.1 General Provisions 9.6
    9.6.2 Occupancy Categories 9.7
    9.6.3 Soil Profile Type 9.7
    9.6.4 Near-Source Factor 9.7
    9.6.5 Seismic Factors 9.8
    9.6.6 Redundancy Factor 9.8
    9.6.7 Design and Analysis 9.9
9.7 Importance of Proper Detailing 9.9

APPENDIX A: REFERENCES   A.1
INDEX   I.1

PREFACE

The main objective of this book is to help civil engineers prepare for the Professional Engineer Examination (PE Exam). The book illustrates the application of the 1997 Uniform Building Code and the current edition of the AASHTO LRFD Bridge Design Specifications to concrete, steel and masonry design. Solved examples are used in the chapters to illustrate the interpretation and application of structural principles and design codes. To keep the size of the book within reason, the authors have elected not to include practice problems within this volume. Numerous practice problems are available in textbooks, manuals and exam guides referenced in Appendix A. The readers are encouraged to apply the principles and procedures learned from this book to solve problems found in the references cited in Appendix A and elsewhere. This is an efficient way to prepare for the PE Exam.

This book also serves as a desk reference for practicing engineers and professionals involved in the structural design and construction of buildings and bridges. The book can also be used effectively for self-study in reviewing the fundamental structural concepts, principles, and procedures in preparation for the Structural Engineering Examination.

Chapters One, Two, Three and Four are devoted to a review of the fundamental principles of engineering mechanics necessary for the sound and reliable design of structures. Chapter Five defines the responsibilities of a structural designer, and emphasizes the need to follow good practices established by the engineering profession and related codes. Chapters Six, Seven, Eight, and Nine deal with the design of concrete, steel, and masonry structures with an introduction to seismic design. Practical problems are used to illustrate the application and interpretation of the controlling codes and specifications. Prospective candidates studying for the PE Exam are strongly encouraged to study these chapters along with the 1997 Uniform Building Code, Volume 2 Structural Engineering Design Provisions, ACI 318-95 Building Code Requirements and Commentary for Reinforced Concrete, the AISC Manual of Steel Construction, Load and Resistance Factor Design, Second Edition, and the ACI 530-95 Building Code Requirements for Masonry Structures.

To the greatest extent feasible, the same equations, notations, and symbols used in the controlling codes and specifications are used in this book.

The authors wish to express their appreciation of the support and patience of their spouses, Juliet, Sheue-Lan Shyu and Mei-Yueh, during the preparation of this book.

In the writing of this book, the authors draw upon their own structural engineering experience and the fine works of many outstanding researchers, teachers, and engineering professionals. The authors are grateful to all the professionals who contribute to engineering knowledge by writing textbooks or reporting on the works they have done for the benefit of the engineering community.

The authors would like to thank Stephanie Law for her careful review and editing of the manuscripts and figures. The authors welcome comments from the readers for improvement in future writings. Comments may be sent to the publisher.

M. Myint Lwin
Chyuan-Shen Lee
J.J. Lee

ABOUT THE AUTHORS

M. MYINT LWIN, PE, SE

M. Myint Lwin, PE, SE, is former Bridge & Structures Engineer of the Washington State Department of Transportation, and is currently a Federal Highway Administration structural engineer based in San Francisco, California. He has 33 years of experience as a practicing and managing engineer, directing the design and construction of bridges and structures incorporating timber, concrete, masonry, and steel.

CHYUAN-SHEN LEE, Ph.D., PE, SE

Chyuan-Shen Lee, Ph.D., PE, SE, contributing author, is a bridge engineer in the Bridge & Structures Office, Washington State Department of Transportation. He has over 12 years of experience as a practicing engineer in the design and construction of bridges and structures incorporating timber, concrete, masonry, and steel.

J.J. LEE, Ph.D, PE, SE

J.J. Lee, Ph.D., PE, SE, contributing author, is Principal with CES, Inc. based in Olympia, Washington. He has over 15 years of experience as a practicing engineer in the design and construction of bridges and structures incorporating timber, concrete, masonry and steel.

CHAPTER 1
PROPERTIES OF MATERIALS

1.1 INTRODUCTION

An understanding of the properties of construction materials is essential for the efficient use and proper selection of materials for safe, economical, and durable designs. This chapter covers the common engineering properties of construction materials usually used in building and bridge design. Properties that are more specific to a particular material will be covered in the specific chapter devoted to the application of the material in design of structures.

Often in engineering practice, it is assumed that two basic characteristics exist in a construction material: (1) a material is *homogeneous,* meaning that the same elastic properties exist at all points in the body, and (2) a material is *isotropic,* meaning that the same elastic properties exist in all directions at any one point of the body. However, not all construction materials are homogeneous or isotropic. When a material does not possess any kind of elastic symmetry, it is called an *anisotropic* material. When the material has elastic symmetry in three mutually perpendicular planes, it is said to be *orthotropic.* A better understanding of these characteristics will be gained after discussion of material properties in this and later chapters.

1.2 STRESS AND STRAIN

Structural engineering is the study and consideration of *stress* and *strain* in individual load-carrying members and in structural systems consisting of load-carrying members. Stress is a measure of the force per unit area (or force divided by area) acting in a member, and strain is a measure of the deformation of a member per unit length (or deformation divided by length). The two are related and are accountable for determining the strength and stiffness of structural members and systems.

A member is in tension when the force causes it to stretch or increase in length. The resulting stress is *tensile stress,* and the unit increase in length is *tensile strain.* A member is in compression when the force causes it to shorten or decrease in length. The resulting stress is *compressive stress,* and the unit decrease in length is *compressive strain.* These definitions assume that the forces act through the centroids of the members. In practice, the forces do not always act through the centroids of members, resulting in the introduction of shearing and bending stresses and strains.

It is necessary to determine the stresses and strains in the structural members and systems to assure that the individual members and the whole structural systems can meet the strength demands and the deflection limitations of the design criteria safely.

1.3 TEST SPECIMENS

Tests are performed in the laboratory to determine the elastic and plastic properties of structural materials. The tests are to provide quality control and assurance in the manufacturing and fabrication processes to ensure that the materials will meet the specifications of a project. The owner or the owner's representative might perform independent testing to assure that the materials furnished are what are specified in the contract. The manufacturers generally provide mill certificates certifying that the products have the chemical and mechanical properties according to standard testing methods.

The American Society for Testing and Materials (ASTM) and the American Association of State Highway and Transportation Officials (AASHTO) issue standard specifications for testing procedures and acceptance criteria for testing materials. ASTM specifications usually are used in building construction, and the AASHTO specifications for bridge construction. ASTM and AASHTO have equivalent standards. However, usually the ASTM will be referred to in this book.

A rectangular tension test specimen is shown in Figure 1.1.

1.4 NORMAL STRESS

If a tension specimen is subjected to a tensile force P acting normal to the longitudinal axis of the specimen as shown in Figure 1.2, the intensity of the normal force P per unit area of the cross section is termed the normal stress σ, and is expressed as

$$\sigma = \frac{P}{A} \tag{1.1}$$

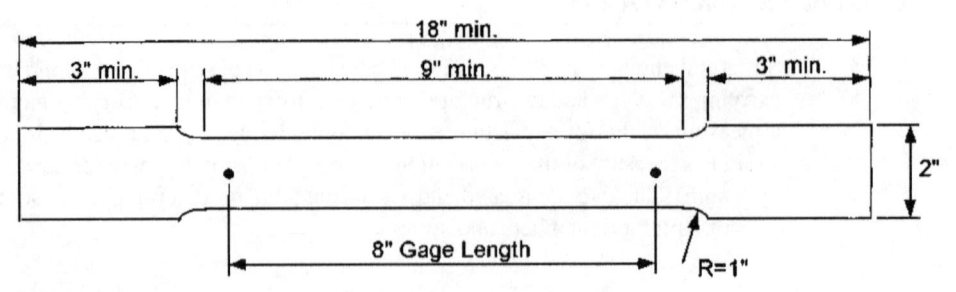

FIGURE 1.1

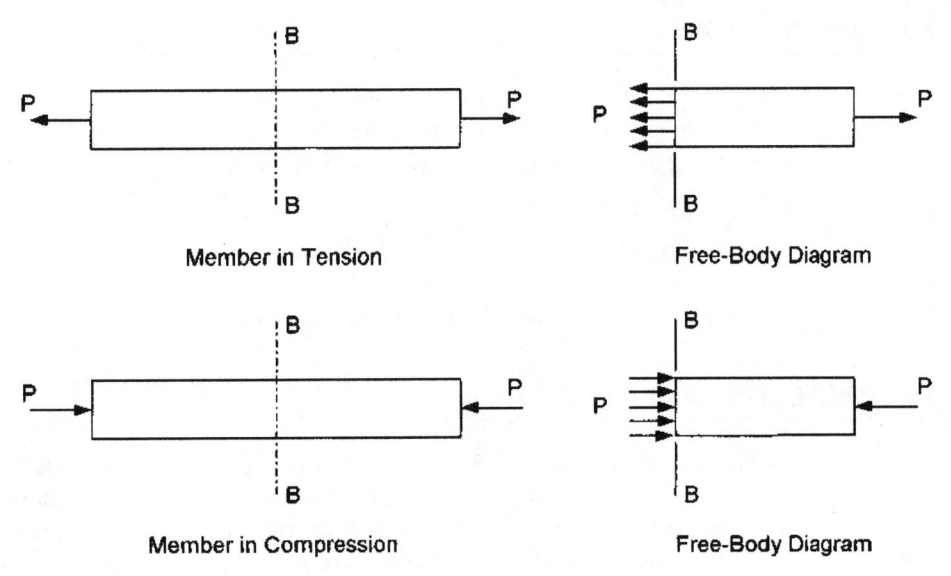

FIGURE 1.2

In Equation 1.1, area A is the original area of the cross-section before the application of the force. This is industry practice. If the specimen is subjected to a compressive force, normal compressive stresses are set up in the specimen.

1.5 NORMAL STRAIN

As the forces in Figure 1.2 are increased gradually, the elongation Δ over the gage length L can be measured for corresponding increase in forces. The values of elongation per unit length may be found by

$$\epsilon = \frac{\Delta}{L} \tag{1.2}$$

This is termed normal strain.

1.6 STRESS-STRAIN DIAGRAMS

The most common type of test is a *tension test,* in which the specimen is stretched by a tensile load. The tensile load is increased in increments gradually from zero until the specimen breaks. The corresponding elongation over the gage length is measured at each increment of load. The normal stress and strain at each load increment can be computed by Equations 1.1 and 1.2, respectively. The values of normal stress and strain then can be plotted in a stress-strain diagram. Strain is plotted horizontally on the x-axis, and stress is plotted vertically on the y-axis. A representative stress-strain diagram for a ductile material is shown in Figure 1.3. The figure shows the behavior of the material at various levels of stress and strain. Figure 1.4 shows typical stress-strain diagrams for some structural materials, such as concrete, timber, and structural steels.

1.7 HOOKE'S LAW

From Figure 1.3, it can be seen that for low values of stress, the curve is a straight line OA. Stress is proportional to strain in this region. If the load is removed in this region, the specimen will return to its original length. The relation between stress and strain is constant and may be expressed as

$$\frac{\sigma}{\epsilon} = E \qquad (1.3)$$

E is a constant and denotes the slope of the straight line *OA*.

1.8 MODULUS OF ELASTICITY

The quantity E in Equation 1.3 is the modulus of elasticity of the material in tension. It is often referred to as Young's modulus. Since the unit for strain ϵ is a pure number, E has the same unit as stress σ. The values of E for engineering materials are found in handbooks. For many engineering materials, the modulus of elasticity in compression is nearly the same as that for tension.

1.9 PROPORTIONAL LIMIT

In Figure 1.3, there is a transition at point *A* from the straight line *OA* to the curve *AB*. Stress no longer is proportional to strain in the region *AB*. The stress at point *A* is the highest stress for which Hooke's law is valid. Point *A* denotes the limit of proportionality of stress to strain. The stress at this point is the *proportional limit*.

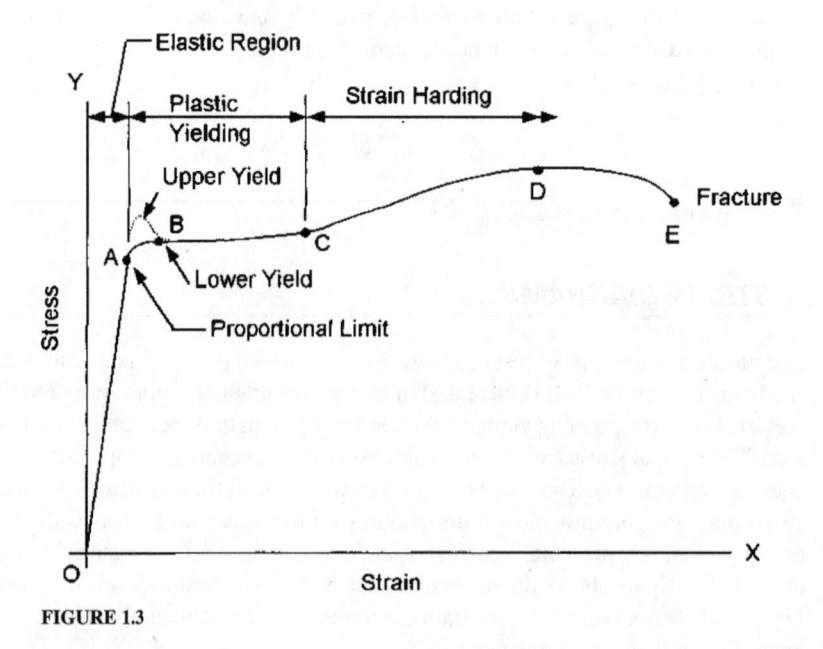

FIGURE 1.3

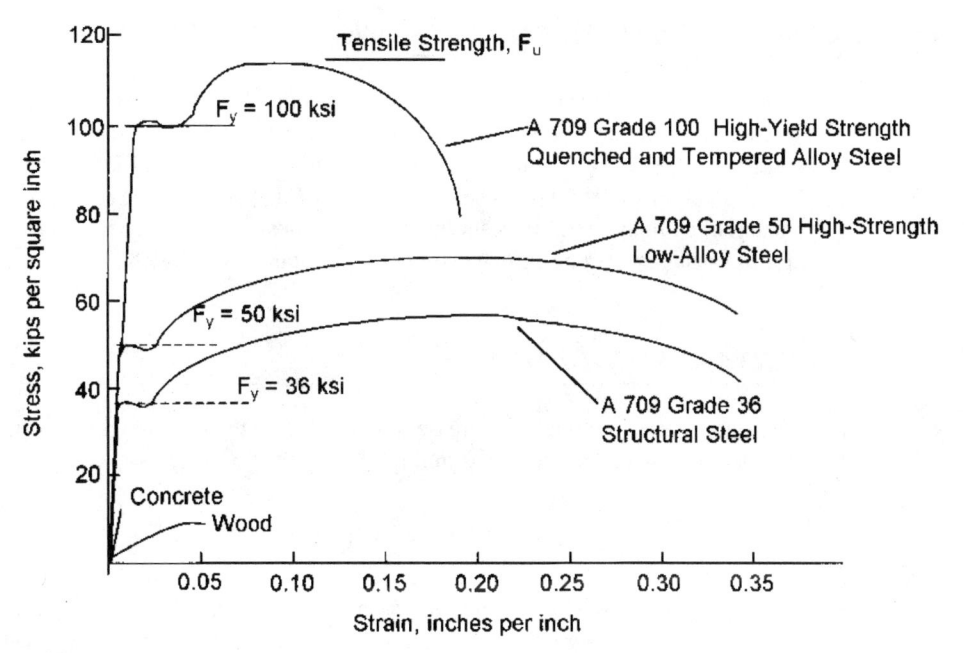

FIGURE 1.4

1.10 YIELD POINT

At point B in Figure 1.3, the curve becomes horizontal. At this point, called the *yield point,* there is an increase in strain without a corresponding increase in stress. This is a very important property of structural materials. Many material specifications and design procedures are based on this value. Some materials exhibit an upper and lower yield point, as shown by the dashed part of the curve in Figure 1.3.

Beyond the yield stress there is a plastic yielding region where an increase in strain occurs without increase in stress. The strain that occurs before the yield stress is referred to as the *elastic strain.* The strain that occurs after the yield stress, with no increase in stress, is referred to as the *plastic strain.* Plastic strains in ductile materials are in the range of 10 to 15 times elastic strains. For brittle materials, there is little or no plastic strain.

For materials that do not exhibit a well-defined yield point, the offset method is used to define a yield point. The common practice is to take an offset of 0.2 percent of strain and draw a line parallel to the straight portion of the initial stress-strain curve, as shown in Figure 1.5. The point of intersection of this line with the stress-strain curve is taken as the yield point of the material at 0.2 percent offset.

1.11 STRAIN HARDENING

A region where additional stress is necessary to produce additional strain follows the plastic strain. This behavior is called strain-hardening and is indicated by the region CD in Figure 1.3.

1.12 ULTIMATE STRENGTH AND BREAKING STRENGTH

The strain-hardening continues to the highest point *D* of the stress-strain diagram. A sharp reduction of the cross-section of the specimen (called *necking*) takes place after point *D* until final fracture at point *E*. The stress at the highest point *D* is known as the *ultimate strength* or *tensile strength*. The stress at point *E* is known as the *breaking strength* of the material. Ultimate strength is another important design characteristic. It is used as the basis for ultimate strength design and strength-limit state design methods.

1.13 PERCENTAGE ELONGATION

Percentage elongation is the increase in length ΔL of the gage length after fracture divided by the initial gage length L_o and multiplied by 100.

$$\text{Percentage elongation} = \frac{\Delta L}{L_o} \times 100 \tag{1.4}$$

Percentage elongation is a measure of the ductility of a material. Structural steel is a ductile material. It has percentage elongation in the range of 8 percent to 12 percent. The values of percentage elongation usually are shown in the mill certificate from the manufacturer.

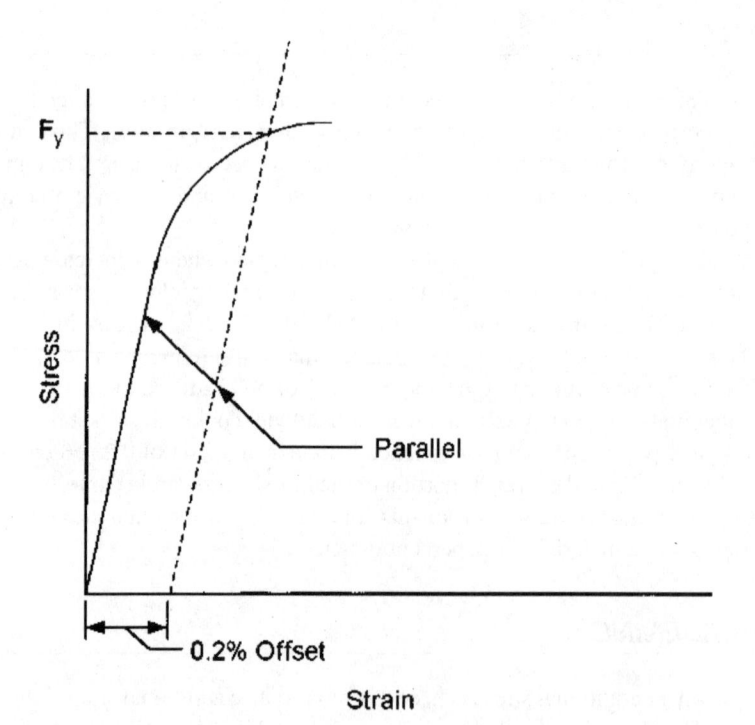

FIGURE 1.5

1.14 PERCENTAGE REDUCTION IN AREA

The percentage reduction in area ΔA is the decrease in cross-sectional area from the original area A_o upon fracture divided by the original area and multiplied by 100.

$$\text{Percentage reduction in area} = \frac{\Delta A}{A_o} \times 100 \tag{1.5}$$

Percentage reduction in area is also a measure of the ductility of a material. An experienced engineer can look at the pieces of a broken tension test specimen and determine the ductility of the material.

1.15 WORKING STRESS

The yield stress, or the ultimate or tensile strength, usually is used to select a working stress for the design of structural members. Frequently, the working stress is determined by dividing the yield stress, or the ultimate strength, by a factor of safety, which is usually provided by design codes or established by the designers. This is the basis for the working stress-design method.

1.16 SECANT MODULUS

The stress-strain diagrams of some materials do not distinctly show a straight portion of elastic behavior. The stress-strain diagrams are curvilinear, even at stresses well below the elastic range. The slope of the secant drawn from the origin to any specified point on the stress-strain curve is the *secant modulus*, as shown in Figure 1.6.

$$E_s = \frac{\sigma}{\epsilon} \tag{1.6}$$

1.17 TANGENT MODULUS

The tangent modulus is the slope of the stress-strain curve at any point in the region of plastic yielding, as shown in Figure 1.6. It is the instantaneous modulus given by

$$E_T = \frac{d\sigma}{d\epsilon} \tag{1.7}$$

1.18 POISSON'S RATIO

When a specimen is subjected to an axial force, it deforms in the direction of the applied force. French mathematician and scientist S. D. Poisson also observed that deformations occurred in the transverse directions perpendicular to the applied force. Poisson's ratio is

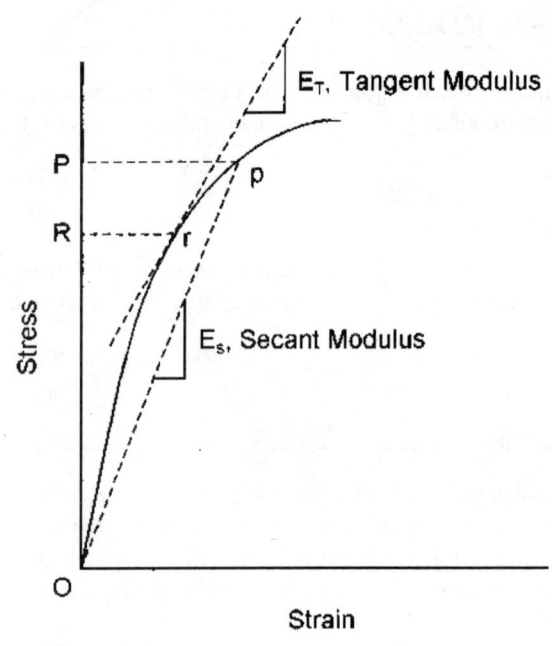

FIGURE 1.6

the absolute value of the ratio of the transverse strain to the axial strain, and is commonly denoted by

$$\mu = \frac{\epsilon_{transverse}}{\epsilon_{axial}} \tag{1.8}$$

Poisson's ratio for structural steels is 0.27, and for concrete is commonly taken as 0.20. The actual value may vary from 0.15 to 0.25, depending on the aggregates, moisture content, age, and compressive strength.

1.19 FATIGUE LIFE

Fatigue is the tendency of materials to crack or fail under many repetitions of a stress considerably less than the ultimate strength. Fatigue is the main cause of cracking or failure of steel members in service. A sound design must address fatigue life to make sure there are no premature cracking or failures.

Fatigue life is the number of stress cycles that cause a structural component or detail to fail at a specified stress range. The fatigue life of a structural component or detail is given in design codes in the form of *S-N* curves, as shown in Figure 1.7. The designers will select the value in accordance with the criteria for the design.

1.20 DUCTILITY

Ductility is the ability of a material to undergo deformation without failure under high tensile stresses. Structures with ductile members and details will be able to sustain large deformation without collapse. This is an important property of structural components and systems, especially in seismic design. Observations of performance of structures in major earthquakes indicate that structures with ductile behavior survived the earthquakes without collapse. This is one of the underlining principles in seismic design and retrofit, which will be discussed in Chapter 10.

1.21 MODULUS OF RESILIENCE

Resilience is the ability of a material to absorb or store energy without permanent deformation. It is measured by the work done on a unit volume of material within the stress-strain curve where Hooke's law applies. This is the region OA shown in Figure 1.8. The quantity of the work done under the line OA is termed the modulus of resilience U_o.

U_o may be calculated as the area under line OA as shown in Figure 1.8.

$$U_o = \frac{1}{2} x \epsilon_y x \sigma_y \tag{1.9}$$

From Young's modulus

$$\epsilon_y = \frac{\sigma_y}{E}$$

Substituting this in Equation 1.9

$$U_o = \frac{\sigma_y^2}{2E} \tag{1.10}$$

The modulus of resilience is a useful value for selecting materials for design where energy must be absorbed by the members.

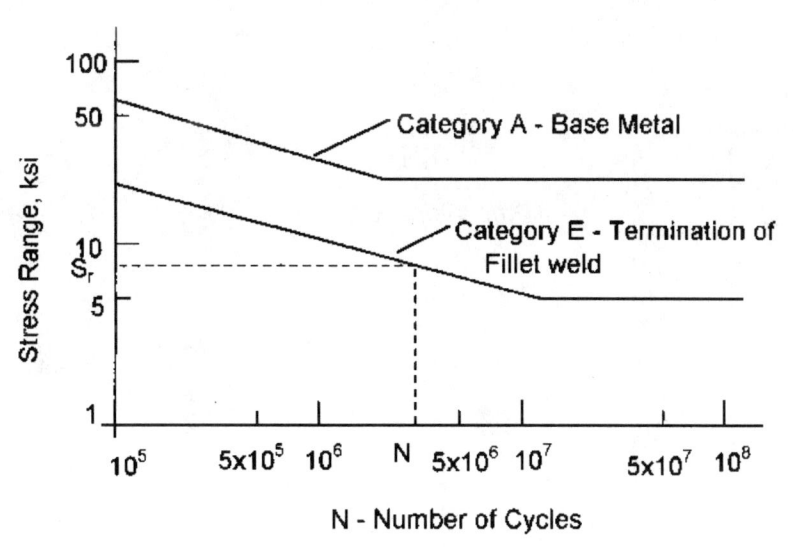

FIGURE 1.7

For example, a structural carbon steel with a proportional limit of 30,000 psi and an E of 30×10^6 psi has a modulus of resilience

$$U_o = \frac{\sigma_y^2}{2E} = \frac{30,000^2}{2(30)10^6} = 15 \text{ in.-lb per cu. in.}$$

A dense structural grade of Douglas fir with a proportional limit of 6500 psi and an E of 1.9×10^6 psi has a modulus of resilience

$$U_o = \frac{\sigma_y^2}{2E} = \frac{6500^2}{2(1.9)10^6} = 11.1 \text{ in.-lb per cu. in.}$$

1.22 HARDNESS

Hardness is a measurement of the resistance of a material to deformation, indentation, or scratching, and can be used to verify the ultimate strength of structural steels after heat treatment. Hardness also can be used in the fabrication shop or in the field to measure the embrittlement of structural steels because of flame cutting. For example, the flame-cut edges of structural low-alloy steel are hardened by the flame-cut process. The excessively hardened edges are highly susceptible to cracking unless the hardened surface is removed by grinding.

There are different methods to test and evaluate hardness, but unfortunately there is no absolute scale for hardness. Each method expresses hardness quantitatively by some arbi-

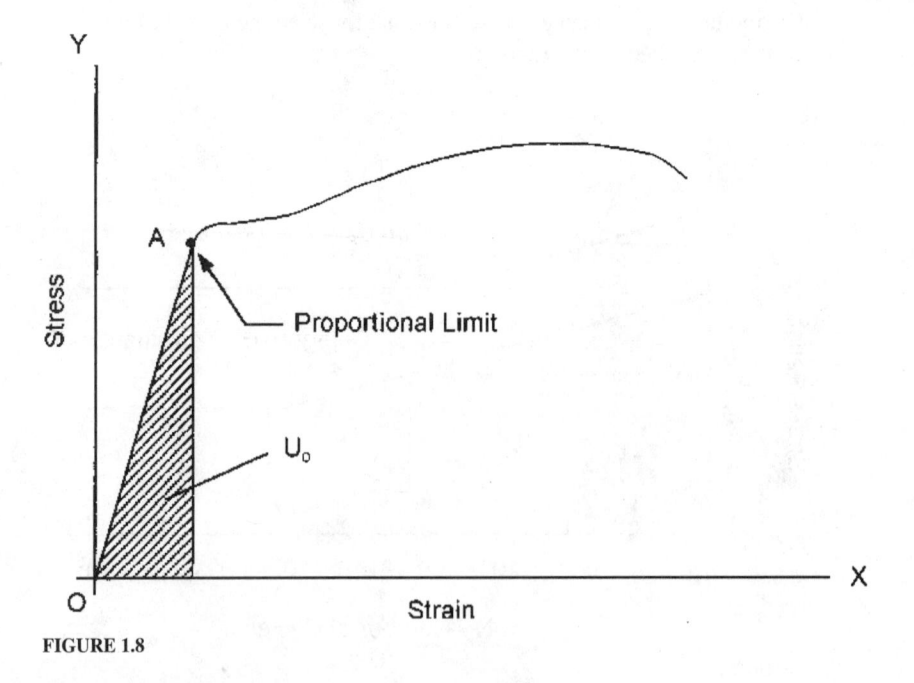

FIGURE 1.8

trarily defined hardness. The two scales commonly used in practice are the Brinell hardness number and the Rockwell hardness number.

The Brinell hardness number HB is a number related to the applied load and to the surface area of the permanent impression made by a ball indenter computed from the equation

$$HB = \frac{2P}{\beta D \left(D - \sqrt{D^2 - d^2}\right)} \quad (1.11)$$

where P = applied load, kgf
D = diameter of ball, mm
d = mean diameter of the impression, mm

The Rockwell hardness number HR is derived from the net increase in the depth indentation as the force on an indenter is increased from a specified preliminary test force to a specified total test force, and then returned to the preliminary test force. The indenters for the Rockwell hardness test include a diamond spero-conical indenter and steel ball indenters of several specified diameters. The Rockwell hardness numbers are quoted with a scale symbol representing the indenter and forces used. The hardness number is followed by the symbol HR and the scale designation. For example, 30 HRC means Rockwell hardness number of 30 on Rockwell C scale. Incidentally, flame-cut edges of structural steels with hardness greater than 30 HRC are highly susceptible to cracking under stress. Such edges should be ground or otherwise treated to reduce the hardness to less than 30 HRC.

1.23 FRACTURE TOUGHNESS

Fracture toughness, or simply *toughness,* is a measure of the ability of a material to withstand impact load without fracturing. It is also a measure of the resistance to extension of a crack in a material. This property is important for crack control and extending the fatigue life of structural steels. When a member or a structural system is subject to impact loading, such as vehicular loading in bridges, it is necessary to specify adequate toughness in the members to avoid premature cracking or failure.

Fracture toughness requirements are provided in building codes and bridge-design specifications for main load-carrying members subjected to tensile stress. The basis for the requirements is to avoid brittle fracture and premature fatigue failure. The fracture toughness is specified in terms of energy absorbed and temperature tested in accordance with ASTM A 673 for the Charpy V-notch Test. A Charpy V-notch impact test is a dynamic test in which a notched specimen is struck and broken by a single blow in a specially designed testing machine. The energy absorbed in such a test is specified as a measure of fracture toughness. For example, for a fracture-critical member of ASTM A 709 Grade 50 steel in a welded structure under a service temperature range of $-1°F$ to $-30°F$, the fracture toughness requirement will be 25 ft lb at 40°F.

Another method of measuring toughness in a material is to determine the *modulus of toughness,* which is defined as the strain energy or work done per unit volume of the material to cause fracture. It is the area under the stress-strain curve *OABCDE* in Figure 1.9. The larger the area, the tougher the material.

1.24 BRITTLE FRACTURE

Under conditions of high restraint, stress concentration, local temperature, fatigue-type loadings, low toughness, improper heat input, and other possibilities, structural materials might lose ductility and toughness, resulting in brittle fracture. Brittle fracture is very sudden, because there is lack of ductility or deformation. There is no warning, or telltale sign. The designers must use good design and detailing practice to avoid brittle fracture in structures.

1.25 CREEP AND SHRINKAGE

Creep and shrinkage are adverse, but very important, properties of concrete. The stress-strain diagram of concrete depends upon the rate of loading and the time history of loading. If the stress is held constant for some length of time, the strain increases. This behavior is known as creep. Concrete loses moisture with time and decreases in volume. This behavior is known as shrinkage.

The amount of creep a particular concrete will exhibit is difficult to estimate accurately. Without specific tests, accuracies of better than 30 percent should be expected. In view of the scatter, it is reasonable to use simple, approximate procedures for estimating creep deformations.

The amount of shrinkage in concrete depends on many factors, such as the composition of the concrete-water, type of cement, quality and gradation of aggregates, and other additives. Curing methods, size of member, and relative humidity also affect the shrinkage properties of concrete.

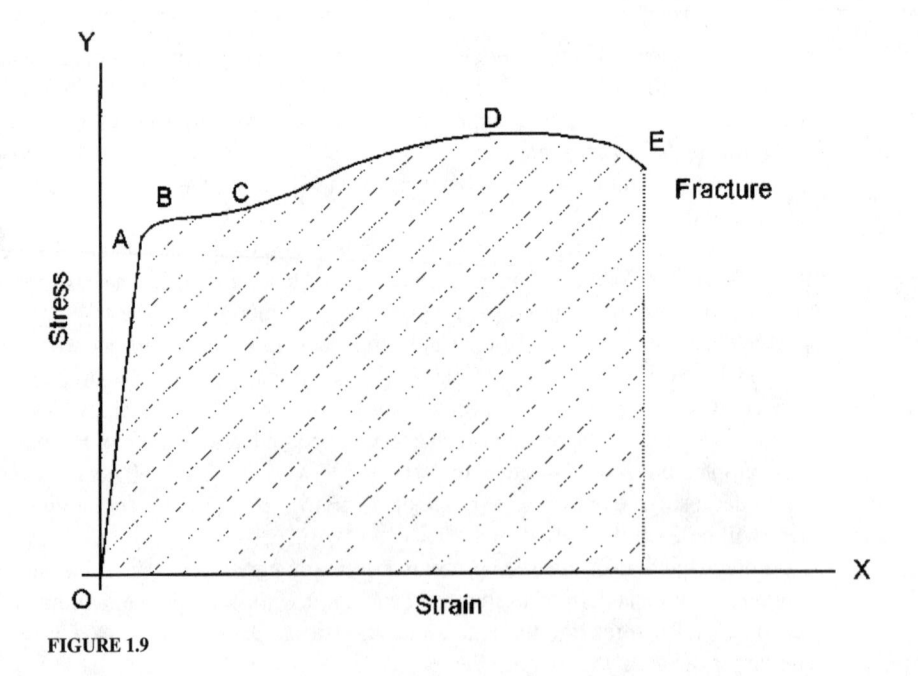

FIGURE 1.9

Creep and shrinkage, as they affect the design of concrete structures, will be covered in Chapter 6.

1.26 RELAXATION

If the strain of a material is held constant for some length of time, the stress will decrease. This behavior is referred to as relaxation. The magnitude of the decrease in stress is small and generally of no significance in structural engineering, except for estimating the prestress loss because of relaxation of the steel prestressing strands in concrete design. This will be discussed in Chapter 6.

1.27 GENERALIZED FORM OF HOOKE'S LAW

The simple form of Hooke's law for an axially loaded tension specimen is given in Equation 1.3. For this case, only the deformation in the direction of load was considered and is given by

$$\frac{\sigma}{\epsilon} = E$$

In the general case an element of material is subjected to three mutually perpendicular normal stresses σ_x, σ_y, σ_z, and the corresponding strains ϵ_x, ϵ_y, ϵ_z. Including the Poisson's effect in the stress-strain relations, the general form of Hooke's law is given by the following equations:

$$\epsilon_x = \frac{1}{E}[\sigma_x - \mu(\sigma_y+\sigma_z)] \qquad (1.12)$$

$$\epsilon_y = \frac{1}{E}[\sigma_y - \mu(\sigma_z+\sigma_x)] \qquad (1.13)$$

$$\epsilon_z = \frac{1}{E}[\sigma_z - \mu(\sigma_x+\sigma_y)] \qquad (1.14)$$

1.28 MATERIAL TESTING

Material testing usually is carried out in accordance with approved ASTM standards. ASTM is a nonprofit organization that provides a forum for producers, users, ultimate consumers, and those having a general interest to meet on common ground and write standards for materials, products, systems, and services. The standards are developed through the voluntary work of 132 standards-writing committees consisting of more than 33,000 technically qualified ASTM members throughout the world. Membership is open to all concerned with the fields in which ASTM is active. Any readers who are interested in participating in the standards-writing committees can obtain information from Member and Committee Services, ASTM, 100 Barr Harbor Drive, West Conshohocken, PA 19428.

Listed below are the ASTM standards that can be used for testing the material properties covered in this chapter. The readers can refer to these standards for further study. They may be available in public libraries or libraries of the state departments of transportation.

A 6—Standard Specification for General Requirements for Rolled Structural Steel Bars, Plates, Shapes and Sheet Piling

A 370—Standard Test Methods and Definitions for Mechanical Testing of Steel Products

A 673—Standard Specification for Sampling Procedure for Impact Testing of Structural Steel

C 39—Standard Test Method for Compressive Strength of Cylindrical Concrete Specimens

C 157—Test Method for Length Change of Hardened Hydraulic-Cement Mortar and Concrete

C 512—Test Method for Creep of Concrete in Compression

C 666—Test Method for Resistance of Concrete to Rapid Freezing and Thawing

C 1202—Test Method for Electrical Indication of Concrete's Ability to Resist Chloride Ion Penetration

E 8—Test Methods for Tension Testing of Metallic Materials

E 9—Test Methods for Compression Testing of Metallic Materials at Room Temperature

E 10—Test Methods for Brinell Hardness of Metallic Materials

E 18—Test Methods for Rockwell Hardness and Rockwell Superficial Hardness of Metallic Materials

E 23—Test Method for Notched Bar Impact Testing of Metal Materials

E 111—Test Method for Young's Modulus, Tangent Modulus, and Chord Modulus

E 132—Test Method for Poisson's Ratio at Room Temperature

E 143—Test Method for Shear Modulus at Room Temperature

E 328—Test Method for Stress Relaxation for Materials and Structures

E 1820—Test Method for Measurement of Fracture Toughness

CHAPTER 2
PROPERTIES OF SECTIONS

2.1 SECTION PROPERTY

Section property is an important factor in the design and analysis of structural members because it controls the efficient use of the material of the structural members. Several section properties commonly are encountered in solving structural design problems and are discussed in the following sections.

2.2 CENTROID OF AN AREA

The area A of the cross section of a member is used directly to compute simple tension, compression, and shear. In engineering practice, it is often necessary to find the centroid of an area—that point in the plane of the area where the moment of the area is zero about any axis passing through the point. Using this definition, the centroid of an area may be determined as follows:

If (x_0, y_0) are the coordinates of the centroid G of the area A shown in Figure 2.1, then

$$x_0 = \frac{\int x dA}{A} \qquad (2.1)$$

$$y_0 = \frac{\int y dA}{A} \qquad (2.2)$$

The quantity $\int x dA$ or $\int y dA$ is the sum of the products obtained by multiplying each element of the area dA by its distance from axis x or y. This quantity is the moment of an area, or first moment of an area, or statitical moment of an area.

2.2 CHAPTER 2

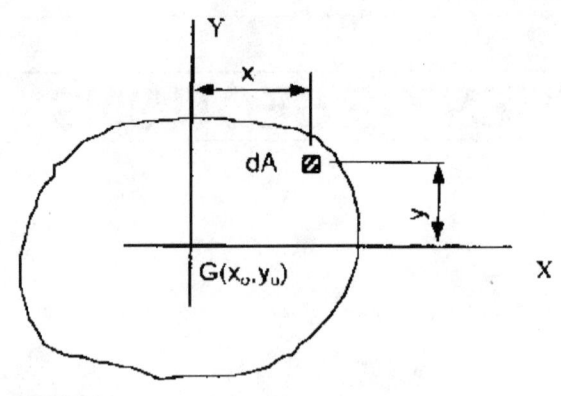

FIGURE 2.1

The centroids of basic shapes are given in design manuals, mathematical handbooks, and books on engineering. For example, Part 7 of the AISC Manual of Steel Construction, Load & Resistance Factor Design (LRFD manual), provides the centroids of a wide variety of basic sections.

For a compound area consisting of basic shapes, the centroid of the area may be determined using the following equations

$$x_0 = \frac{\sum_i A_i x_i}{\sum_i A_i} \tag{2.3}$$

$$y_0 = \frac{\sum_i A_i y_i}{\sum_i A_i} \tag{2.4}$$

EXAMPLE 2.1
Find the centroid of the area shown in Figure 2.2.

Solution
The area is divided into basic shapes of one 24 in. × 6 in. rectangle, two 8 in. × 12 in. rectangles, and two 6 in. × 6 in. triangles as shown in Figure 2.2b. Next, find the areas of the basic sections

$A_1 = 24$ in. $\times 6$ in. $= 144$ in^2

$A_2 = 8$ in. $\times 12$ in. $= 96$ in^2

$A_3 = \frac{1}{2}$ in. $\times 6$ in. $\times 6$ in. $= 18$ in^2

(a)

(b)

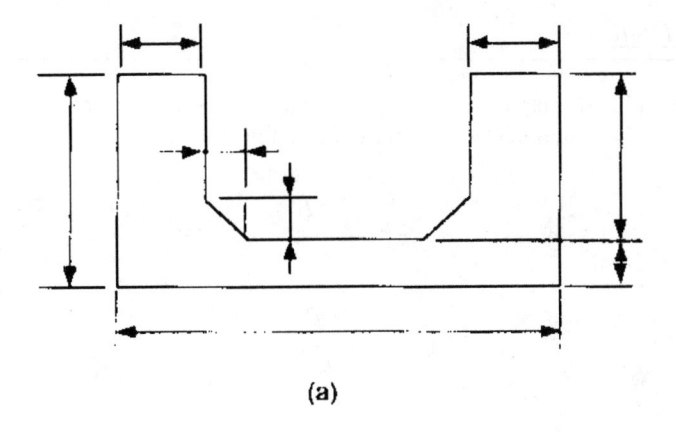

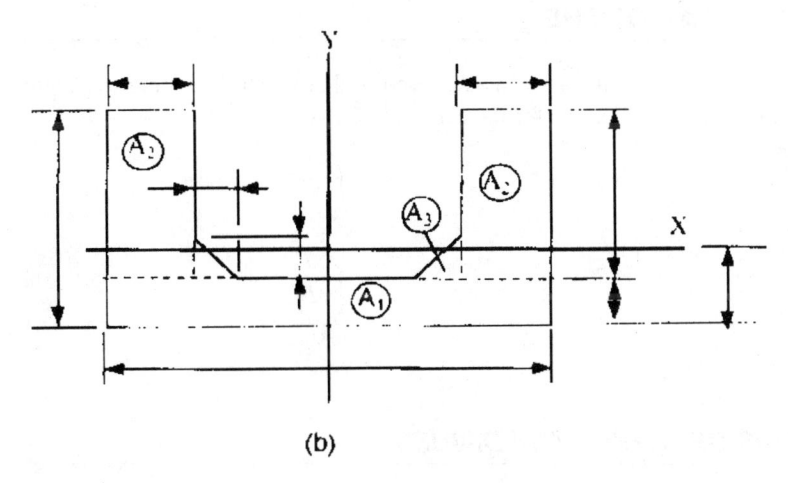

FIGURE 2.2

Next, find the y coordinates of the centroids of the basic shapes.

$y_1 = 3$ in.

$y_2 = 12$ in.

$y_3 = 8$ in.

Using Equation 2.4,

$$y_0 = \frac{144 \times 3 + 2 \times 96 \times 12 + 2 \times 18 \times 8}{144 + 96 + 18} = 11.72 \text{ in. (from the bottom)}$$

By symmetry about the y axis, $x_0 = 0$.

2.3 CENTROID OF A LINE

The centroid of a line L may be derived in a similar way as that for determining the centroid of an area. The equations for the centroid of a line are

$$x_0 = \frac{\int x \, dL}{L} \qquad (2.5)$$

$$y_0 = \frac{\int y \, dL}{L} \qquad (2.6)$$

2.4 CENTROID OF A VOLUME

The centroid of a volume V can be derived similarly to the way for determining the centroid of an area. The equations for the centroid of a volume are

$$x_0 = \frac{\int x \, dV}{V} \qquad (2.7)$$

$$y_0 = \frac{\int y \, dV}{V} \qquad (2.8)$$

2.5 THEOREMS OF PAPPUS-GULDINUS

The theorems of Pappus-Guldinus provide very simple methods for calculating the surface of an area generated by revolving a plane curve and the volume generated by revolving an area about a nonintersecting line in its plane.

In the case of a surface generated by a plane curve, consider the line segment of length L in the x-y plane generating a surface when revolved about the x axis as shown in Figure 2.3. An element of this surface is the ring generated by dL. The area of this ring is

$$dA = 2\pi \, dL$$

and the total area is

$$A = 2\pi \int y \, dA$$

Since

$$y_0 L = \int y \, dL$$

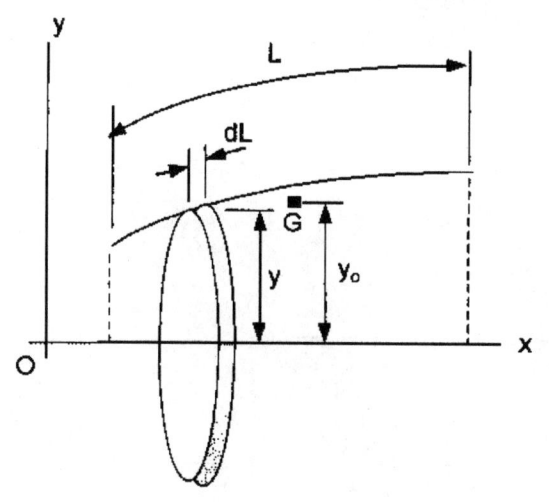

FIGURE 2.3

the area becomes

$$A = 2\pi y_0 L \tag{2.9}$$

In the case of a volume generated by revolving an area, consider the area A in the x-y plane generating a volume when revolved about the x axis, as shown in Figure 2.4.

An element of this volume is the ring of cross section dA and radius y. The volume of the element is

$$dV = 2\pi y \, dA$$

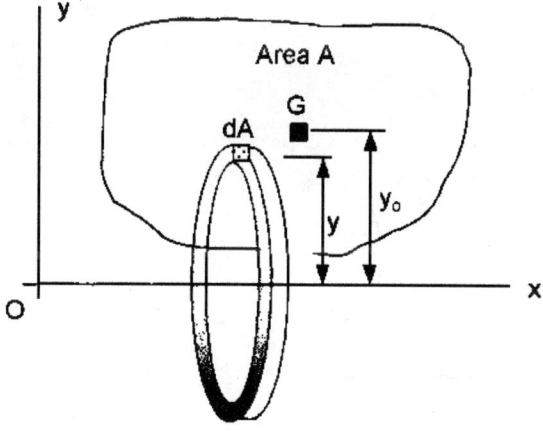

FIGURE 2.4

and the total volume is

$$V = 2\pi \int y \, dA$$

Since

$$y_0 = \int y \, dA$$

the volume becomes

$$V = 2\pi y_0 A \tag{2.10}$$

2.6 MOMENT OF INERTIA

The moment of inertia of the cross section of a structural member is a measure of the resistance to bending, rotation, and buckling by virtue of the geometry and size of the section. The moment of inertia is an important property in solving design problems of beams and long columns.

The moment of inertia of an area (also known as the second moment of an area) with respect to an axis is the sum of the products obtained by multiplying each element of the area dA by the square of its distance from the axis.

Consider the area A in the x-y plane as shown in Figure 2.5. By definition, the moment of inertia of the element dA about the x and y axes are $dI_x = y^2 dA$ and $dI_y = x^2 dA$, respectively (Fig. 2.5). Therefore the moments of inertia of area A about the x and y axes are

$$I_x = \int y^2 \, dA \tag{2.11}$$

$$I_y = \int x^2 \, dA \tag{2.12}$$

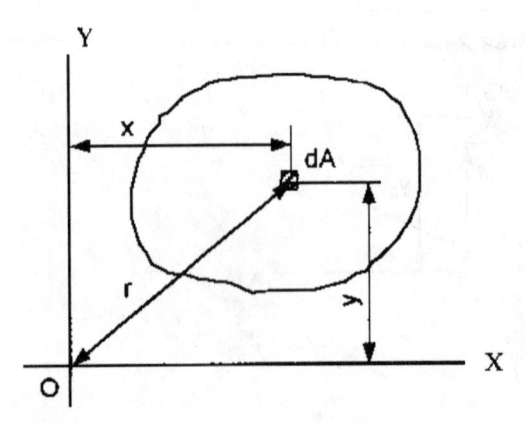

FIGURE 2.5

By similar definition, the moment of inertia of the area A about the pole 0 (z axis) is given by

$$J_z = \int r^2 \, dA \tag{2.13}$$

The expressions of Equations 2.11 and 2.12 are known as the rectangular moment of inertia, and the expression of Equation 2.13 is known as the polar moment of inertia. From the relationship $x^2 + y^2 = r^2$, the moments of inertia are related by the expression

$$J_z = I_x + I_y \tag{2.14}$$

2.7 TRANSFER OF AXES

The moment of inertia of an area about a noncentroidal axis can be expressed in terms of the moment of inertia about a parallel centroidal axis. In Figure 2.6, the x_0 and y_0 axes pass through the centroid G of the area. Let it be desired to find the moments of inertia of the area about the parallel x and y axes.

By definition, the moment of inertia of the element dA about the x axis is

$$dI_x = (y_0 + y_1)^2 dA \tag{2.15}$$

Expanding and integrating give

$$I_x = \int y_0^2 \, dA + 2y_1 \int y_0 \, dA + y_1^2 \int dA \tag{2.16}$$

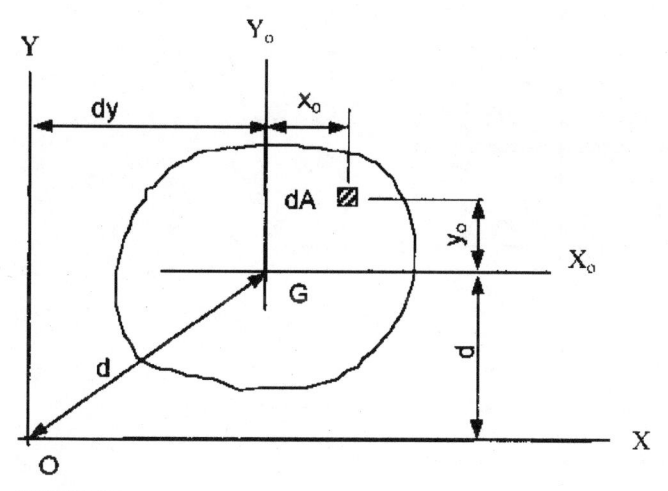

FIGURE 2.6

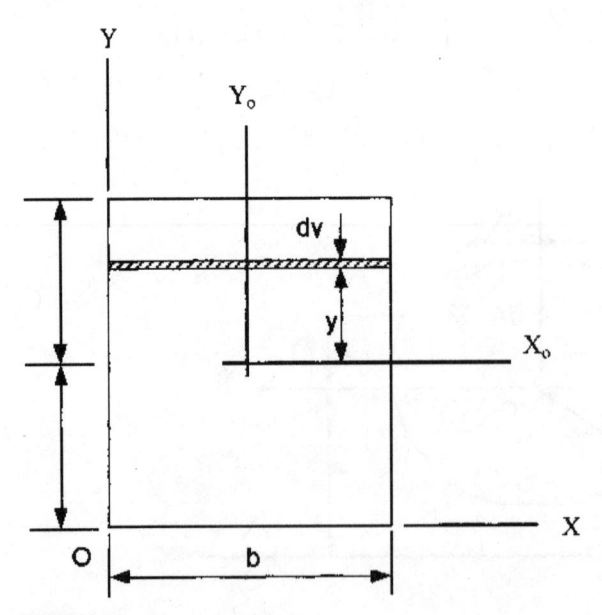

The first integral is the moment of inertia I_x about the centroidal x_0 axis. The second integral is zero, because $y_0 = 0$ about the centroidal axis. The third integral is Ad_x^2. Thus the moments of inertia I_x and I_y become

$$I_x = I_0 + Ay_1^2 \qquad (2.17)$$
$$I_y = I_0 + Ax_1^2 \qquad (2.18)$$

By Equation 2.14,

$$J_z = J_0 + Ar^2 \qquad (2.19)$$

Equations 2.17, 2.18, and 2.19 are the parallel axis theorem, which states that the moment of inertia of an area about any axis is equal to the moment of inertia about a parallel centroidal axis, plus the product of the area and the square of the distance between the two axes. The two axes must be parallel, and one of the axes must pass through the centroid. If a transfer is made between two parallel axes, neither of which passes through the centroid, it is first necessary to transfer from one axis to the parallel centroidal axis and then to transfer from the centroidal axis to the second axis.

EXAMPLE 2.2
In Figure 2.7, determine the moment of inertia of the rectangular area about the centroidal x_0-y_0 axes and the x axis.

Solution
For determining the moment of inertia about the x_0 axis, consider a horizontal strip of area $b\,dy$ at distance y from the x_0 axis.

FIGURE 2.7

From Equation 2.11, the moment of inertia I_{x0} about the centroidal x_0 axis is given by

$$I_{x0} = \int_{-d/2}^{d/2} y^2 b \, dy = \frac{1}{12} bd^3 \qquad \text{Answer}$$

Similarly, the moment of inertia about the centroidal y_0 axis is

$$I_{y0} = \int_{-d/2}^{d/2} x^2 d \, dx = \frac{1}{12} db^3 \qquad \text{Answer}$$

By the parallel axis theorem, the moment of inertia about the x axis is

$$I_x = \frac{1}{12} bd^3 + bd\left(\frac{d}{2}\right)^2 = \frac{1}{3} bd^3 \qquad \text{Answer}$$

EXAMPLE 2.3
In Figure 2.8, determine the moment of inertia of the triangular area about its base and about a parallel axis through the centroid.

Solution
Consider a horizontal strip of area parallel to and at distance y from the base, as shown in Figure 2.7. The area of the strip

$$dA = x \, dy = [(d - y)b/d]dy$$

From Equation 2.11, the moment of inertia I_x about the base is given by

$$I_x = \int y^2 \, dA = \int_0^d y^2 \frac{d - y}{d} b \, dy = b\left[\frac{y^3}{3} - \frac{y^4}{4h}\right]_0^d = \frac{bd^3}{12} \qquad \text{Answer}$$

By the parallel axis theorem, the moment of inertia I_o about the centroidal axis is given by (the centroid is at a distance $d/3$ above the base),

$$I_0 = I_x - A(\text{distance})^2 = \frac{bd^3}{12} - \left(\frac{bd}{2}\right)\left(\frac{d}{3}\right)^2 = \frac{bd^3}{36} \qquad \text{Answer}$$

The $A(\text{distance})^2$ term is negative because the axis is transferred from the base to a parallel centroidal axis.

EXAMPLE 2.4
In Figure 2.9, determine the moments of inertia of the area of a circle about the polar axis and about a diametral x axis or y axis through the center.

Solution
Consider a circular strip of area $2\pi r_0 \, dr_0$ at distance r_0 from the center.

2.10 CHAPTER 2

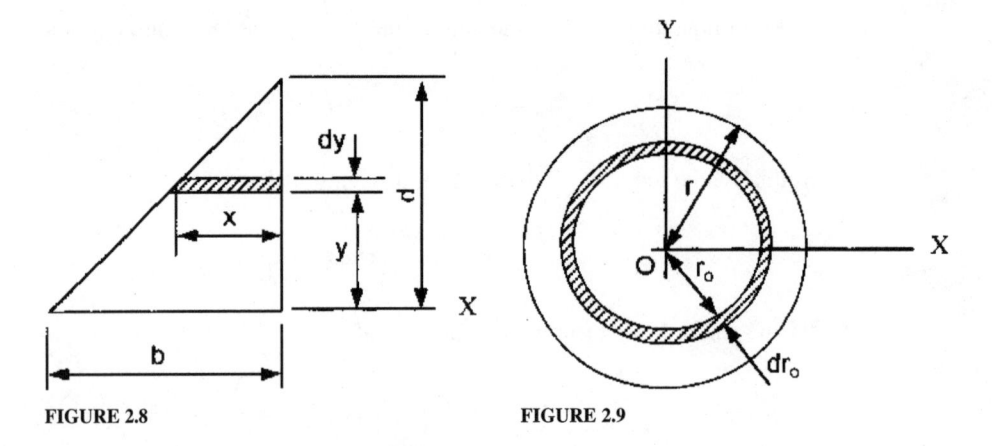

FIGURE 2.8

FIGURE 2.9

From Equation 2.13, the moment of inertia about the polar z axis through the center O is given by

$$J_z = \int_0^r r_0^2 (2\pi r_0 dr_0) = \frac{\pi r^4}{2} \qquad \text{Answer}$$

From Equation 2.14 and by symmetry $I_x = I_y$,

$$I_x = I_y = \frac{1}{2} J_z = \frac{\pi r^4}{4} \qquad \text{Answer}$$

2.8 METHODS FOR DETERMINING MOMENT OF INERTIA

Moment of inertia I is an important property in solving stiffness or deflection problems in beams and long columns. Many simplified formulas for computing I of typical sections are given in design handbooks and the LRFD manual. However, in engineering practice, the sections often consist of typical and nontypical shapes. It is necessary to know several methods for determining I. Some practical methods are presented in the following subsections.

2.8.1 Method 1: Moment of Inertia for Typical Sections

For finding moments of inertia of typical sections, such as rectangles, triangles, and circles, the formulas derived in Examples 2.2, 2.3, and 2.4, and those given in engineering handbooks and manuals can be used.

For example, the moment of inertia for a rectangle about its neutral axis is

$$I_0 = \frac{bd^3}{12}$$

whereas the moment of inertia for a rectangle about the base is

$$I_b = \frac{bd^3}{3}$$

where b = width of rectangle and
       d = depth of rectangle

2.8.2 Method 2: Moment of Inertia by Elements

In this method, the section is broken into elements of typical shapes, such as rectangles and triangles, where the moments of inertia can be found in the LRFD manual or other engineering handbooks and manuals. In accordance with Equations 2.17 and 2.18 of the parallel axis theorem, each element has a moment of inertia about its own centroidal axis (neutral axis) plus the moment of inertia resulting from transferring the centroidal axis of the element to the centroidal axis of the full section. This method is illustrated by the following example.

EXAMPLE 2.5
Find the moment of inertia about the centroidal or neutral axis of the compound section shown in Figure 2.10a.

Solution
The compound section is divided in three rectangular elements, as shown in Figure 2.10b.
Step 1. Locate the neutral axis.

The neutral axis n-a is given by taking moment about the base of the section:

$$n - a = \frac{\sum M}{\sum A} = \frac{(3 \times 6 \times 13.5) + (8 \times 2 \times 8) + (4 \times 8 \times 2)}{(3 \times 6) + (8 \times 2) + (4 \times 8)} = 6.59 \text{ in.}$$

Step 2. Find the moment of inertia of the section.

Having found the neutral axis, the moment of inertia of the section can be calculated by applying the parallel axis theorem to each element. The elements and their centroidal distances are shown in Figure 2.10b.

Element I: $\quad I = \dfrac{6 \times 3^3}{12} + 6 \times 3 \times 6.91^2 = 873.0 \text{ in}^4$

Element II: $\quad I = \dfrac{2 \times 8^3}{12} + 2 \times 8 \times 1.41^2 = 117.1 \text{ in}^4$

Element III: $\quad I = \dfrac{8 \times 4^3}{12} + 8 \times 4 \times 4.59^2 = 716.8 \text{ in}^4$

Hence, total moment of inertia about the neutral axis of whole compound section is

$$I_0 = 873.0 + 117.1 + 716.8 = 1706.9 \text{ in}^4 \qquad \textit{Answer}$$

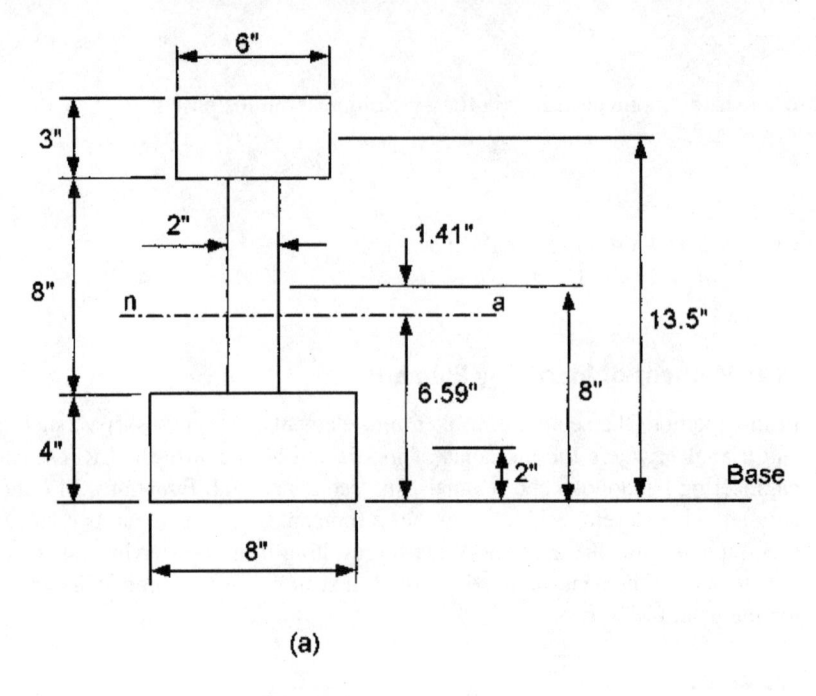

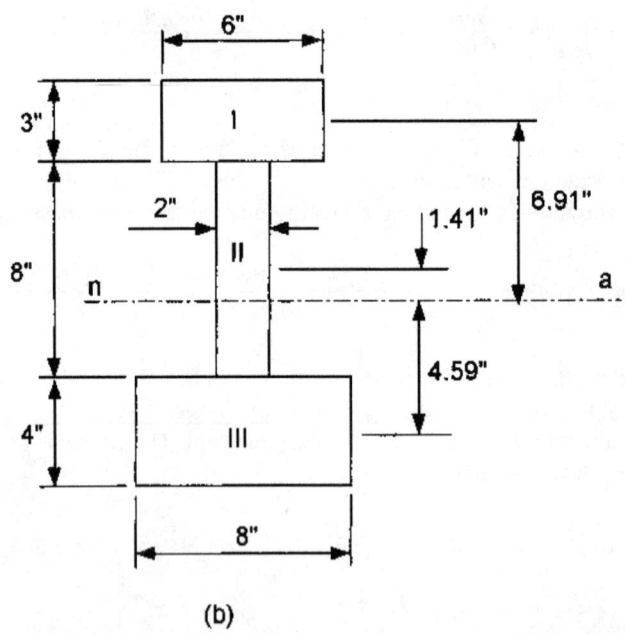

FIGURE 2.10

2.8.3 Method 3: Moment of Inertia by Areas

This method is used to compute moment of inertia without first calculating the neutral axis. It is a useful and efficient method in the preliminary design of beams or columns when the section is subject to change. Areas can be added to increase the moment of inertia of the section or subtracted when the section is oversized. The moment of inertia of the new areas or the subtracted areas can be computed and added to or subtracted from the previous values to compute the new moment of inertia. The method is straightforward, and its application will be illustrated by a couple of solved examples. First, the background of the method is given below.

Based on the parallel axis theorem, the moment of inertia of the whole section about the x axis or the base is given by

$$I_x = I_0 + Ay^2 \tag{2.20}$$

or

$$I_0 = I_x - Ay^2 \tag{2.21}$$

Since

$$y = \frac{\text{total moments about the x-axis or base}}{\text{total area}} = \frac{M}{A}$$

Therefore

$$y^2 = \frac{M^2}{A^2} \tag{2.22}$$

Substituting the value of y^2 in Equation 2.21

$$I_0 = I_x - \frac{M^2}{A} \tag{2.23}$$

where I_0 = moment of inertia of the whole section about the centroidal or neutral axis
I_x = sum of the moments of inertia of all the elements about a common reference axis, which may be the base or the top or somewhere in between the section
M = sum of the moments of all the elements about the same common reference axis
A = total area or sum of the areas of all the elements of the section

Each element has its own moment of inertia I_g about its centroidal or neutral axis. This moment of inertia I_g must be added to Equation 2.23.

$$I_0 = I_x + I_g - \frac{M^2}{A} \tag{2.24}$$

EXAMPLE 2.6

Using Method 3, find the moment of inertia and the neutral axis of the section of Example 2.5 (See Figure 2.11).

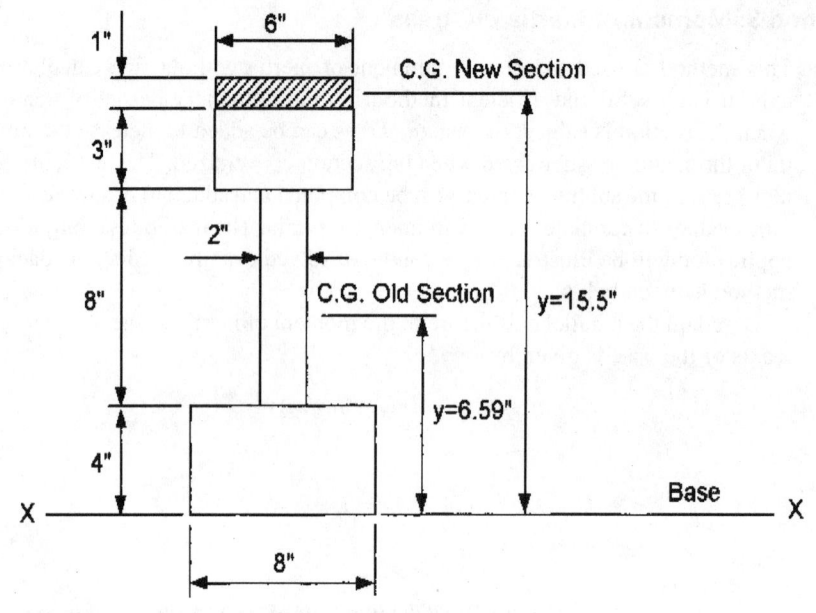

FIGURE 2.11

Solution
The base of the section is used as the common reference axis *x-x*. The computation is best carried out in a tabular form.
Step 1. Compute moment of inertia about the base.

Element	Size	Area A in.^{2}	y in.	$M = Ay$ in.^{3}	Ay^2 in.^{4}	I_g in.^{4}
III	8 in. × 4 in.	32	2	64.0	128.0	42.7
II	2 in. × 8 in.	16	8	128.0	1024	85.3
I	6 in. × 3 in.	18	13.5	243.0	3280.5	13.5
	Total =	66		435	4432.5	141.5

Step 2. Compute moment of inertia and find the neutral axis of the section.
From Equation 2.24, we have

$$I_0 = I_x + I_g - \frac{M^2}{A}$$

$$= 4432.5 + 141.5 - \frac{435^2}{66}$$

$$= 2980.0 \text{ in.}^4 \qquad \textit{Answer}$$

Neutral axis *n-a* is given by

$$y_0 = \frac{M}{A} = \frac{435}{66} = 6.59 \text{ in.} \qquad \text{Answer}$$

EXAMPLE 2.7
The designer wants to add a 1-inch-thick element to the top flange of Example 2.6. Find the moment of inertia and the neutral axis of the new section as shown in Figure 2.12.

Solution
The calculations from Example 2.6 can be brought forward. Only the section properties of the new element need to be computed.

Element	Size	Area A in.^{2}	y in.	$M = Ay$ in.^{3}	Ay^2 in.^{4}	I_g in.^{4}
Properties of Old Sec.		66	6.59	435.0	4432.5	141.5
New Element	1 in. × 6 in.	6	15.5	93.0	1441.5	0.5
Total =		72		528.0	5874.0	142.0

$$I_0 = 5874.0 + 142.0 - \frac{528.0^2}{72} = 2144.0 \text{ in.}^4 \qquad \text{Answer}$$

$$y_0 = \frac{528}{72} = 7.3 \text{ in.} \qquad \text{Answer}$$

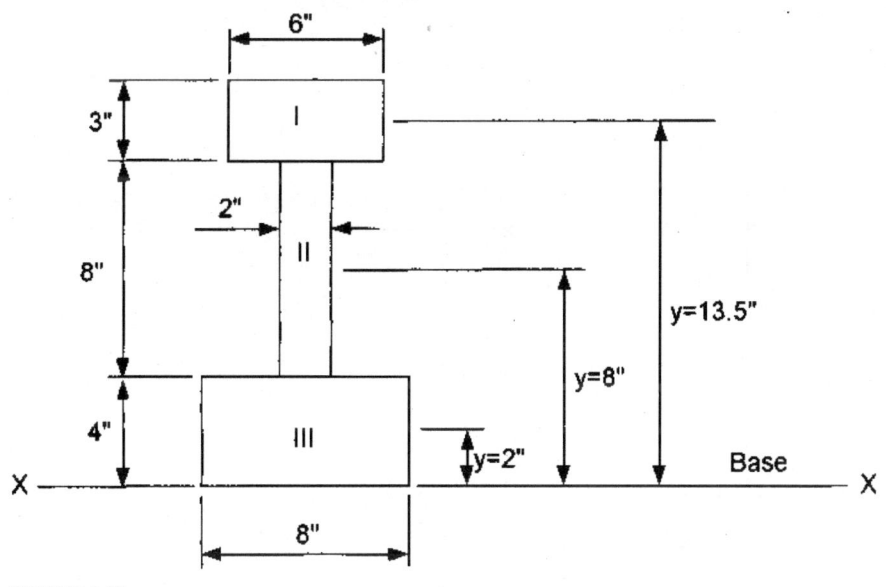

FIGURE 2.12

2.9 PRODUCT OF INERTIA

The product of inertia I_{xy} of an area with respect to a pair of rectangular axes is defined as the sum of the products obtained by multiplying each element of the area dA by its coordinates with respect to the x-y axes, as shown in Figure 2.5.

By definition

$$I_{xy} = \int xy \, dA \qquad (2.25)$$

The product of inertia can be positive or negative. For an axis of symmetry, the product of inertia about the axis of symmetry is zero.

EXAMPLE 2.8

Determine the product of inertia of a rectangle with respect to the x and y axes shown in Figure 2.13.

From Equation 2.25, we have

$$I_{xy} = \int\int xy \, dxdy = \int \left[\frac{x^2}{2}\right]_0^b y \, dy$$

$$= \frac{b^2}{2}\left[\frac{y^2}{2}\right]_0^h = \frac{b^2 d^2}{4} \qquad \textit{Answer}$$

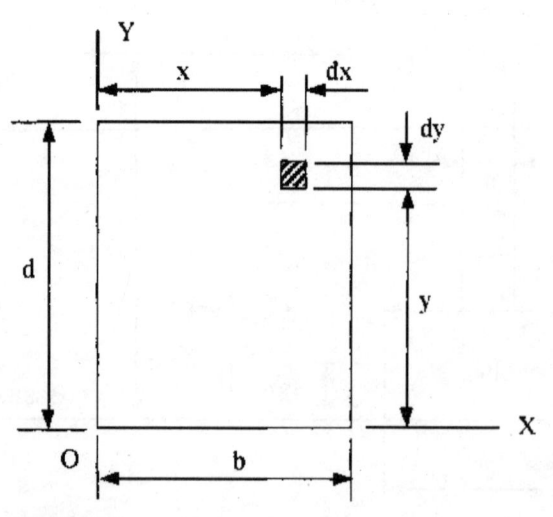

FIGURE 2.13

2.10 TRANSFER OF AXES FOR PRODUCT OF INERTIA

In Figure 2.6, x_0 and y_0 axes are the centroidal axes passing through the centroid G of the area A. The x and y axes are located at distances y_1 and x_1 from the x_0 and y_0 axes respectively. From Equation 2.25, we have

$$I_{xy} = \iint (x_0 + x_1)(y_0 + y_1) dx\, dy$$

$$= \iint x_0 y_0\, dxdy + \iint x_0 y_1\, dxdy + \iint x_1 y_0\, dxdy + \iint x_1 y_1\, dxdy$$

The first integral is the product of inertia about the centroidal axes and is equal to $I_{x_0 y_0}$. The second and third integrals are equal to zero because x_0 and y_0 are measured from the centroidal axes. The fourth integral equals $x_1 y_1 A$. Hence

$$I_{xy} = I_{x_0 y_0} + x_1 y_1 A \tag{2.26}$$

2.11 INCLINED AXES

In the design and analysis of columns, it is often necessary to determine the maximum and minimum moments of inertia of an area about inclined axes.

In Figure 2.14, the moments of inertia of the area about the x' and y' axes are

$$I_{x'} = \int y'^2\, dA = \int (y\cos\theta - x\sin\theta)^2\, dA$$

$$= \int (y^2 \cos^2\theta - 2xy \sin\theta \cos\theta + x^2 \sin^2\theta)\, dA$$

$$= \cos^2\theta \int y^2\, dA + \sin^2\theta \int x^2\, dA - 2\sin\theta \cos\theta \int xy\, dA$$

By definition, the first two integrals are the moments of inertia and the third integral is the product of inertia of the area about the x and y axes. Hence

$$I_{x'} = I_x \cos^2\theta + I_y \sin^2\theta - 2I_{xy} \sin\theta \cos\theta$$

$$\sin^2\theta = \frac{1 - \cos 2\theta}{2}, \quad \cos^2\theta = \frac{1 + \sin 2\theta}{2}, \quad 2\sin\theta \cos\theta = \sin 2\theta$$

Substituting and simplifying, we have

$$I_{x'} = \frac{I_x + I_y}{2} + \frac{I_x - I_y}{2}(\cos 2\theta) - I_{xy} \sin 2\theta \tag{2.27}$$

In a similar way, we have

$$I_{y'} = \frac{I_x + I_y}{2} - \frac{I_x - I_y}{2}(\cos 2\theta) + I_{xy} \sin 2\theta \tag{2.28}$$

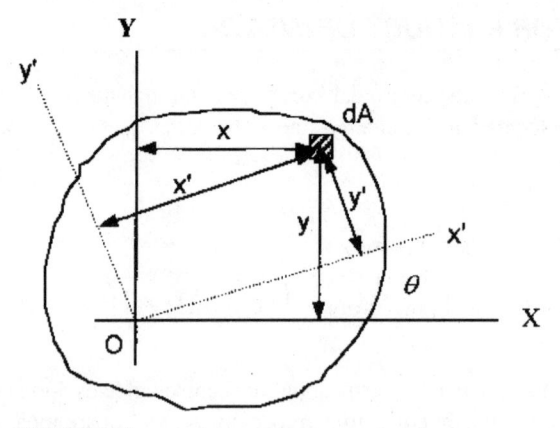

FIGURE 2.14

and

$$I_{x'y'} = \int x'y' \, dA$$

$$= \frac{I_x - I_y}{2} \sin 2\theta + I_{xy} \cos 2\theta \qquad (2.29)$$

Adding Equations 2.27 and 2.28, we have

$$I_{x'} + I_{y'} = I_x + I_y = J_z \qquad (2.30)$$

This is the polar moment of inertia about O, which is the same as Equation 2.14.

The value of θ that makes $I_{x'}$ a maximum or a minimum can be determined by setting the derivative of $I_{x'}$ with respect to θ to zero. Thus

$$\frac{dI_{x'}}{d\theta} = (I_y - I_x) \sin 2\theta - 2I_{xy} \cos 2\theta = 0$$

Solving the equation gives

$$\tan 2\theta = \frac{2I_{xy}}{I_y - I_x} = \frac{-2I_{xy}}{I_x - I_y} \qquad (2.31)$$

Equation 2.31 gives two values of 2θ differing by π, since $\tan 2\theta = \tan(2\theta + \pi)$. Hence the two solutions for θ differ by $\pi/2$ or $90°$. One solution defines the axis of maximum moment of inertia and the other value defines the axis of minimum moment of inertia. These two rectangular axes are known as the principal axes of inertia, and the maximum and minimum moment of inertia are termed the principal moments of inertia, which can be expressed as

$$I_{max,min} = \frac{1}{2}(I_x + I_y) \pm \sqrt{\frac{1}{4}(I_x - I_y)^2 + I_{xy}^2} \qquad (2.32)$$

EXAMPLE 2.9

A structural steel $\angle\, 5 \times 3 \times \frac{1}{2}$ section has the nominal dimensions shown in Figure 2.15. The values of I_x and I_y about the centroidal x and y axes are given in the LRFD manual. Find I_{xy} and determine the maximum and minimum values of the moments of inertia of the angle.

Solution

From the LRFD manual, the centroid G is located as shown in the Figure 2.15, and the area and moments of inertia of the angle are

$$A = 3.75 \text{ in}^2$$
$$I_x = 9.45 \text{ in}^4$$
$$I_y = 2.58 \text{ in}^4$$

By symmetry, the product of inertia for each rectangle about its own centroidal axes is zero. Hence, from Equation 2.26

$$I_{xy} = I_{x_0 y_0} + x_1 y_1 A$$

we have

For Rectangle 1
$$I_{xy} = 0 + (1.5 - 0.75)(-1.25 - 0.25)(3 \times 0.5)$$
$$= (+0.75)(-1.5)(1.5) = -1.688 \text{ in}^4$$

For Rectangle 2
$$I_{xy} = 0 + (-0.25 - 0.25)(2.5 - 1.75)(4.5 \times 0.5)$$
$$= (-0.5)(+0.75)(2.25) = -0.844 \text{ in}^4$$

For the complete angle $\quad I_{xy} = -1.688 - 0.844 = -2.532 \text{ in}^4 \qquad$ *Answer*

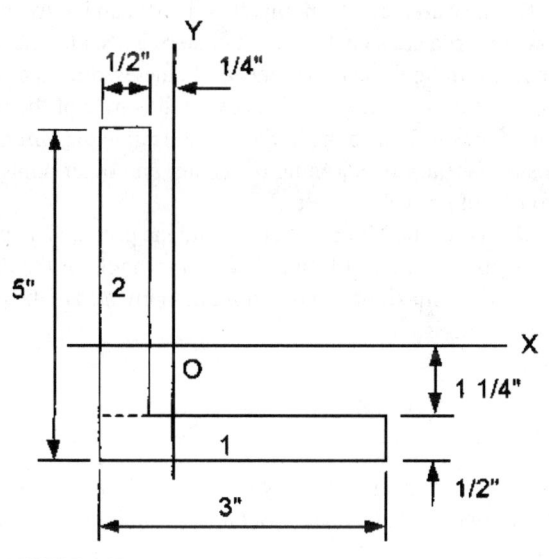

FIGURE 2.15

The angle of inclination of the principal axes of inertia is given by Equation 2.31. Hence

$$\tan \theta = \frac{2(-2.532)}{2.58 - 9.45} = 0.737$$

$$2\theta = 36°23'', \theta = 18°12''$$

$$\cos 2\theta = \cos 36°23'' = 0.805$$

$$\sin 2\theta = \sin 36°23'' = 0.593$$

From Equations 2.27 and 2.28

$$I_{max} = \frac{9.45 + 2.58}{2} + \frac{9.45 - 2.58}{2}(0.805) - (-2.532)(0.593) = 10.281 \text{ in}^4$$

$$I_{max} = \frac{9.45 + 2.58}{2} - \frac{9.45 - 2.58}{2}(0.805) + (-2.532)(0.593) = 1.749 \text{ in}^4$$

I_{max} and I_{min} may also be computed from Equation 2.32 as follows:

$$I_{max} = \tfrac{1}{2}(9.45 + 2.58) + \sqrt{\tfrac{1}{4}(9.45 - 2.58)^2 + (-2.532)^2} = 10.281 \text{ in}^4$$

$$I_{min} = \tfrac{1}{2}(9.45 + 2.58) - \sqrt{\tfrac{1}{4}(9.45 - 2.58)^2 + (-2.532)^2} = 1.749 \text{ in}^4$$

2.12 THE MOHR'S CIRCLE

The Mohr's circle might be used graphically to find the principal axes of inertia and determine the principal moments of inertia, i.e., the values of maximum and minimum moments.

A horizontal axis for the measurement of moments of inertia and a vertical axis for the measurement of products of inertia are selected first. On these axes, plot the points (I_x, P_{xy}) and ($I_{y'} - P_{xy}$). With the line joining these two points as diameter, draw a circle. The angle of the diameter to the horizontal axis is 2θ, which is twice the angle of the principal axes. The points of intersection of the circle with the horizontal axis give the minimum and maximum principal moments of inertia. Lines joining the center and other points on the circle define moments of inertia on other inclined axes.

Figure 2.16 illustrates the use of the Mohr's circle to find the principal axes and the maximum and minimum moments of inertia of Example 2.9. The angle of the diameter AB to the horizontal x axis is 2θ, and the maximum and minimum moments are the points of intersection of the circle with the x axis.

2.13 RADIUS OF GYRATION

The radius of gyration r of an area is an imaginary distance from an axis where the entire area of the section could be concentrated and still obtain the same moment of inertia about the same axis.

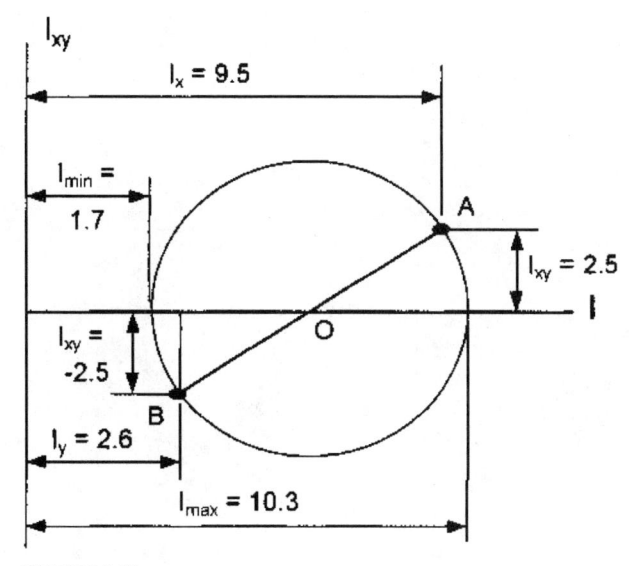

FIGURE 2.16

Hence, if r_x and r_y are the radii of gyration from the x and y axes, then

$$I_x = r_x^2 A \quad \text{and} \quad I_y = r_y^2 A$$

or

$$r_x = \sqrt{\frac{I_x}{A}} \quad \text{and} \quad r_y = \sqrt{\frac{I_y}{A}} \qquad (2.33)$$

This property is used in solving buckling problems in beams and columns.

CHAPTER 3
STRENGTH OF MATERIALS

3.1 INTRODUCTION

In structural design, the engineer must consider both dimensions and material properties to satisfy requirements of strength and rigidity. When loaded, a structure or member must carry loads safely (not break) and also meet the performance requirement (not deform excessively). This chapter provides the engineer with fundamentals about relationships between the loads applied to a nonrigid body and the resulting internal forces and deformations induced in the body.

3.2 TENSION AND COMPRESSION

A member is in *tension* when a force causes it to stretch, or increase, and is in *compression* when a force causes it to shorten. When a prismatic bar is loaded by a pair of axial force P either in tension or compression, as shown in Figure 3.1, the internal tensile stress or compression stress on a plane lying normal to its axis can be expressed by the following equation:

$$\sigma = \frac{P}{A} \tag{3.1}$$

The axial deformation of the bar in elongation or shortening can be expressed as

$$\delta = \frac{PL}{EA} \tag{3.2}$$

where δ = Total axial deformation
        P = Axial load acting through the centroid of the cross section

CHAPTER 3

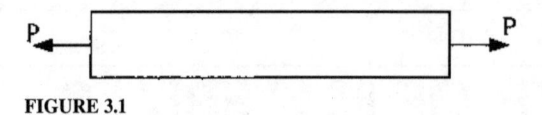

FIGURE 3.1

L = Total length of the bar
E = Modulus of elasticity of material
A = Cross-sectional area of the bar.

Equation 3.2 is subject to the following restrictions:

1. The load must be axial.
2. The bar must have a constant cross section.
3. The material must be homogeneous.
4. The stress must not exceed the proportional limit.

When the axial force or the cross-sectional area of material varies along the axis of the bar, Equation 3.2 no longer is suitable. However, it can be used to find the elongation (or shortening) of each part of bar for which the axial force and the cross-sectional area are constant and the material is homogeneous. Then, the total elongation (or shortening) can be obtained from the equation

$$\delta = \sum_{i=1}^{n} \frac{P_i L_i}{E_i A_i} \tag{3.3}$$

where i = Index identifying the various parts of the bar
 n = Total number of parts.

EXAMPLE 3.1
Compute the total elongation of a composite bar, shown in Figure 3.2.

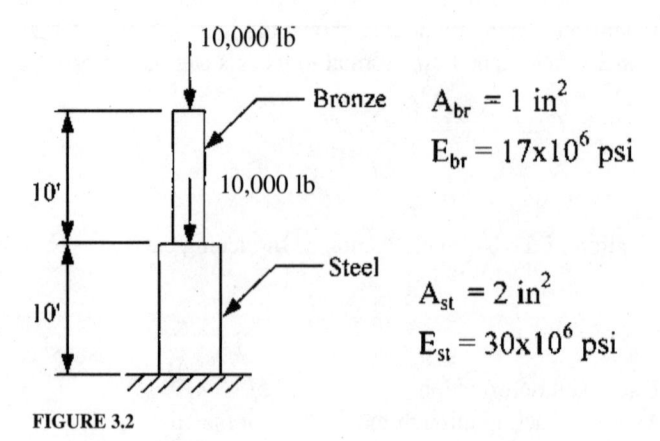

10,000 lb
Bronze $A_{br} = 1 \text{ in}^2$
 $E_{br} = 17 \times 10^6 \text{ psi}$
10,000 lb
10'
Steel
10' $A_{st} = 2 \text{ in}^2$
 $E_{st} = 30 \times 10^6 \text{ psi}$

FIGURE 3.2

Solution

The total axial shortening of the bar is equal to the sum of the shortening of the bronze and the steel bar. Using Equation 3.3, we can write the following equation:

$$\delta_{total} = \left(\frac{PL}{EA}\right)_{br} + \left(\frac{PL}{EA}\right)_{st}$$

$$= \frac{(-10{,}000)[12(10)]}{(17 \times 10^6)(1.0)} + \frac{(-20{,}000)[12(10)]}{(30 \times 10^6)(2.0)} = -0.11 \text{ in.} \qquad Answer$$

3.3 DETERMINATE FORCE SYSTEM

A structure that is initially at rest and remains at rest when acted on by a system of force is said to be in a state of *static equilibrium*. For a planar structure or member to remain in static equilibrium, the following three conditions must be fulfilled simultaneously by the applied loads and reactions or by the internal resisting forces over a cross section:

$$\Sigma F_x = 0 \qquad \Sigma F_y = 0 \qquad \Sigma M = 0 \qquad (3.4)$$

These are the equations of static equilibrium of a planar structure subjected to a general system of forces.

If the supports are replaced by the reactions they supply to the structure, the structure will be acted on by a general system of forces consisting of the known applied loads and the unknown reactions. The equations of static equilibrium can be written in terms of these known applied loads and the unknown reactions. They also can be written in terms of the unknown internal resistance forces and known applied load. Under certain conditions, the simultaneous solution of the equations of static equilibrium is sufficient to determine the magnitude of the unknown reactions or unknown internal resistance forces. In such a case, the structure is said to be *statically determinate*.

3.4 INDETERMINATE FORCE SYSTEM

There are other instances in which the equations of static equilibrium are not sufficient to determine all the internal resistance forces and reactions. Such cases are called *statically indeterminate* and require the use of the additional relations that depend on displacement of the structure or deformations in the members. The degree of indeterminacy for a structure is the number of unknowns above the number of equations of static equilibrium available for solution. The following outline of steps is the analysis procedure involving indeterminate structures.

Step 1. Draw a free-body diagram.

Step 2. Note the number of unknowns involved (magnitudes and positions).

Step 3. Recognize the type of force on the free-body diagram, and note the number of independent equations of equilibrium available.

Step 4. Write a deformation equation for each unknown that exceeds the number of equilibrium equations.

Step 5. When the number of independent equilibrium equations and deformation equations equal the number of unknowns, the equations can be solved simultaneously.

EXAMPLE 3.2

The composite bar shown in Figure 3.3a is attached at both ends to rigid supports and loaded axially by the force P of 10,000 lbs at an intermediate point C. Determine the reactions at supports A and B.

$$A_{br} = 1 \text{ in}^2 \quad A_{st} = 2 \text{ in}^2 \quad E_{br} = 17 \times 10^6 \text{ psi} \quad E_{st} = 30 \times 10^6 \text{ psi}$$
$$a = 10 \text{ ft} \quad b = 10 \text{ ft}$$

Solution

The free-body diagram in Figure 3.3b shows two unknown reactions. From equilibrium, we have only one equation available: $R_a + R_b = P$ which contains both unknown reactions. No other equations of equilibrium are available to indicate in what portion the load is distributed to each support. These reactions cannot be found by static alone. The structure is statically indeterminate. A second relation between these forces must be obtained from the elastic deformation of the bar.

As the load is applied at point C, it moves downward an amount δ_c. The forces R_a and R_b in the top and bottom part can be expressed in terms of δ_c, as follows:

$$R_a = \frac{(E_{br})(A_{br})}{a}\delta_c = \frac{(17 \times 10^6)(1.0)}{(12)(10)}\delta_c$$

$$R_b = \frac{(E_{st})(A_{st})}{b}\delta_c = \frac{(30 \times 10^6)(2.0)}{(12)(10)}\delta_c$$

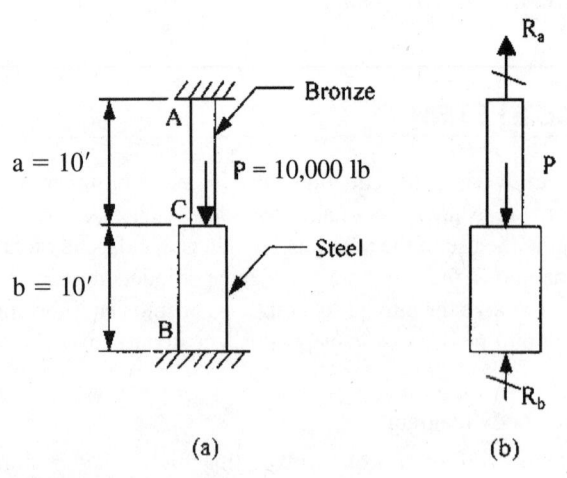

FIGURE 3.3

From the equilibrium equation, we obtain

$$\frac{(17 \times 10^6)(1.0)}{(12)(10)} \delta_c + \frac{(30 \times 10^6)(2.0)}{(12)(10)} \delta_c = 10,000$$

which yields $\delta_c = 0.0156$ in.

Knowing δ_c, we can now find the reactions R_a and R_b

$$R_a = \frac{(17 \times 10^6)(1.0)}{(12)(10)} \delta_c = \frac{(17 \times 10^6)(1.0)}{(12)(10)} (0.0156) = 2208 \text{ lbs}$$

and

$$R_b = \frac{(30 \times 10^6)(2.0)}{(12)(10)} \delta_c = \frac{(30 \times 10^6)(2.0)}{(12)(10)} (0.0156) = 7792 \text{ lbs} \qquad Answer$$

3.5 ELASTIC ANALYSIS

Equation 3.2 is formulated to determine the total elongation (or shortening) δ in a homogeneous prismatic member subjected to axial load P. It is based on simplified assumptions related to the stress-strain diagrams shown in Figure 3.4. For sufficiently small strains, these diagrams indicate that the stresses and strains are related linearly. If the external loads producing strain do not exceed a certain limit, the strain disappears with the removal of the load. The linear relation between stress and strain are known generally as Hooke's law (Section 1.7). When a material behaves elastically and exhibits a linear relationship between stress and strain, it is said to be linearly elastic, and the principle of superposition is applied. This is an extremely important property of many engineering materials, including structural steel, plastic, timber, and concrete. An analysis made with these assumptions is called an elastic analysis of the structure.

3.6 PLASTIC ANALYSIS

As noted in Figure 3.4, if the load is sufficiently small, the relation between stress and strain is essentially linearly elastic; that is, the stress-strain curve is a straight line, and loading and unloading proceed along this line. As the load is increased enough, the stress-strain curve becomes nonlinear. Depending on the material, the loading-unloading process can be elastic (reversible) or plastic (nonreversible). If the path of unloading coincides with the path of loading, we say the material is nonlinearly elastic. If the unloading path does not follow the loading path, we say the behavior is nonelastic or plastic. Since an actual stress-strain curve, shown in Figure 3.4, is difficult to use in mathematical solutions of complex problems, ideal stress-strain curves usually are used in analysis. For a few materials, notably structural steel,

FIGURE 3.4

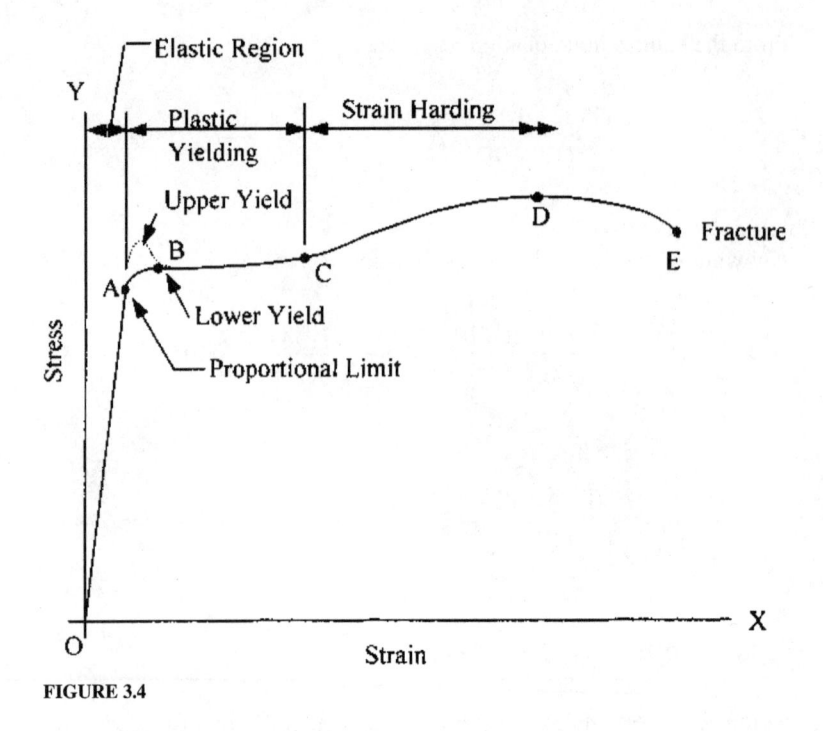

the stress-strain can be idealized, shown in Figure 3.5. The stress-strain curve is idealized by two straight lines—a linear elastic region up to the yield point followed by a region of considerable yielding under constant stress. An analysis made with these assumptions is called a plastic analysis of the structure. The concept of plastic design was developed in the early 20th century. The first published papers on the possibility of the application of plastic analysis to the structural design were by Kazinczy in Hungary in 1914 and by Kist in Holland in 1917. The basic physical property used in the plastic methods is ductility, a property associated with behavior outside the elastic range.

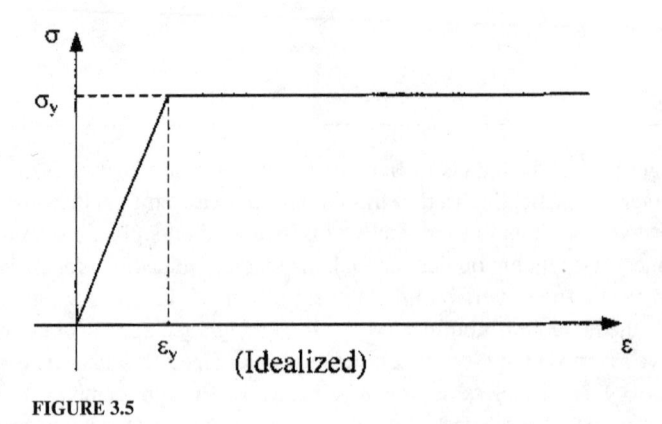

FIGURE 3.5

3.7 STRAIN ENERGY IN TENSION AND COMPRESSION MEMBERS

Consider a force gradually applied to a bar—the load increases from zero to its applied value as the bar undergoes deformation δ. As long as the material follows Hooke's law, a linear relationship exists between the load and the deflection, as shown in Figure 3.6.

During loading of the bar, force P works on the bar and is stored as strain energy. The total work done by the applied load during this period is given by

$$U = W = \int_0^\delta F\,ds = \int_0^\delta \left(\frac{Ps}{\delta}\right)ds = \frac{1}{2}P\delta = \frac{P^2 L}{2EA} \tag{3.5}$$

If the bar varies in cross section or material, this result may be applied to segments of length dx and integrated over the length of the bar to obtain

$$U = \int_0^L \frac{P_x^2\,dx}{2E_x A_x} \tag{3.6}$$

3.8 SHEAR STRESS

This type of stress differs from tensile and compressive stress in that it is caused by forces acting along, or parallel to, the area resisting the forces rather than perpendicular to it, as in the case of tensile and compressive stresses. A shearing stress is produced whenever the applied loads cause one section of a body to tend to slide past its adjacent section. An example is shown in a bolted lap connection in Figure 3.7: The bolt resists shear across its cross-sectional area. This type of shear stress can be called *direct shear*. The exact distribution

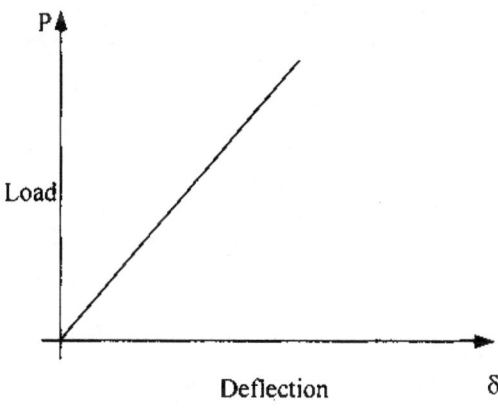

FIGURE 3.6

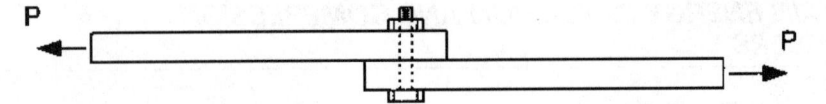

FIGURE 3.7

of shear stress is not easily determined, but we can obtain the average shear stress by dividing the total shear force V by the area A over which it acts:

$$\tau_{aver} = \frac{V}{A} \tag{3.7}$$

In contrast to the direct shear, induced shear stress that can be produced by other means, such as by loading beams, twisting shafts, bending, direct forces, and so forth, will be discussed in detail in subsequent sections.

3.9 SHEAR STRAIN

Shearing forces cause a shearing deformation, just as axial forces cause elongation, but with an important difference. An element subject to tension undergoes an increase in length; an element subject to shear does not change the length of its sides—instead it distorts its shape from a rectangle into a parallelogram as shown by the dashed lines in Figure 3.8.

The shearing strain is the angular change between two perpendicular faces of a differential element subjected only to pure shear stress. In Figure 3.8, this defines $\tan\gamma = \delta_s/L$, which is equal to the total shearing deformation δ_s where the upper edge of the element slides horizontally with respect to the lower side divided by the height of the element. However, since the angle γ usually is very small, $\tan\gamma \approx \gamma$ and we obtain

$$\gamma = \frac{\delta_s}{D} \tag{3.8}$$

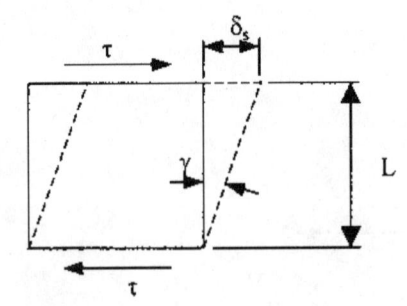

FIGURE 3.8

3.10 SHEAR MODULUS

If the material has a linear elastic region, then the relation between shearing stress and shearing strain is

$$\tau = G\gamma \tag{3.9}$$

The constant G is called the shear modulus of elasticity, or the modulus of rigidity. If Poisson's ratio μ and the modulus of elasticity E are known, the shear modulus G is expressed by

$$G = \frac{E}{2(1 + \mu)} \tag{3.10}$$

This equation shows that the relationships of E, G, and μ are not independent properties of the material.

3.11 SHEAR DEFORMATION AND STRAIN ENERGY IN PURE SHEAR

Assuming that the material follows Hooke's law, the load-deflection diagram for shear (V versus δ_s) is analogous to the diagram shown in Figure 3.6 for a bar in tension. The work done by the force V and stored in the form of elastic strain energy is

$$U = \frac{V\delta_s}{2} \tag{3.11}$$

Combining Equation 3.7, $\tau = V/A$, Equation 3.8, $\gamma = \delta_s/L$, and Equation 3.9, $\tau = G\gamma$, we obtain the following expression for shear deformation:

$$\delta_s = \frac{VL}{GA} \tag{3.12}$$

Substituting this result for δ_s into Equation 3.11, we obtain the following equation for the strain energy:

$$U = \frac{V^2 L}{2GA} \tag{3.13}$$

3.12 TORSION

Torsional stresses occur in members, such as power transmission shafts, coupling, and other machine parts that transmit a torque (a couple) from one plane to a parallel plane.

3.10 CHAPTER 3

The most efficient shape for carrying a torque is a hollow circular shaft. The stress formulas for torsion are based on the following assumptions:

1. A plane section before twisting remains plane after twisting.
2. The diameter of the circular cross section of the member remains a straight line.
3. The member is loaded by twisting couples acting in planes that are perpendicular to the axis of the member.
4. The torsional stresses remain within the elastic range.

The first two assumptions apply only to shafts of circular section.

3.13 TORSIONAL SHEARING STRESS AND STRAIN

Consider a torsional moment T acting on a solid circular shaft of homogeneous material and uniform cross section, as shown in Figure 3.9a. During torsion, there will be rotation about the longitudinal axis of one end of the bar with respect to the other. At the same time, longitudinal lines on the outside face of the bar, such as fiber AB and CD, will rotate through a small angle to the position AB' and CD'. If we develop the curve surface as shown in Figure 3.9b, a rectangular element on the surface of the bar $ABCD$ is distorted into a rhomboid $AB'CD'$ because of this rotation. The angle γ represents the magnitude of the shearing strain. Using Figure 3.9b, we can write the following relations:

$$\gamma \approx \tan\gamma = \frac{BB'}{L} \tag{3.14}$$

As shown in Figure 3.9c, the distance BB' is the length of a small arc of radius r subtended by the angle ϕ, which is the angle of rotation about the longitudinal axis of one end of bar with respect to the other. The angle ϕ, is called the angle of twist. Thus, we find that $BB' = r\phi$. From Equation 3.14, we have

$$\gamma = \frac{r\phi}{L} \tag{3.15}$$

Now, consider any internal fiber located at radius ρ from the axis of the shaft. From assumption 2 in Section 3.12, the radius of such a fiber also rotates through the angle ϕ. The length of this deformation is the arc of radius ρ subtended by the angle ϕ. Thus, the shearing strain at any radial distance ρ from center O is expressed by

$$\gamma_\rho = \frac{\rho\phi}{L} \tag{3.16}$$

From Equations 3.15 and 3.16, we obtain

$$\frac{\gamma L}{r} = \frac{\gamma_\rho L}{\rho} \quad \text{and} \quad \gamma_\rho = \frac{r}{\rho}\gamma \tag{3.17}$$

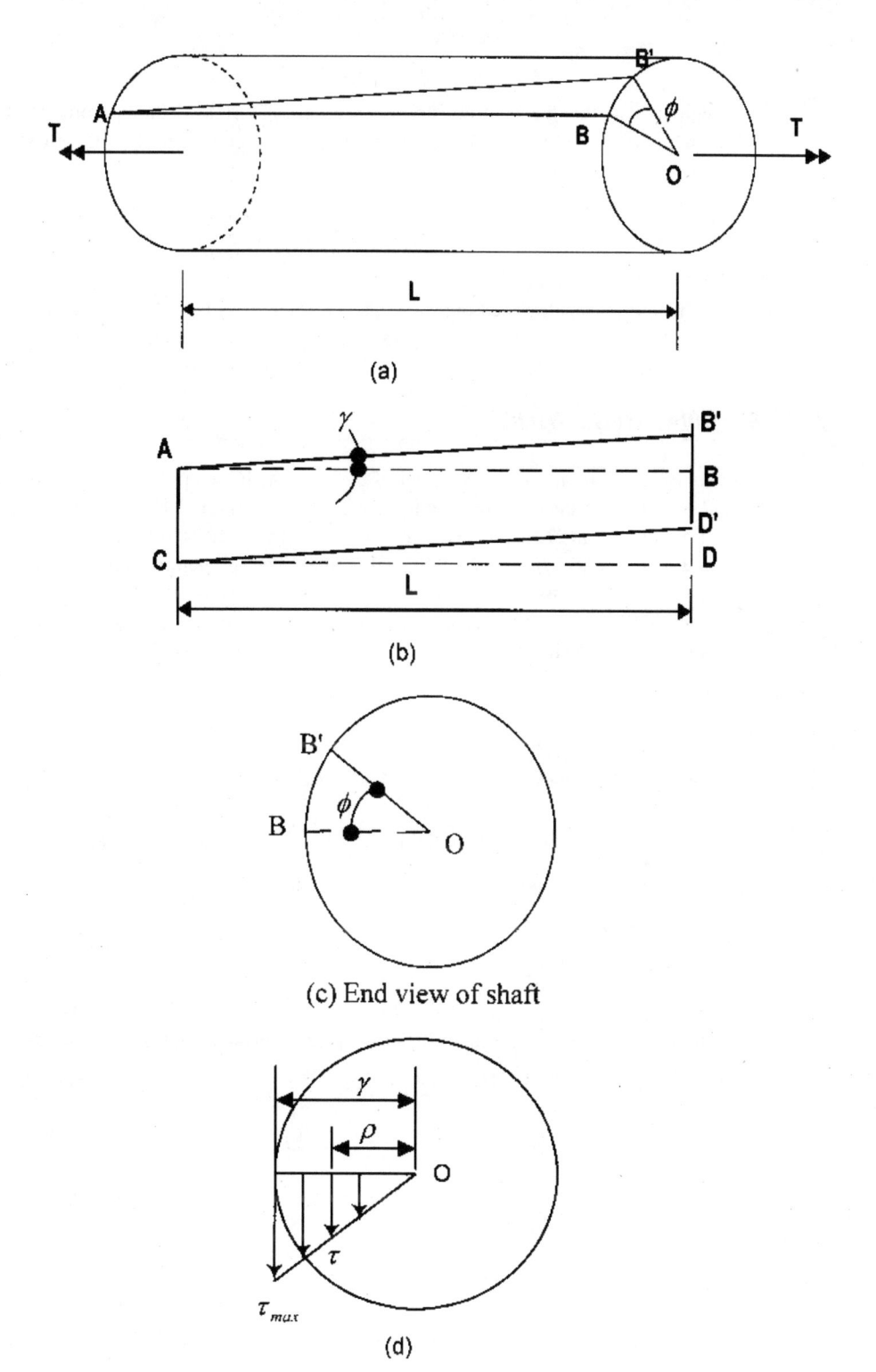

FIGURE 3.9

3.11

which indicate that the shearing strain is proportional to the distance from the axis of shaft. From Equation 3.9 for a linear elastic material, the magnitude of the shearing stress at a distance ρ is

$$\tau = G\gamma_\rho = G\frac{\rho\phi}{L} = \left(\frac{G\phi}{L}\right)\rho \qquad (3.18)$$

which indicates that the shearing stress also varies linearly with the radial distance ρ from the axis of the shaft and has the maximum value at the outer surface, as shown in Figure 3.9d.

3.14 TORSIONAL RESISTANCE

A free-body diagram of the segment of the shaft with the applied torque T on the left end and resisting torque T_r on the right end is shown in Figure 3.10.

Resisting torque T_r is the resultant of the differential shear forces, τdA, acting on element of area dA at a distance ρ from the shaft center. The moment of this force about the shaft axis is $\tau\rho dA$. To satisfy the condition of static equilibrium, the resisting torque T_r equals the applied torque T. The resisting torque T_r is the sum of the resisting torque developed by all differential shear forces τdA:

$$T = T_r = \int_{area} \rho\tau dA \qquad (3.19)$$

Substituting Equation 3.18 into Equation 3.19, we obtain

$$T = \int_{area} \rho\left(\frac{G\phi}{L}\right)\rho dA = \left(\frac{G\phi}{L}\right)\int_{area} \rho^2 dA = \left(\frac{G\phi}{L}\right)J \qquad (3.20)$$

where

$$J = \int_{area} \rho^2 dA \qquad (3.21)$$

is the polar moment of inertia of the cross section. From Equation 3.20, we obtain the total angle of twist

$$\phi = \frac{TL}{GJ} \qquad (3.22)$$

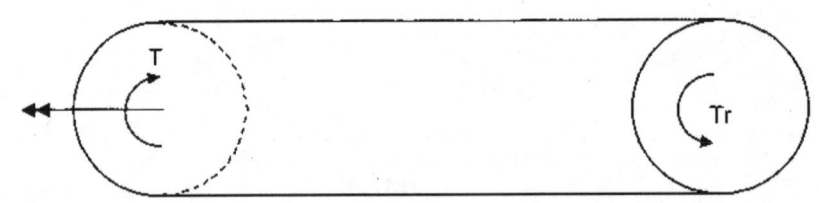

FIGURE 3.10

The product *GJ* is known as the torsional rigidity of the shaft.

By replacing ϕ in Equation 3.18 with TL/GJ, the magnitude of the shearing stress at a distance ρ is

$$\tau = \frac{T\rho}{J} \tag{3.23}$$

This equation is known as the elastic torsion formula. The maximum shear stress at the outer surface of the shaft, distance r from center, is

$$\tau_{max} = \frac{Tr}{J} \tag{3.24}$$

3.15 SHEARING FORCE AND BENDING MOMENT

A beam is a structural member subjected to loads acting transversely to its longitudinal axis, causing the member to bend. Common beams encountered in practice include girders; floor beams; stringers in bridges, joists, purlins, and rafters; and lintels in buildings.

Figure 3.11a shows a simple beam that carries a concentrated load *P*. Assume that a cutting plane *A-A* at a distance *x* from left divides the beam into two segments. The left segment in Figure 3.11b shows that the only externally applied load is R_1. By definition of Newton's Second Law, the segment can be prevented from experiencing a vertical acceleration and a clockwise angular acceleration by having a transverse force V_r, and a couple M_r presented

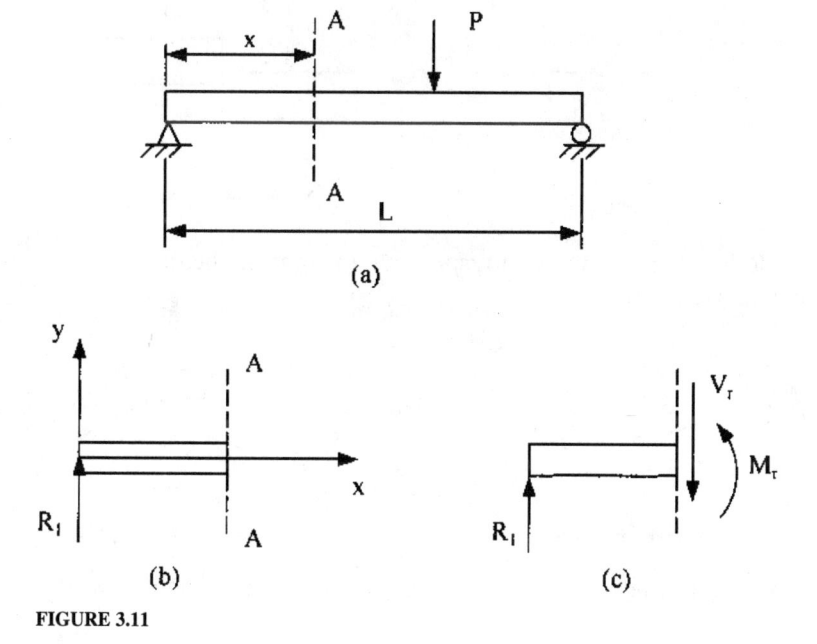

FIGURE 3.11

at section *A-A*, as shown in the free-body diagram of Figure 3.11c. The force V_r is the resultant of the shearing stresses at section *A-A* and is called the *resisting shearing force*. The couple M_r is the resultant of the normal stresses at the section *A-A* and is called the *resisting moment*.

The magnitudes and senses of V_r and M_r can be determined from the equations of equilibrium $\Sigma F_y = 0$ and $\Sigma M_o = 0$ where o is any axis perpendicular to xy plane at selected section *A-A*. This may be summarized mathematically as

$$V_r = V = (\Sigma F_y)_L \tag{3.25}$$

and

$$M_r = M = (\Sigma M)_L = (\Sigma M)_R \tag{3.26}$$

in which V is shearing force and M is bending moment. The subscript L indicates that the summation includes only the external loads acting on the beam segment to the left of the section considered, and the subscript R refers to loads to the right of the section. The resisting shear V_r is always equal, but oppositely directed, to the shearing force V. A sign convention is necessary for the correct interpretation of results obtained from equations or diagrams for shear and moment. The following convention will give consistent results regardless of whether one proceeds left to right or right to left. As shown in Figure 3.12, a positive shear force at a section tends to move the left segment upward with respect to the right segment, and vice versa. The sign convention for moment is shown in Figure 3.13. The bending moment in a horizontal beam is positive at sections for which the top of the beam is in compression and the bottom is in tension.

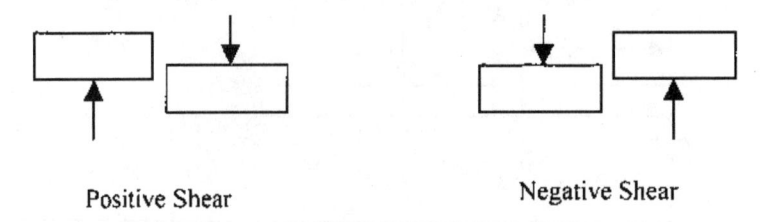

Positive Shear   Negative Shear

(Relative movements corresponding to signs of shearing)
FIGURE 3.12

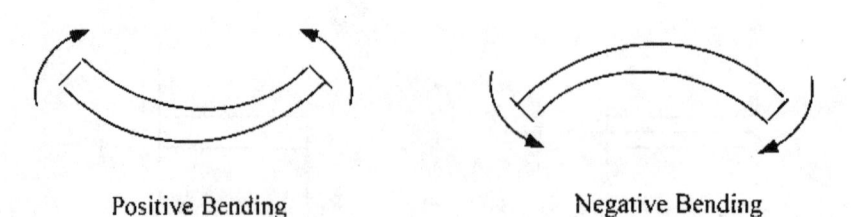

Positive Bending   Negative Bending

(Curvatures corresponding to signs of bending moment)
FIGURE 3.13

3.16 LOAD, SHEAR, AND MOMENT RELATIONSHIPS

In this section, we discuss the relationships between loads, shears, and bending moments in a beam. These relationships will be used to facilitate construction of shear and moment diagrams, as demonstrated in Section 3.17. Consider a segment of a beam of length dx, as shown in Figure 3.14, in which the upward direction is positive for the applied load w, and the shears and moment are shown as positive, according to the sign convention established in Section 3.15.

Applying the equilibrium equation, $\Sigma F_y = 0$, gives

$$V + wdx - (V + dV) = 0$$

from which

$$dV = wdx \quad \text{or} \quad w = \frac{dV}{dx} \tag{3.27}$$

From a moment summation about point O, the equation $\Sigma M_o = 0$, gives

$$M + Vdx + wdx\left(\frac{dx}{2}\right) - (M + dM) = 0$$

from which

$$dM = Vdx + \frac{w(dx)^2}{2}$$

The second term of right side in this equation is the square of a differential that is negligible in comparison with the other terms. This equation reduces

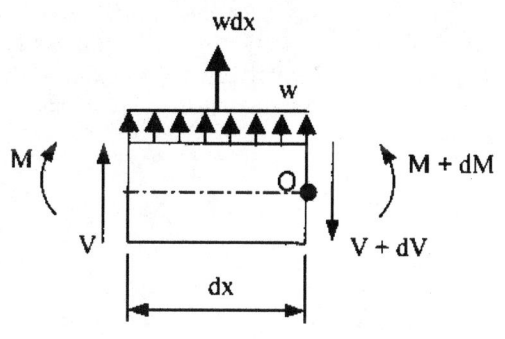

$$dM = Vdx \quad \text{or} \quad V = \frac{dM}{dx} \tag{3.28}$$

FIGURE 3.14

3.17 SHEAR AND BENDING MOMENT DIAGRAM

The shear force V and bending moment M in a beam usually will vary with the distance x, defining the location of the cross section at which they occur. When designing a beam, it is desirable to know the values of V and M at all cross sections of beam. Shear and moment diagrams are the graphical visualization of the shear and moment plotted on V-x and M-x axes, showing how they vary along the axis of the beam. A convenient arrangement for constructing shear and moment diagrams is to draw a free-body diagram of the entire beam and construct the shear and moment diagram directly below.

Calculating values of shear and moment at various sections along the beam, and plotting enough points to obtain a smooth curve, we can draw shear and moment diagrams. This procedure, however, is time-consuming, and two other methods may prove to be quicker.

The first method consists of writing algebraic equations for the shear V and moment M and constructing curves from the equations. This disadvantage of this is that unless the load is uniformly distributed or varies according to a known equation along the entire beam, no single expression can be written for the shear V or the moment M, which applies to the entire length of the beam. Instead, it is necessary to divide the beam into intervals bounded by the abrupt changes in the loading. An origin should be selected (different origins may be used for different intervals); positive directions should be shown for the coordinate axes, and the limits of the abscissa (usually x) should be indicated for each interval. Complete shear and moment diagrams should indicate values of shear and moment at each section where the load changes abruptly and at sections where they are maximum or minimum. Sections where the shear and moment are zero also should be located.

The second method is to draw the shear diagram from the load diagram and the moment diagram from the shear diagram by means of the mathematical relationship developed in Section 3.16. This method may not produce a precise curve, but is less time-consuming than the first method, and it does provide the information usually required.

As stated in Equation 3.27, $dV/dx = w$, the slope of the shear diagram is equal to the intensity of loading at any section in the beam. Integrating Equation 3.27, we obtain

$$\int_{V_1}^{V_2} dV = \int_{x_1}^{x_2} w\, dx = V_2 - V_1 \tag{3.29}$$

in which V_1 is the shear at position x_1 and V_2 is the shear at x_2. Equation 3.29 implies that the change in shearing force $V_2 - V_1$ between section x_1 and x_2 is equal to the area under the load diagram between the two sections. Similarly, Equation 3.28, $dM/dx = V$, indicates that the slope of the moment diagram is equal to the shear at any section in the beam. The integration of Equation 3.28 gives

$$\int_{M_1}^{M_2} dM = \int_{V_1}^{V_2} V\, dx = M_2 - M_1 \tag{3.30}$$

Equation 3.30 shows that the change in moment between any two sections equals to the area under the shear diagram for this interval.

Equations 3.29 and 3.30 provide useful means of computing the changes in shear and moment and the numerical values of shear and moment at various sections along a beam, while Equations 3.27 and 3.28 enable us to sketch the shapes of the shear and moment diagrams.

The equations in Section 3.16 were derived with the x positive to the right, the applied load positive upward, and the shear and moment signs as shown in Figures 3.12 and 3.13. Positive slopes are directed up to the right, and negative slopes are directed down to the right. A positive distributed load will result in a positive slope on the shear diagram, and a negative load will give a negative slope. The change of shear at a concentrated force is equal to the concentrated force. Similarly, a positive shear represents a positive slope, and a negative shear represent a negative slope. A couple applied to a beam will cause the moment to change abruptly by an amount of the couple. Figure 3.15 shows possible slopes of curves. Equations for area under the curves most commonly encountered are shown in Figure 3.16.

The following procedure can be used for the construction of shear and moment diagrams:

Step 1. Draw the free-body diagram.
Step 2. Compute the reactions.
Step 3. Compute values of shear at the change of load points, using either

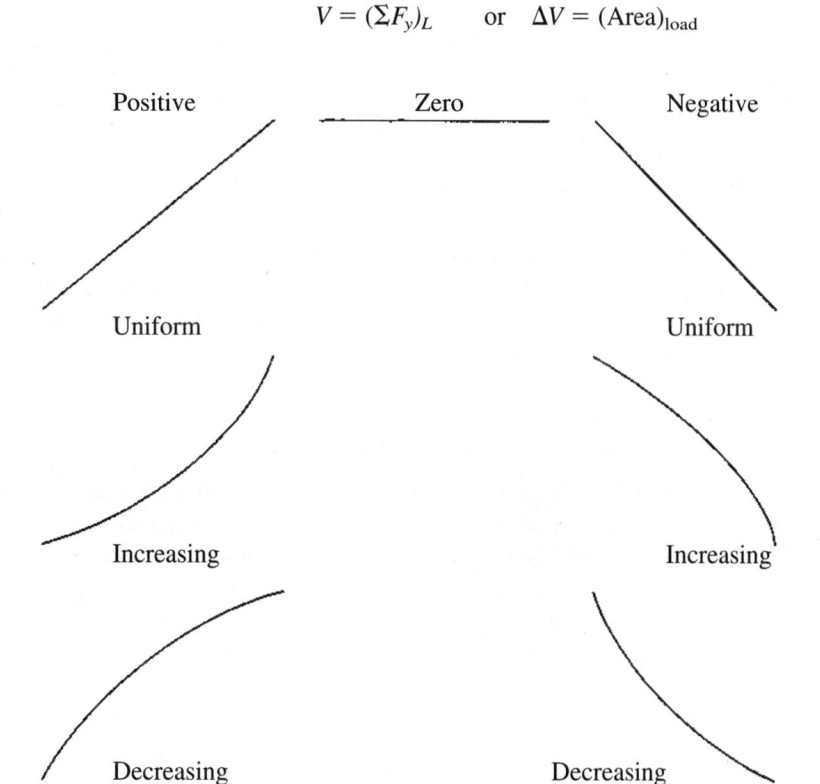

$$V = (\Sigma F_y)_L \quad \text{or} \quad \Delta V = (\text{Area})_{\text{load}}$$

Positive — Zero — Negative

Uniform — Uniform

Increasing — Increasing

Decreasing — Decreasing

Seven possible slopes of curves

FIGURE 3.15

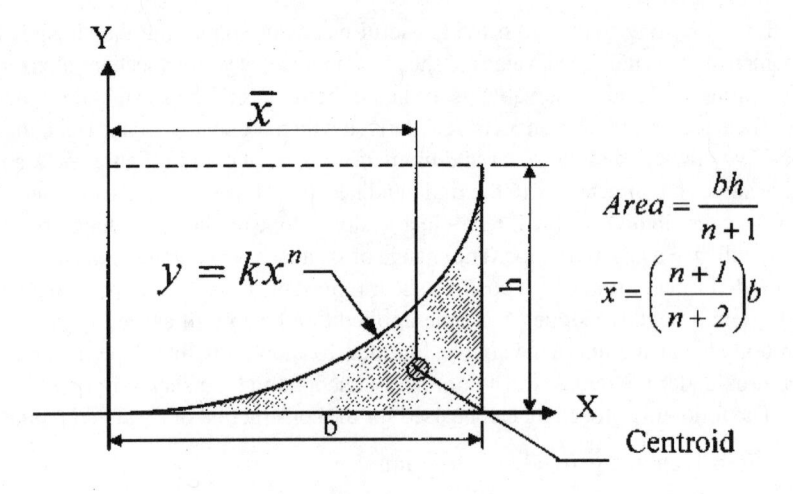

FIGURE 3.16 Areas under curve

$$\text{Area} = \frac{bh}{n+1}$$

$$\bar{x} = \left(\frac{n+1}{n+2}\right)b$$

Step 4. Sketch the shear diagram by plotting the values of shear computed in Step 3 on an axis equal to the length of beam, determining the shape from Equation 3.27, where the slope of the shear diagram equals the intensity of load.

Step 5. Locate the points of zero shears.

Step 6. Compute values of bending moment at the change of load points and point of zero shear, using either $M = (\Sigma M)_L = (\Sigma M)_R$ or $\Delta M = (\text{Area})_{\text{shear}}$.

Step 7. Sketch the moment diagram through the ordinates of the bending moment computed in Step 6. The shape of the diagram is determined from Equation 3.28, in which the slope of the moment diagram equals the intensity of corresponding ordinate of the shear diagram.

3.18 STRESSES IN BEAMS

To design a beam properly, the designer must anticipate and control the maximum stresses developed at critical sections. As shown in the shear and bending moment diagram, we have observed that, with the exception of those cases where the moment or shear is zero, an internal shear force V and an internal bending moment M act on every section of beam. A beam section must be strong enough to withstand the maximum bending moment; it also must have satisfactory shear resistance. Fortunately, the maximum stresses caused by the bending moment and the shearing force, which control the design, act at different locations and develop with no interference with each other. The stresses caused by the bending moment are known as *bending* or *flexure* stresses. The intensity of the flexure stress at a distance y from the neutral axis is given by the flexure formula

$$\sigma_b = \frac{My}{I} \tag{3.31}$$

where M = Bending moment at the section being considered
  I = Moment of inertia of the section about the neutral axis
  y = Distance from the neutral axis.

The intensity of the transverse shear stress is given by the equation

$$\tau = \frac{VQ}{Ib} \tag{3.32}$$

where V = Shear at the section being considered
  Q = First moment of the cross-sectional area with respect to the neutral axis lying above or below the horizontal shear plane to be investigated
  I = Moment of inertia of the section about the neutral axis
  b = Thickness of shear plane.

3.19 LOADS ACTING ON BEAMS

In the design of a structure, loading conditions must be established before the stress analysis can be done. General information about the loads for which a given structure should be designed usually is given in the building codes and specifications, such as the Uniform Building Code, the Basic Building Code, the Standard Building Code, and AASHTO Standard Specifications for Highway Bridges. The loads acting on a structure can be classified into two general types: dead and live loads.

Dead loads are those that are applied continuously and never change magnitude or position, such as the weight of structure or any permanent attachments to the structure, including walls, floors, roofs, and fixed-service equipment in buildings, and roadway slab, curbs, sidewalks, and railings on bridges.

Live loads are movable or moving loads and can vary in magnitude. Movable loads are loads that can be moved from one location to another on a structure, such as people and furniture on a building floor. Moving loads are loads that move continuously over the structure, such as vehicles, trains, and pedestrians on a bridge, and crane trolleys on a runway girder. Loads of this type usually are applied rapidly and have a much greater effect than that generated by the same loads considered fixed in position. Live loads, when applied to a structure, should be positioned to give the most severe possible effect. To obtain this, we must know not only what loading is imposed, but also where it is placed, including full loading, partial loading, and alternate span loading.

Loads also can be further classified with respect to the area over which they are applied. Loads such as those exerted by columns on a supported beam, wheel loads on a supported bridge deck, and longitudinal girders are considered concentrated loads. Any load applied to a relatively small area compared with the size of the loaded member is assumed to be a concentrated load. On the other hand, loads distributed either in part or in full along a length or over an area, such as weight of structure and floor load on a supported beam, are called uniform loads.

3.20 NEUTRAL AXIS

As stated in Section 3.15, transversely applied loads cause a beam to bend. Figure 3.17 shows a segment of bended beam with the distortion greatly exaggerated. The longitudinal fibers on the convex side of the beam elongated, while those on the concave side shortened. Somewhere between the convex and concave sides, there is a surface in which the longitudinal fibers do not change in length. This surface is called the *neutral surface* of the beam. The intersection of the neutral surface with any cross section is called the *neutral axis* of the section.

3.21 SECTION MODULUS

In deriving the flexure formula, Equation 3.31, we make the following assumptions:

1. The beam is initially straight and of constant cross section.
2. The plane sections of the beam remain plane after the beam is bent.
3. The material in the beam is homogeneous, and the moduli of elasticity in tension and compression are equal.
4. The stresses are within the range of the proportional limit of the material.
5. The applied load act in a plane containing the axis of symmetry of the cross section and the loads must be perpendicular to the longitudinal axis of the beam.

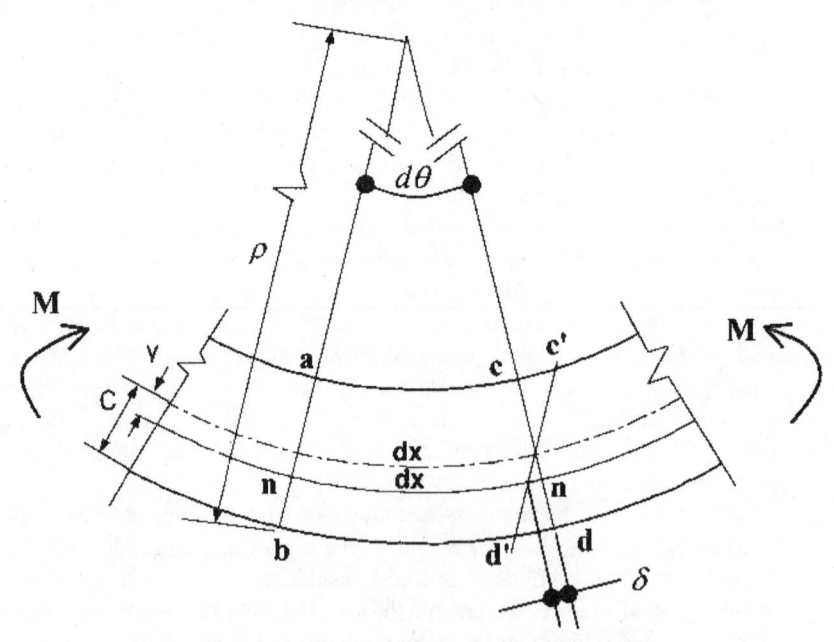

Deformations of a beam in pure bending

FIGURE 3.17

From assumptions 1 and 2 and the geometry shown in Figure 3.17, we see that the deformation of a typical fiber at a distance y from the neutral axis is given by $\delta = y\,d\theta$.

If we denote the radius of curvature of the neutral surface by ρ, the original length of the member is equal to $\rho\,d\theta$. Since all elements had the same initial length dx (assumption 1), the strain of any element can be determined by dividing the deformation by the original length of the element dx. From that, the strain becomes

$$\epsilon_x = \frac{\delta}{dx} = \frac{y\,d\theta}{\rho\,d\theta} = \frac{y}{\rho} = \kappa y \tag{3.33}$$

Equation 3.33 indicates that the strains ϵ_x are directly proportional to the distance from the neutral axis.

Assuming that the material is homogeneous and the stresses are within the range of the proportional limit of the material (assumptions 3 and 4), the stress is given by

$$\sigma_x = E\epsilon_x = \left(\frac{E}{\rho}\right)y = \kappa E y \tag{3.34}$$

Since the radius of curvature ρ of the neutral surface is independent of the location y of fiber, and the modulus of elasticity E is equal in tension and compression (assumption 3), Equation 3.34 indicates that the stress in any fiber varies directly with its location y from the neutral surface. Figure 3.18 is a free-body diagram showing the stress distribution in the beam. There are tensile stresses below the neutral axis and compressive stresses above.

For a beam in pure bending, to satisfy the condition of equilibrium $\Sigma F_x = 0$, we must have $\int \sigma_x\,dA = 0$.

Replacing σ_x with $\kappa E y$, we obtain

$$\int \kappa E y\,dA = \kappa E \int y\,dA = 0.$$

Because the curvature κ and the modulus of elasticity E are constants, we conclude that $\int y\,dA = 0$.

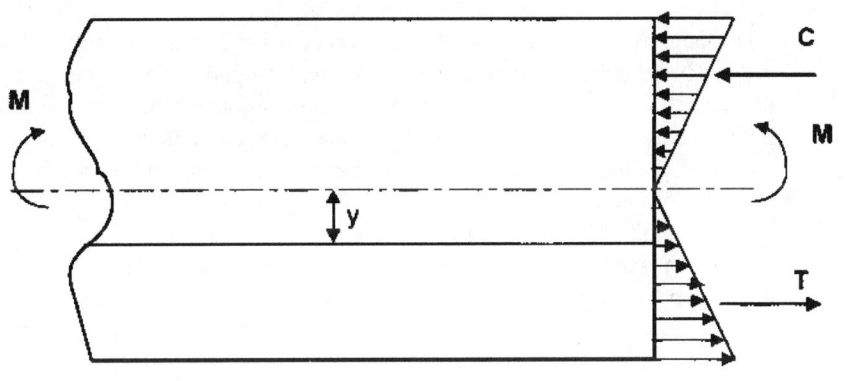

FIGURE 3.18

This equation states that the first moment of the area of the cross section with respect to the neutral axis is equal to zero; that is, the neutral axis must coincide with the centroidal axis.

The condition of equilibrium $\Sigma M_z = 0$ requires that the bending moment be balanced by the resisting moment. From the stress-distribution diagrams of Figure 3.18, the resisting moment about the neutral axis is equal to $\int y(\sigma_x dA)$. By replacing σ_x with $\kappa E y$, we obtain

$$M = \int y(\kappa E y \, dA) = \kappa E \int y^2 \, dA \tag{3.35}$$

Since the term $\int y^2 dA$ is moment of inertia I of the cross-sectional area with respect to the neutral axis, Equation 3.35 can be rewritten as

$$M = \kappa E I \quad \text{or} \quad \kappa E = \frac{M}{I} \tag{3.36}$$

Combining Equations 3.34 and 3.36, we obtain the flexure formula (Equation 3.31)

$$\sigma_b = \frac{My}{I} \tag{3.37}$$

This equation indicates that the flexure stress in any section varies directly with the distance from the neutral axis. Thus, the maximum stress in the beam will occur at the surface farthest from the neutral axis. Denoting the distance from the neutral axis to the extreme fiber by c, the maximum flexure stress in any section is given by

$$\sigma_{mas} = \frac{Mc}{I} = \frac{M}{S} \tag{3.38}$$

where $S = I/c$ is called the section modulus of the beam.

3.22 PLASTIC MOMENT

It has long been known that under a given set of loads, a simply supported beam and a continuous beam behave quite differently. A simply supported beam has one point where the moment is a maximum. As the loading increases, this moment increases proportionately until the extreme fiber stresses equal the yield point of the material. If the loading is further increased, the material will deform more rapidly and the deflection will increase at a greater rate. Therefore, the load that causes the first yield is considered to be the critical load for a simple beam. On the other hand, an indeterminate beam has more than two points of peak moment. As the loading increases, the cross section at the greatest of these peaks will reach the yield point. As more load is added, a zone of yielding develops there, but elsewhere the structure is still elastic, and this controls the total deflection. Therefore, for an indeterminate beam, the stresses that are calculated for elastic design purposes often are not true maximum stresses at all.

With the knowledge that an indeterminate beam possesses a greater load-carrying capacity than indicated by the elastic design concept, we will determine the *true maximum moment of resistance* of an indeterminate beam. The assumptions made in Section 3.21 still apply, except that the stresses need not be proportional to strain. Plane cross sections remain plane under pure bending, a condition which is just as valid for nonlinear inelastic materials as for linear elastic materials. Thus, the strains in the beam are proportional to their distances from the neutral axis. If the stress-strain diagram is known, with the aid of equilibrium equations, it is always possible to determine stresses, strains, deflections, and the ultimate load-carrying capacity of the beam.

A typical stress-strain curve for an elastic-plastic material is shown in Figure 3.19. The stress-strain relationship consists of two straight lines. Such a material follows Hooke's law up to the yield stress, and then stress remains constant at yield stress σ_y wherever the strain exceeds the yield strain ϵ_y. Figure 3.20 shows the successive stages of strains and stress distribution as moment is increased. At stage 1, the extreme fiber strains just reach the yield strain and the stress distribution is still elastic. Applying the flexure formula, we find that the resisting moment at this section is $M_y = \sigma_y S$.

When more moment is applied, say to stage 2, the strains at the extreme fiber of the cross section will continue to increase, and the maximum strain will exceed the yield strain ϵ_y. From the stress-strain curve, the maximum stresses, however, will remain constant and equal to σ_y. We can obtain the moment at this stage by integrating.

With additional increase in the bending moment, the plastic region extends farther inward toward the neutral axis until the stress distribution at stage 3 is reached. At this stage, the beam has reached the maximum moment that can be sustained by a beam of elastic-plastic material. The bending moment corresponding to this stress distribution is called the *plastic moment* M_p.

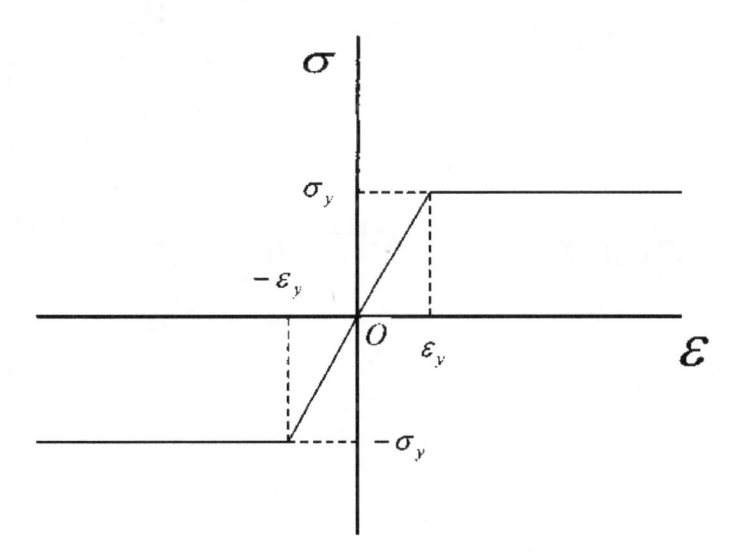

Stress-Strain curve for an elastic plastic material
FIGURE 3.19

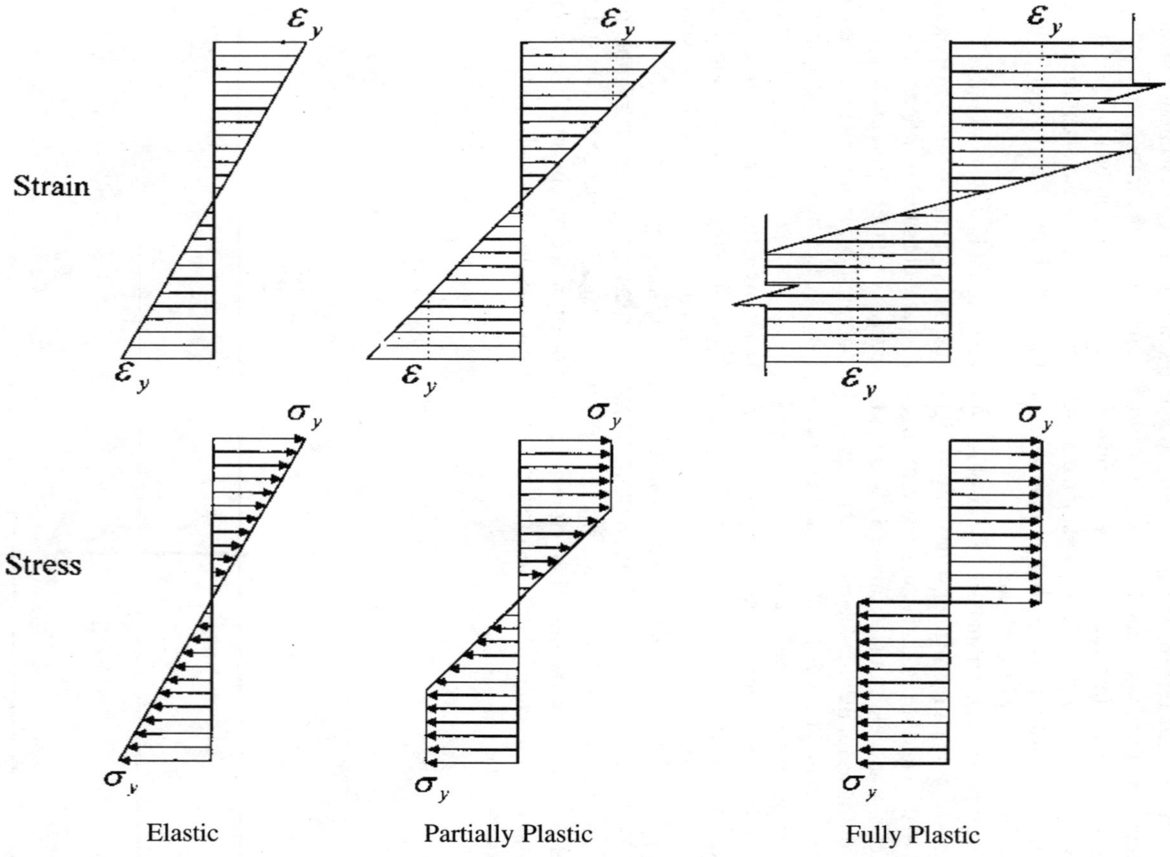

Successive stages of strain and stress

FIGURE 3.20

3.23 PLASTIC SECTION MODULUS

In order to find M_p, it is necessary to locate the neutral axis of the cross section first. From the equilibrium equation $\Sigma F_x = 0$, the sum of the force in the x direction on the cross section must be zero, which gives $T = C$ $\sigma_y A_t = \sigma_y A_c$, or $A_t = A_c$.

Thus we conclude that the neutral axis divides the cross section into two equal areas.

The magnitude of the moment M_p can be found simply by taking the moments about the neutral axis of the tensile and compressive forces $\sigma_y A/2$. Therefore, we obtain

$$M_p = \sigma_y A \frac{(y_1 + y_2)}{2} \tag{3.39}$$

where y_1 and y_2 are the distances from the neutral axis to the centroids of area above and below the neutral axis.

The quantity $A(y_1 + y_2)/2$ is called the *plastic modulus* for the cross-section and given the symbol Z. This yields

$$M_p = \sigma_y Z \tag{3.40}$$

where

$$Z = \frac{A(y_1 + y_2)}{2}$$

The procedure in each instance is to divide the cross section into two equal areas, locate the centroid of each half, and then use Equation 3.39 to calculate M_p.

3.24 DEFLECTION OF BEAMS

As described in Section 3.1, a structure or member must carry loads safely (not break) and meet the performance requirement (not deform excessively). Frequently, the design of a beam is governed by the amount of its permissible deflection. For example, the live-load deflection of beams that carry plaster ceiling usually are limited to a deflection of $\frac{1}{360}$ of their length to minimize the cracking of plaster ceilings. Another application is that the roof members shall be cambered to prevent problems from ponding of water because of dead-load deflection. The bridge girders shall be cambered to match the design profile grade elevation when supporting full dead load. In these cases, deflections must be computed to determine the amount of camber built into the roof members and bridge girders. Probably most important for the structural engineer's interest in deflection evaluation is that the calculation of deflection is essential to the analysis of statically indeterminate structures. A structure is statically indeterminate when the equations of static equilibrium available are not sufficient to determine all the internal resistance forces and reactions. Deflection condition equations representing consistent deflections can be used for this purpose.

3.25 METHODS FOR DETERMINING BEAM DEFLECTION

There are numerous methods of computing deflections in a structure. The most common and useful in structural analysis are:

1. Double integration
2. Moment area
3. Elastic weight
4. Superposition

3.25.1 The Double-Integration Method

This method usually starts with the relationship between the curvature d^2y/dx^2 and the bending moment M. The procedure consists of two successive integrations to get the equation for deflection y of the elastic curve. The constants of integration can be evaluated from the applicable boundary or matching conditions of the beam. The relationships between deflection, slope, moment, shear, and load intensity are as follows:

Deflection $\quad y \quad$ (3.41)

Slope $\quad \dfrac{dy}{dx} \quad$ (3.42)

Curvature (moment) $\quad \dfrac{d^2y}{dx^2} = -\dfrac{M}{EI} \quad$ (3.43)

Shear $\quad \dfrac{d^3y}{dx^3} = -\dfrac{V}{EI} \quad$ (3.44)

Load intensity $\quad \dfrac{d^4y}{dx^4} = +\dfrac{w}{EI} \quad$ (3.45)

Calculating the deflection of a beam by the double-integration method generally involves the following steps:

1. Use equilibrium equations to determine the reactions.
2. Establish the equations for bending moments.
3. Use $d^2y/dx^2 = -M/(EI)$ and integrate once to obtain the slope equations.
4. Integrate the slope equations once to obtain the deflection equations.
5. Apply boundary conditions to evaluate the constants of integration.

This method is useful when the complete elastic curve equations are required. It can be cumbersome if the loading condition changes along the beam, since this requires that a separate moment equation be written between each change-of-load point and two integrations be made for each such moment equation.

EXAMPLE 3.3
Determine the deflection curve of a simply supported beam AB subjected to the concentrated loads P as shown in Figure 3.21

FIGURE 3.21

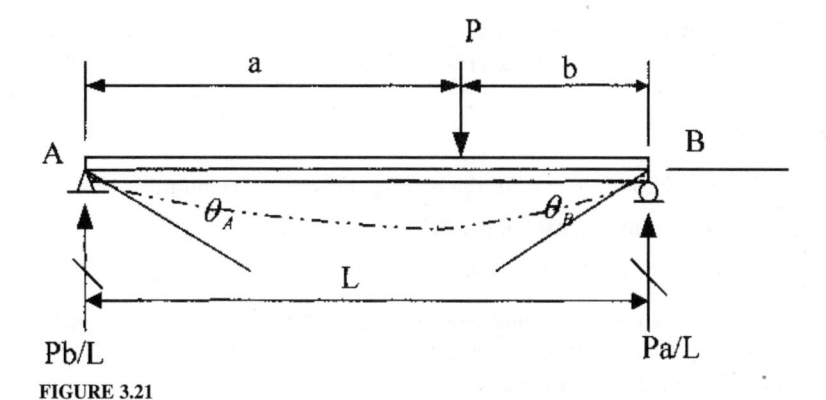

Solution

From a free-body diagram of the beam and the equations of equilibrium, the left reaction is found to be *Pb/L* upward and the right reaction is *Pa/L* upward. As shown in Figure 3.21, the x-y coordinates system is introduced, where the *x* axis coincides with the original unbent position of the beam, and the origin of the coordinates is selected at the left support. In Figure 3.21 there are two intervals to be considered—one for the part of the beam to the left of the load and one for the right-hand part. Therefore, two different equations are required to describe the bending moment in the beam. One equation is valid to the left of the load *P*, the other holds to the right of this force.

In the region to the left of the load *P* we have the bending moment $M = (Pb/L)x$ for $0 < x < a$. The differential equation of the bent beam for this part of the beam becomes

$$EI\frac{d^2y}{dx^2} = -\frac{Pbx}{L} \quad (0 \le x \le a) \tag{a}$$

The first integration gives $EI\dfrac{dy}{dx} = -\dfrac{Pbx^2}{2L} + C_1 \quad (0 \le x \le a)$ (b)

The next integration gives $EIy = -\dfrac{Pbx^3}{6L} + C_1 x + C_2 \quad (0 \le x \le a)$ (c)

In the region to the right of the force *P*, the bending moment is $M = (Pb/L)x - P(x - a)$ for $a < x < L$. Thus

$$EI\frac{d^2y}{dx^2} = -\frac{Pbx}{L} + P(x - a) \quad (a \le x \le L) \tag{d}$$

The first integration gives

$$EI\frac{dy}{dx} = -\frac{Pbx^2}{2L} + \frac{P(x-a)^2}{2} + C_3 \quad (a \le x \le L) \tag{e}$$

Now performing the second integration, we obtain

$$EIy = -\frac{Pbx^3}{6L} + \frac{P(x-a)^3}{6} + C_3 x + C_4 \quad (a \le x \le L) \tag{f}$$

The integration of each equation gives rise to two constants of integration, so there are four constants of integration to be determined. These can be found from the following conditions:

1. Under the load P, at $x = a$, the slopes for the two portions of the beam must be equal
2. The deflection at $x = a$ for the two portions of the beam must be equal
3. At the left support, $x = 0$, the deflection is zero
4. At the right support, $x = L$, the deflection is zero.

Four conditions (two matching and two boundary) are sufficient for the evaluation of the four constants integration, and the problem can be solved.

For clarity, these four conditions are listed as follows:

Available matching conditions: dy/dx from the left equation = dy/dx from the right equation at $x = a$ and y from the left equation = y from the right equation at $x = a$

Available boundary conditions: $y = 0$ when $x = 0$ and $y = 0$ when $x = L$

Substituting the first condition values into Equations (b) and (e) gives

$$-\frac{Pba^2}{2L} + C_1 = -\frac{Pba^2}{2L} + C_3 \text{ from which } C_1 = C_3.$$

Substituting the second condition values into Equations (c) and (f) gives

$$-\frac{Pba^3}{6L} + C_1 a + C_2 = -\frac{Pba^3}{6L} + C_3 a + C_4 \text{ from which } C_2 = C_4.$$

Applying the boundary conditions to equations (c) and (f), we obtain $C_2 = 0$ and

$$-\frac{PbL^2}{6} + \frac{Pb^3}{6} + C_3 L = 0$$

Thus

$$C_2 = C_4 = 0 \quad \text{and} \quad C_1 = C_3 = \frac{Pb(L^2 - b^2)}{6L}$$

The equations for the deflection curve are:

$$EIy = -\frac{Pbx^3}{6L} + \frac{Pb(L^2 - b^2)}{6L} x = \frac{Pbx}{6L}(L^2 - b^2 - x^2) \quad (0 \le x \le a) \tag{g}$$

$$EIy = -\frac{Pbx^3}{6L} + \frac{P(x-a)^3}{6} + \frac{Pb(L^2 - b^2)}{6L} x$$

$$= \frac{Pbx}{6L}(L^2 - b^2 - x^2) + \frac{P(x-a)^3}{6} \quad (a \le x \le L) \tag{h}$$

These two equations are necessary to describe the deflection curve of the beam. Each equation is valid only in the region indicated.

The slopes for the two portions of the beam are found by substituting the values of C_1 and C_3 into Equations (b) and (e), and we obtain

$$EI\frac{dy}{dx} = -\frac{Pbx^2}{2L} + \frac{Pb(L^2 - b^2)}{6L} = \frac{Pb}{6L}(L^2 - b^2 - 3x^2) \quad (0 \leq x \leq a) \tag{i}$$

$$EI\frac{dy}{dx} = -\frac{Pbx^2}{2L} + \frac{P(x-a)^2}{2} + \frac{Pb(L^2 - b^2)}{6L}$$

$$= \frac{Pb}{6L}(L^2 - b^2 - 3x^2) + \frac{P(x-a)^2}{2} \quad (a \leq x \leq L) \tag{j}$$

From these equations, the slope at any point of the deflection curve can be calculated. Thus, substituting $x = 0$ into Equation (i) and $x = L$ into Equation (j), we obtain the angles of rotation at the ends of the beam:

$$\theta_A = \frac{Pb}{6LEI}(L^2 - b^2) = \frac{Pab(L + b)}{6LEI}$$

$$\theta_B = -\left[\frac{Pb}{6LEI}(L^2 - b^2 - 3L^2) + \frac{P(L-a)^2}{2EI}\right] = \frac{Pab(L + a)}{6LEI}$$

The maximum deflection of the beam occurs at the point where the deflection curve has a horizontal tangent. If $a > b$, the maximum deflection occurs in the left-hand portion of the beam (between $x = 0$ and $x = a$), and we can locate this point by equating the slope dy/dx from Equation (l) to zero.

k. Solving Equation (i) for this value of x,

$$x = \sqrt{\frac{L^2 - b^2}{3}} \quad (a \geq b) \tag{k}$$

At this point the deflection is found by substituting x from Equation (k) into Equation (g), yielding

$$y_{max} = \frac{Pb}{6LEI}\sqrt{\frac{L^2 - b^2}{3}}\left(L^2 - b^2 - \frac{L^2 - b^2}{3}\right) = \frac{Pb(L^2 - b^2)^{3/2}}{9\sqrt{3}LEI} \quad (a \geq b)$$

If the load P acts at the middle of the beam ($a = b = L/2$), the deflection at the midspan and the angles at the ends, by substitution, are found to be

$$\theta_A = \theta_B = \frac{PL^2}{16EI} \qquad y_{max} = \frac{PL^3}{48EI}$$

3.25.2 The Moment-Area Method

This method is especially useful when applied in the determination of slope or deflection at a specified point. Its name comes from the fact that it utilizes the area of the bending moment diagram.

The method is based on two theorems, known as Moment Area Theorems I and II:

Theorem I: The change in slope expressed in radians between tangents drawn to any two points on the elastic curve is equal to the area of the M/EI diagram between these two points.

Theorem II: The deviation of any point B on a continuous elastic curve to a tangent drawn to any other point A on the elastic curve, measured in a direction perpendicular to the original axis of the beam, is equal to the moment of the area of the M/EI diagram between points A and B about point B.

Referring to Figure 3.22, moment area Theorem I gives the angle θ_{AB} between tangents to the elastic curve at A and B on the elastic curve as

$$\theta_{B/A} = \int_{X_A}^{X_B} \frac{M}{EI} dx \qquad (3.46)$$

In Figure 3.22 the distance $\Delta_{B/A}$ from point B on the elastic curve to a tangent drawn at point A, measured in a direction perpendicular to the original axis of the beam is

$$\Delta_{B/A} = \int_{X_A}^{X_B} \frac{M}{EI} dx \qquad (3.47)$$

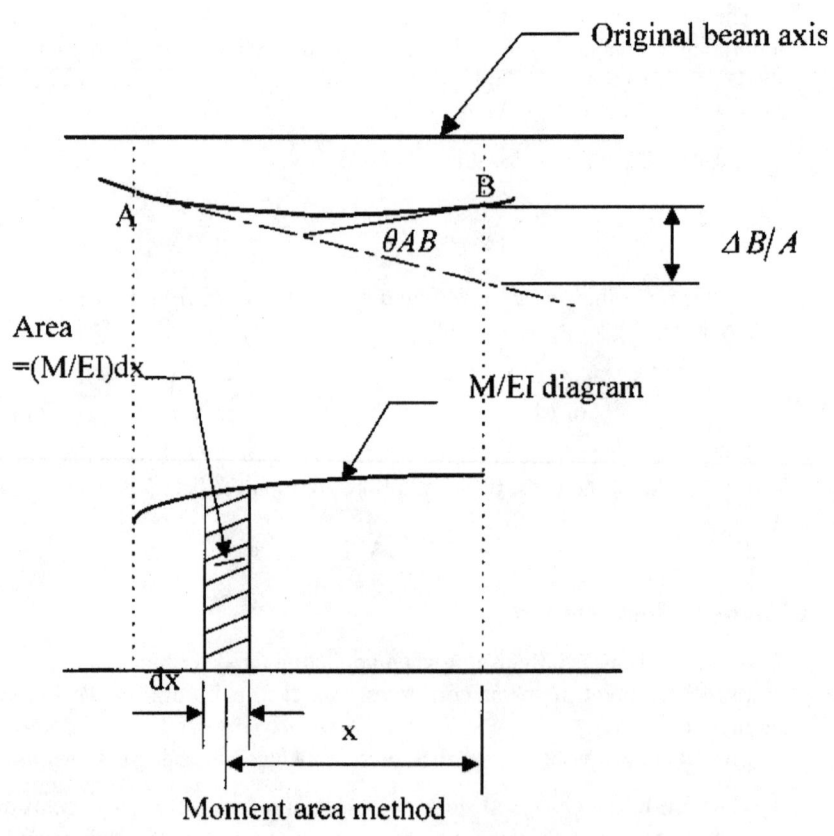

FIGURE 3.22 Moment area method

where x is the distance between the area $(M/EI)\,dx$ and point B. The geometric significance of the integral $\int x\,(M/EI)\,dx$ is that the integral is equivalent to the sum of moments of the individual $(M/EI)\,dx$ area between A and B about the ordinate at B.

EXAMPLE 3.4
Two concentrated loads P act on a simple beam AB at points as shown in Figure 3.23a. Find the angle of rotation θ_a of the deflection curve at point A, and the maximum deflection of the beam at midspan.

Solution
To find the angle at A, we noted that θ_a is equal to the displacement of point B from the tangent at A, $\Delta B/A$, divided by the length L. According to the second moment-area principle, this displacement $\Delta B/A$ equals the moment about B of the moment-area divided by EI between A and B.

Therefore, to evaluate $\Delta B/A$ by the moment-area method, we first draw the moment diagram for the beam as shown in Figure 3.23b. The M/EI diagram is developed from this moment diagram, with the values of the moment diagram divided by the corresponding values of EI as shown in Figure 3.23c. Using the triangular and rectangular components of the M/EI diagram and taking the moments of these areas about B, we obtain for the displacement at B; $\Delta B/A = A_1X_1 + A_2X_2 + A_3X_3$ where X_1, X_2, and X_3 are the distances from B to the centroids of the respective areas. The value of deflection is found to be

$$\frac{\Delta B}{A} = \left(\frac{1}{18}\frac{PL^2}{EI}\right)\left(\frac{7}{9}L\right) + \left(\frac{1}{9}\frac{PL^2}{EI}\right)\left(\frac{L}{2}\right) + \left(\frac{1}{18}\frac{PL^2}{EI}\right)\left(\frac{2}{9}L\right) = \frac{1}{9}\frac{PL^3}{EI}$$

from which

$$\theta_A = \frac{\Delta B/A}{L} = \frac{1}{9}\frac{PL^2}{EI}$$

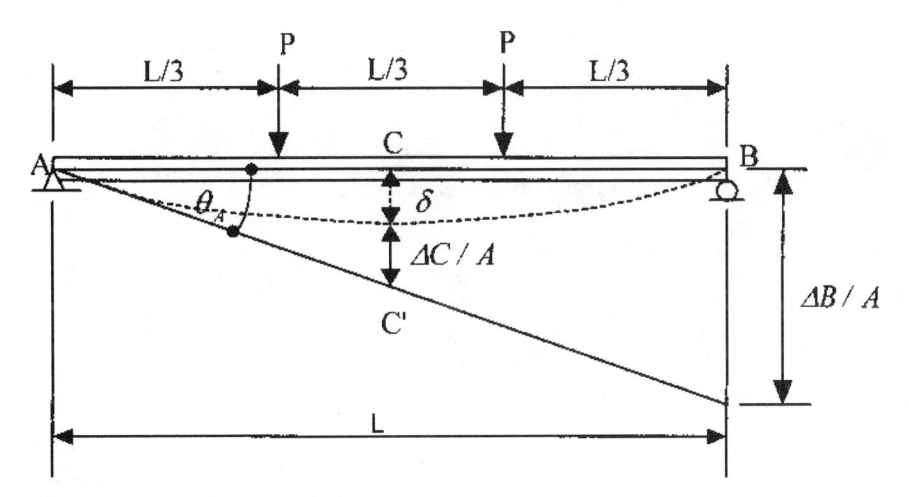

FIGURE 3.23a

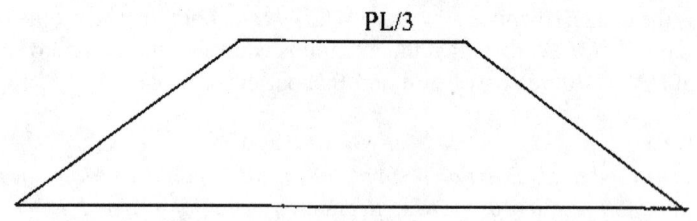

Moment diagram - Moment area between A and B

FIGURE 3.23b

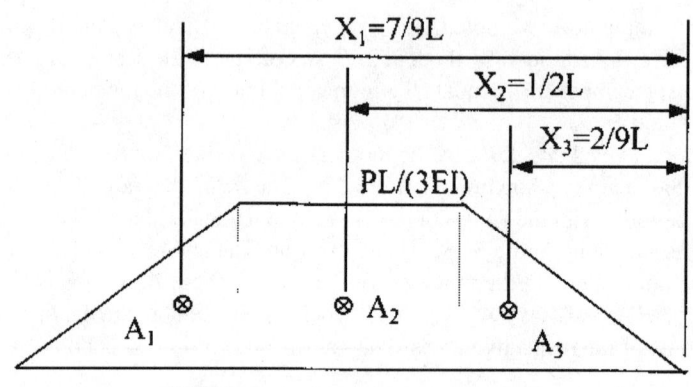

M/EI diagram - between A and B

FIGURE 3.23c

M/EI diagram - between A and C

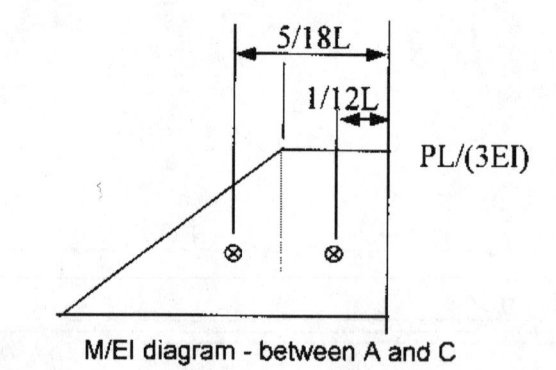

FIGURE 3.23d

The magnitude of the maximum deflection δ at midspan is found as the distance CC' minus the displacement $\Delta C/A$. The distance CC' is equal to $(L/2)\theta_a$. To evaluate displacement $\Delta C/A$ we take the moment of the portion of the M/EI between A and C about C

$$\frac{\Delta C}{A} = \left(\frac{1}{18}\frac{PL^2}{EI}\right)\left(\frac{5}{18}L\right) + \left(\frac{1}{18}\frac{PL^2}{EI}\right)\left(\frac{L}{12}\right) = \frac{13}{648}\frac{PL^3}{EI}$$

The maximum deflection δ at midspan can be found as follows:

$$\delta = \left(\frac{L}{2}\right)\theta_a - \Delta C/A = \left(\frac{L}{2}\right)\left(\frac{1}{9}\frac{PL^2}{EI}\right) - \frac{13}{648}\frac{PL^3}{EI} = \frac{23}{648}\frac{PL^3}{EI}$$

3.25.3 The Elastic Weight Method

This method, also known as the conjugate-beam method, was developed by Otto Mohr in 1868. Its purpose is to transform the problem of solving the slopes and deflections of a beam resulting from the actual applied loads to a problem of solving the shears and moments of a conjugate beam because of the elastic load.

Relationships between the real beam and the conjugate beam are as follows:

1. The span of the conjugate beam is equal to the span of the real beam.
2. The load of the conjugate beam is the *M/EI* diagram of the real beam.
3. The slope at any section on the real beam is equal to the shear at the corresponding section of the conjugate beam.
4. The deflection at any section on the real beam is equal to the moment at the corresponding section of the conjugate beam.

In addition, the setup of the supports and connections of the conjugate beam must induce shear and moment in the conjugate beam in conformity to the slope and deflection induced by the counterparts in the real beam. The rules for selecting the supports and other details of conjugate beams are summarized as

Fixed end	↔	Free end
Simple end	↔	Simple end
Interior support	↔	Interior connection

This method is easy to apply when one must determine the deflection and the slope at any section of a simply supported beam. General procedure for the application of the conjugate-beam method for simply supported beams can be summarized in the following steps:

1. Draw the shear and moment diagram for the real beam.
2. Establish the conjugate beam.
3. Apply the moment diagram, divided by the flexural rigidity *EI*, as a load to the conjugate beam.
4. Calculate the reactions of the conjugate beam.
5. Determine the shear *V* on the conjugate beam where the slope of the beam is desired; and determine the moment *M* on the conjugate beam where the deflection of the beam is desired.

EXAMPLE 3.5

Use the conjugate-beam method to compute the end slopes θ_A and θ_B and the maximum deflection for the beam of Figure 3.24a.

Solution

This beam was analyzed previously by the double-integration method in Example 3.3. To solve by the conjugate-beam method, the following steps apply:

1. Establish the conjugate beam.

   To do this, we first plot the moment diagrams, as shown in Figure 3.24b. The conjugate beam loaded with the *M/EI* diagram is given in Figure 3.24c.

2. Compute θ_A and θ_B.

   The slopes at supports in the real beam are determined by evaluating the reactions R'_A and R'_B in the conjugate beam. The centroid of a typical triangle is at $(L + a)/3$ from the left support and at $(L + b)/3$ from the right support. Thus we find the slopes at ends to be

   $$\theta_A = R'_A = \frac{Pab}{2EI}\frac{(L+b)}{3}\left(\frac{1}{L}\right) = \frac{Pab(L+b)}{6LEI} \quad (a)$$

   $$\theta_B = R'_B = \frac{Pab}{2EI}\frac{(L+a)}{3}\left(\frac{1}{L}\right) = \frac{Pab(L+a)}{6LEI} \quad (b)$$

3. Compute maximum deflection Δ_{max}.

   The maximum deflection in the real beam occurs where the shear in the conjugate beam is zero. Assume Δ_{max} occurs at a distance x from the left support $(a > b)$.

   $$V'_x = R'_A - \frac{1}{2}\frac{Pbx}{LEI}x = \frac{Pab(L+b)}{6LEI} - \frac{1}{2}\frac{Pbx}{LEI}x = \frac{Pb[a(L+b) - 3x^2]}{6LEI} = 0$$

   $$3x^2 = a(L+b) = (L-b)(L+b) = L^2 - b^2$$

   $$x = \sqrt{\frac{L^2 - b^2}{3}} \quad (a \geq b) \quad (c)$$

   $$\Delta_{max} = M'_x = R'_A x - \frac{1}{2}\frac{Pbx}{LEI}x\left(\frac{x}{3}\right) = \frac{Pab(L+b)}{6LEI}x - \frac{Pbx^3}{6LEI} = \frac{Pb(L^2 - b^2)^{3/2}}{9\sqrt{3}LEI}$$

   $$(a \geq b)$$

3.25.4 The Method of Superposition

Superposition is applicable whenever the effects of the loads to be determined, such as stresses and deflections, are a linear function of the applied loads. This assumption is based on the following conditions:

1. The material of the structure must be elastic and follows Hooke's law at all points and throughout the range of loading considered; that is, the stress is proportional to the strain.

STRENGTH OF MATERIALS

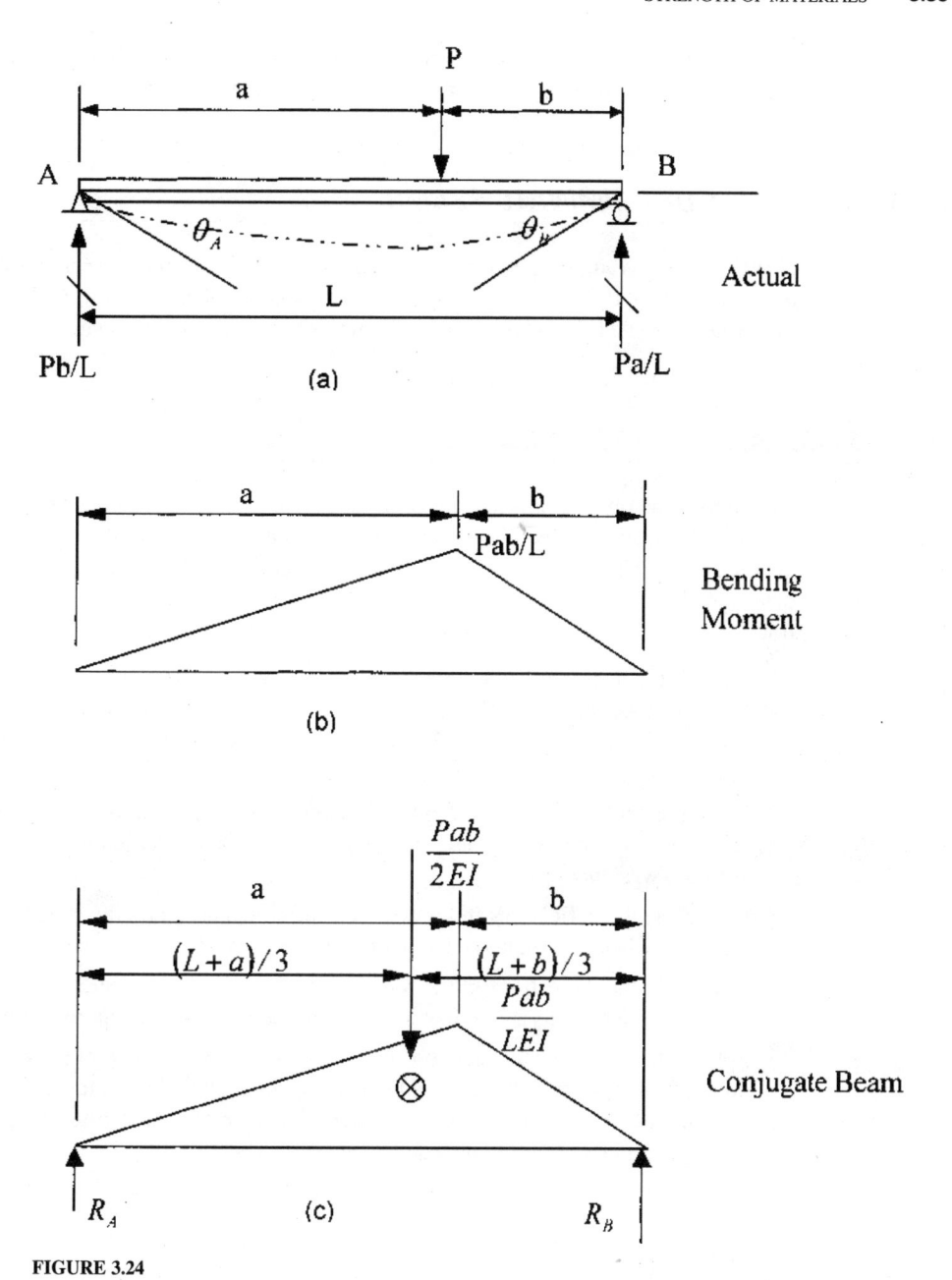

FIGURE 3.24

(a) Actual

(b) Bending Moment

(c) Conjugate Beam

Labels on figure:
- (a): P, a, b, A, B, θ_A, θ_B, L, Pb/L, Pa/L
- (b): a, b, Pab/L
- (c): a, b, $(L+a)/3$, $(L+b)/3$, $\dfrac{Pab}{2EI}$, $\dfrac{Pab}{LEI}$, R_A, R_B

2. The changes in the geometry of the structure are so small that they can be neglected, and the computations can be based on the original geometry of the structure.

Under such conditions, the desired effects can be found due to each load acting separately, and then the results may be superimposed to obtain the total value due to all loads acting simultaneously. The results for the separate loads frequently are available in various handbooks or readily determined by one of the methods already presented.

3.26 STATICALLY DETERMINATE BEAMS

Cantilever beams and simple beams have only two reactions, and they can be obtained from a free-body diagram of the beam by applying the equations of equilibrium. Such beams are statically determinate because the reactions can be obtained from the equations of equilibrium.

3.27 STATICALLY INDETERMINATE BEAMS

Continuous and other beams that are continuous over two or more supports, where there are not enough equations of equilibrium to determine the reactions, are statically indeterminate. When a beam is statically indeterminate, the evaluation of the redundant reactions can be determined by the method of superposition. The following is a general outline of the solution for a statically indeterminate beam, using the method of superposition:

1. Identify the degree of static indeterminacy of the beam, which is the number of reactions in excess of the number that can be found from equations of static equilibrium. These quantities can be viewed as redundant (or excess) forces that are not needed to maintain an immovable structure in static equilibrium.

2. Make the beam statically determinate by removing the redundant supports. The statically determinate structure left after the redundants are released is called the *released structure*.

3. Determine the deflections in the released structure caused by the actual loads at the locations where the redundant supports were removed.

4. Determine the deflections in the released structure caused by each of the unknown redundants separately at the locations where the redundant supports were removed.

5. Formulate simultaneous equations by using conditions of compatibility and the principle of superposition. There will be as many such equations as there are unknown redundants. These equations can be solved for the redundants, after which all other reactions and stress resultants can be found from static.

3.28 SHEAR CENTER

When the theory of bending of beam was developed in Section 3.21, one of the assumptions was that the loads were applied in a plane containing the axis of symmetry of the cross section. When this assumption is not satisfied, the beam usually will be subjected to twisting as well as bending. However, it is possible to place the loads in such a plane that the beam will not twist. The *shear center* (also known as the flexural center or the center of twist) is defined as the point in the cross section that the transverse shear force must pass so that the effects of the twisting are eliminated. So determining the position of shear center is of great interest.

For a cross section having two axes of symmetry, or antisymmetry, such as a circular, rectangular, wide-flange section or a Z section, the shear center coincides with the centroid of the cross section. For the cross-sections with one axis of symmetry, such as a channel section, the shear center always will be on the axes of symmetry. For a section consisting of two intersecting rectangular elements, such as an angle section, a tee section or a V section, the shear center is at the junction of the two legs. For a general cross section, the shear center can be determined by considering the equilibrium of the twisting moment of the applied loads and the moment due to the shear stresses.

3.29 UNSYMMETRIC BENDING

The flexure formula developed in Section 3.21 was restricted to loads lying in a plane that contained an axis of symmetry of the cross section. From symmetry it can be concluded that the neutral axis will pass the centroid of the section and will be perpendicular to the plane of loading. In this section, we will extend the work to cover the cases of beams with a plane of symmetry, but with load (couple) applied not in, or parallel to, the plane of symmetry.

We consider first the case shown in Figure 3.25, in which a beam has two planes of symmetry subjected to loads inclined to axes of symmetry.

Each load can be resolved into two components acting in the two planes of symmetry. Each component produces flexure stresses that are normal to the cross section and can be solved directly by the flexure formula. After solving the bending problem for each of these conditions, the resultant stresses are obtained by superposition; that is

$$\sigma = \frac{M_z y}{I_z} + \frac{M_y z}{I_y} \tag{3.48}$$

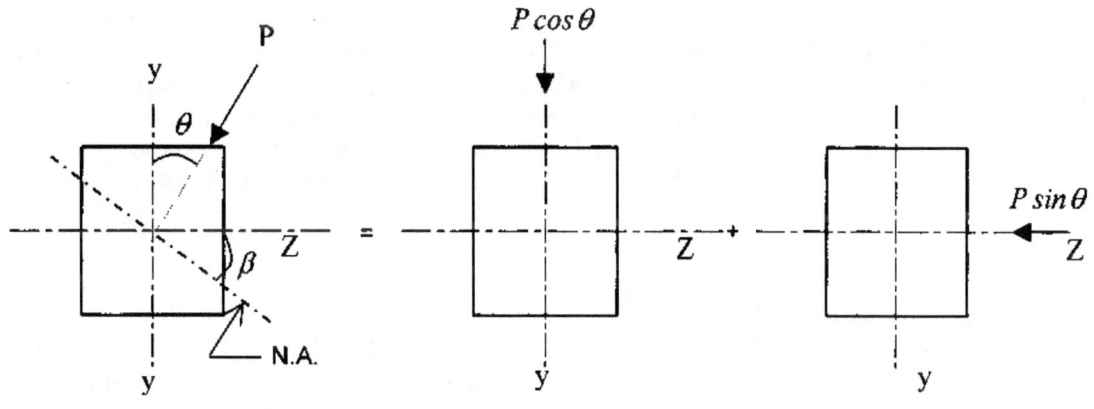

FIGURE 3.25

where M_z = Bending about the z axis caused by $P\cos\theta$
M_y = Bending moment about y axis due to $P\sin\theta$.

The bending moments M_z and M_y can be written in terms of total moment M as follows:

$$M_z = M\cos\theta \quad \text{and} \quad M_y = M\sin\theta \tag{3.49}$$

Substituting these expressions for M_z and M_y into Equation 3.48 gives the elastic flexure formula for unsymmetrical bending

$$\sigma = \frac{M\cos\theta y}{I_z} + \frac{M\sin\theta z}{I_y} \tag{3.50}$$

The neutral axis of the cross section of a beam is defined to be the axis in the cross section for which $\sigma = 0$. From which, by canceling the common term M and rearranging, we obtain

$$\frac{\cos\theta y}{I_z} = -\frac{\sin\theta z}{I_y}$$

Thus, the equation of the neutral axis of the cross section in the yz plane

$$y = -\left(\frac{I_z}{I_y}\tan\theta\right)z \tag{3.51}$$

The slope of the line is $dy/dz = \tan\beta$, where β is the angle between the neutral axis of bending and the z axis. Therefore,

$$y = -\left(\frac{I_z}{I_y}\tan\theta\right)z \tag{3.52}$$

From this, we see that the neutral axis is, in general, not perpendicular to the load unless the angle $\theta = 0$ or $I_y = I_z$. In the first case, the plane of loads is in the principal plane xy and the neutral axis coincides with the principal axis z. In the second case, the two principal moments of inertia of the cross section are equal; this reduces to the special kind of symmetry where all centroidal moments are equal (circle, square, etc.).

Once the neutral axis is located on the cross section at angle β as indicated in Figure 3.25, the maximum tensile and compressive flexure stresses in the cross section are easily determined.

3.30 CURVED BEAMS

One of the assumptions in the development of the flexure formula in Section 3.21 was that the beam is initially straight and of constant cross section, from which all longitudinal elements of a beam are seen to have the same length. Strains and stresses are directly proportional to the distance from the neutral axis. However, members subjected to bending are not always straight, such as in the case of crane hooks—they are curved before a bending moment is applied. When the member is sharply curved, the unstrained lengths of the longitudinal elements no longer are equal, and the linear variation of strain over the cross section no longer is valid. In this section, a theory will be developed to determine the flexure stresses in a given curved beam subjected to pure bending.

Using the same assumption as in the case of straight beams, the transverse cross sections that were originally plane and normal to the centerline remain plane after bending. We can obtain the equations for finding the stress distribution of a curved beam of constant cross section subjected to pure bending M applied in the plane of bending. In accordance with this assumption, the longitudinal deformation of any element will be proportional to the distance of the element from the neutral surface. In Figure 3.26a, the total elongation of a fiber at a distance y from the neutral axis is $yd\phi$. The unstrained length of this fiber is $(R + y)d\theta$. Hence the longitudinal strain is

$$\epsilon = \frac{\delta}{L} = \frac{yd\phi}{(R + y)d\theta} \tag{3.53}$$

Then, by Hooke's law, the stress is

$$\sigma = E\epsilon = \frac{Ed\phi}{d\theta} \frac{y}{R + y} \tag{3.54}$$

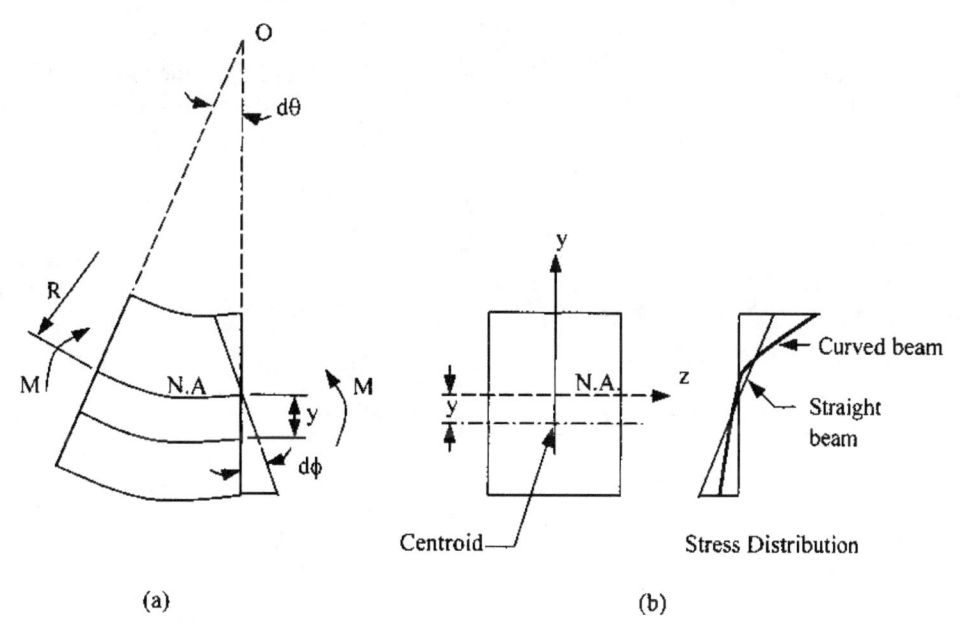

FIGURE 3.26

The plot of Equation 3.54 is shown in Figure 3.26b. The stress distribution is no longer linear, and the neutral surface is shifted from the centroid of the section toward the center of curvature.

In the case of pure bending, the conditions of equilibrium require that the sum of the normal forces distributed over a cross section equal zero and the moment of these forces equal the applied bending moment M. This statement can be expressed by the following two equations:

$$\int_A \sigma dA = \int_A \frac{Ed\phi}{d\theta} \frac{y}{(R+y)} dA = \frac{Ed\phi}{d\theta} \int_A \frac{y}{(R+y)} dA = 0 \tag{3.55}$$

$$\int_A \sigma y dA = \int_A \frac{Ed\phi}{d\theta} \frac{y^2}{(R+y)} dA = \frac{Ed\phi}{d\theta} \int_A \frac{y^2}{(R+y)} dA = M \tag{3.56}$$

Since $Ed\phi/d\theta$ cannot be zero, we obtain

$$\int_A \frac{y}{(R+y)} dA = 0 \tag{3.57}$$

The radius of the neutral surface, R, can be obtained by solving Equation 3.57 for each specific problem.

The integral of Equation 3.56 can be rewritten in the following form:

$$\int_A \frac{y^2}{(R+y)} dA = \int_A y dA - R \int_A \frac{y}{(R+y)} dA \tag{3.58}$$

The first integral on the right side of Equation 3.58 is the moment of the entire cross-sectional area about the neutral axis. The second integral, from Equation 3.57, was shown to equal zero. Equation 3.58 can now be rewritten as

$$\int_A \frac{y^2}{(R+y)} dA = \int_A y dA = \bar{y} A \tag{3.59}$$

Substituting this simplified expression in Equation 3

$$\frac{Ed\phi}{d\theta} \bar{y} A = M \tag{3.60}$$

from which Equation 3.54 may be written

$$\sigma = \frac{M}{\bar{y} A} \frac{y}{R+y} \tag{3.61}$$

This is a general expression for the elastic flexural stress in an initially curved beam.

3.31 PLASTIC DEFORMATIONS OF BEAMS

As described in previous sections, when the stress exceeds the elastic limit, a portion of the deformation can remain after the load is removed. The deformation remaining after an applied load is removed is called *plastic deformation*.

3.32 PLASTIC HINGE

To illustrate the concept of a plastic hinge, consider the behavior of a simple beam of elastic-plastic material subjected to a steadily increasing concentrated load P at the midspan, as shown in Figure 3.27.

The bending moment diagram is triangular, with the maximum moment M_{max} equal to $PL/4$. When the bending moment is increased beyond the yield moment M_y, the strain in the outermost fibers increases beyond the yield strain ϵ_y so that a region of contained plastic flow will exist in the central part of the beam. A further increase of bending moment causes yield to spread inward toward the neutral axis in the central part of beam. Ultimately, the two zones of yield meet. The curvature at the center of beam then becomes extremely large, and no

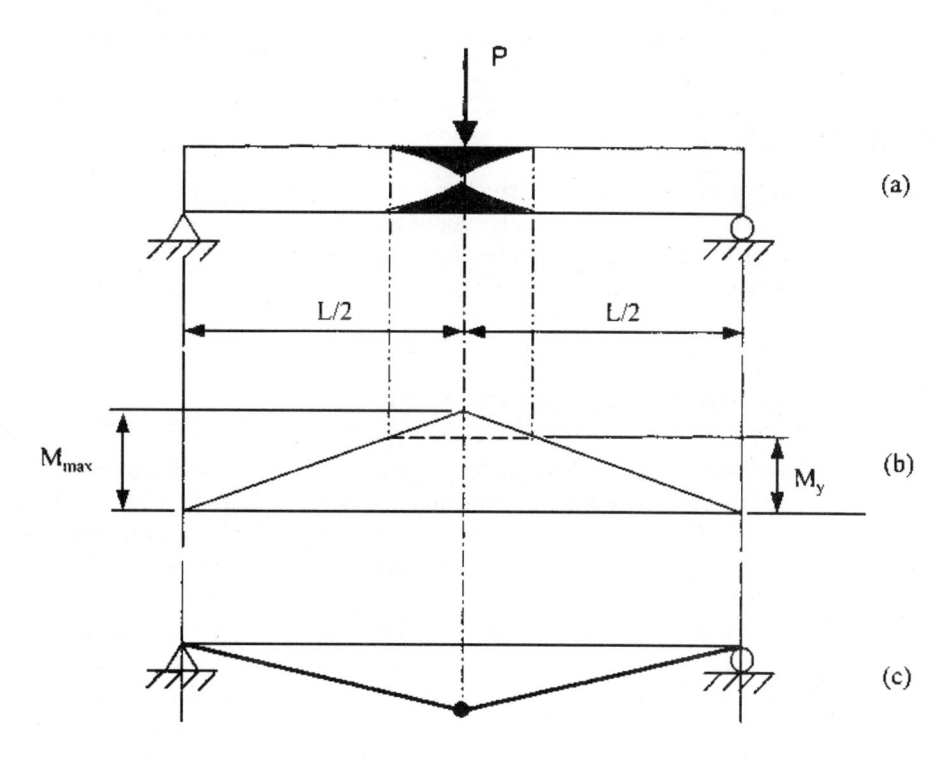

(Partially plastic beam with central concentrated load)

FIGURE 3.27

further increase in the maximum moment can occur. The beam fails by excessive rotations at the middle cross section, while the two halves of the beam remain comparatively rigid. Thus, the beam behaves like two rigid bars linked by a plastic hinge that permits the bars to rotate relative to one another under the action of a constant moment M_p. Therefore, the definition of a plastic hinge is a zone of yielding due to flexure in a structural member. At those sections where plastic hinges are located, the member acts as though it were hinged, except with a constant restraining moment M_p.

3.33 COLLAPSE MECHANISM

It has been shown that if the loads applied to a statically determined structure, such as a simply supported beam, steadily increase, collapse occurs as soon as a single plastic hinge forms. A beam in this condition forms a mechanism that can continue to deflect under the ultimate load. The terms *failure mechanism* and *collapse mechanism* often are used to describe this condition. In contrast, a statically indeterminate structure, such as a fixed-ended beam, generally does not collapse when the first plastic hinge forms at the most highly stressed section of beam. Instead, the beam will behave as a statically determinate simple beam. As the load further increases, the rotation takes place at the first plastic hinge, and the bending moment at this hinge stays constant at the fully plastic moment M_p. Eventually, another plastic hinge forms, the structure has formed a mechanism, and no further increase in the load is possible.

3.34 COLUMNS

A column is a compression member whose unbraced length is at least 10 times its lateral dimension. Columns are usually subdivided into three groups: short, intermediate, and long columns. The distinction between the groups is determined by their mode of failure. In general, long columns fail by buckling or excessive lateral bending, intermediate columns fail by a combination of crushing and buckling, and short columns fail by crushing or yielding.

3.35 CRITICAL BUCKLING LOAD OF A COLUMN

If a column is perfectly straight, the material is perfectly homogeneous, and the compressive loads are truly axial, the column will remain straight under any value of loads, and it will never bend. However, actual columns are never perfectly straight—they always contain small imperfections of material, unavoidable flaws of fabrication, and are rarely perfectly axially loaded. The initial crookedness of the column and the placement of the lateral loads cause it to deform laterally. As the load is increased, the deflection can become large and can lead to a catastrophic failure. This situation is called *buckling* and is defined as a sudden large deformation of a structure because of slight increase of an existing load under which the structure had exhibited little deformation before the load was increased. The *critical load,* or *buckling load,* is the maximum load a column can support and still remain straight. A small increase above this load will increase the deflection significantly, with possibility of stability failure.

3.36 SLENDERNESS RATIO

Swiss mathematician Leonhard Euler published the first solution of the critical load for long, slender columns, known as Euler's column formula, in 1757. The Euler's formula for a hinge-end column is given by

$$P_{cr} = \frac{\pi^2 EI}{L^2} \tag{3.62}$$

where P_{cr} = Critical column load, also called the Euler buckling load
E = Modulus of elasticity
I = Moment of inertia
L = Column length.

Euler's formula shows that the critical load for a column is independent of the strength of material. Therefore, using high-strength steel in lieu of ordinary structural steel will not increase the critical load of a slender steel column. The critical load, however, can be increased by increasing the moment of inertia I of its cross section. For a given area, this result can be accomplished by distributing the material as far as possible from the centroid of the cross section.

The critical compressive stress in a centrally loaded column is founded by dividing the critical load by the cross-sectional area. Thus, the critical stress is

$$\sigma_{cr} = \frac{P_{cr}}{A} = \frac{\pi^2 EI}{AL^2} = \frac{\pi^2 E}{(L/\gamma)^2} \tag{3.63}$$

in which we have replaced I by $A\gamma^2$ and γ is the least radius of gyration of the cross section, i.e.,

$$\gamma = \sqrt{\frac{I}{A}} \tag{3.64}$$

The ratio L/γ is called the slenderness ratio of the column.

The critical load for columns with other end conditions can be expressed in terms of the critical load for a hinge-end column. Equation 3.62 becomes

$$P_{cr} = \frac{\pi^2 EI}{(KL)^2} \tag{3.65}$$

where KL is the effective column length.

3.37 EFFECTIVE LENGTH OF A COLUMN

The effective column length represents the distance between adjacent points of inflection in the buckled column or in the imaginary extension of the buckled column. The effective length factor K is the ratio of the effective length of an idealized hinge-end column to the actual length of a column with various other end conditions. Theoretical values of K for some idealized column end conditions are given in the following table:

Case	End Conditions	Theoretical Effective Length Factors, K	Recommended Effective Length Factors, K
1	Both ends hinged	1.0	1.0
2	One end hinged, one end fixed	0.7	0.8
3	Both ends fixed	0.5	0.65
4	One end fixed, one end free	2.0	2.1
5	One end fixed, one end rotation fixed and translation free	1.0	1.2
6	One end hinged, one end rotation fixed and translation free.	2.0	2.0

The theoretical K values in the above table are for six idealized conditions in which joint rotation and translation are either free or fully fixed. However, for most real situations because of friction or the restraining influence of foundations, a hinge-end column is never completely free to rotate. On the other hand, in buildings, bridges, and similar structures, columns are framed into horizontal beams, walls, or adjacent structures, and joints are seldom fully fixed because of the flexibility of the supporting members. Therefore, the effective length factor is dependent on the amount of bending stiffness and lateral support supplied at column ends. Methods for estimating the effective length can be found in *Manual of Steel Construction—Allowable Stress Design* by the American Institute of Steel Construction.

3.38 BEAM COLUMNS

Beam columns are members subjected to both axial compression and bending. The bending can be caused either by moments applied to the ends of member or by transverse loads acting directly on the member. In the previous sections, we have discussed the effect of the small amounts of bending caused by unavoidable imperfections in columns. In this section, we will learn how the members behave when both the bending and the axial compression are due to intentionally applied loads. In order to understand this behavior, we begin by considering a simply supported member subjected a transverse concentrated load Q at midspan and axial force P, as shown in Figure 3.28a. The free-body diagram is shown in Figure 3.28b. Equating the external moment at a distance x from the left support to the internal resisting moment $-EIy''$ gives

$$EI\frac{d^2y}{dx^2} + Py = -\frac{Qx}{2} \qquad (3.66)$$

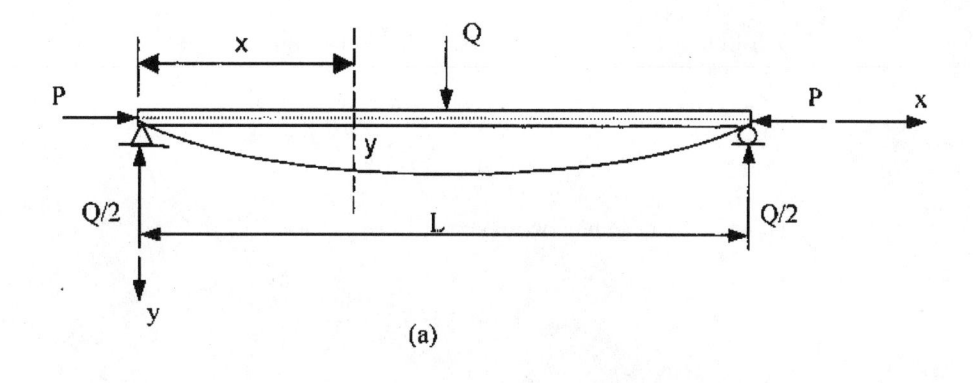

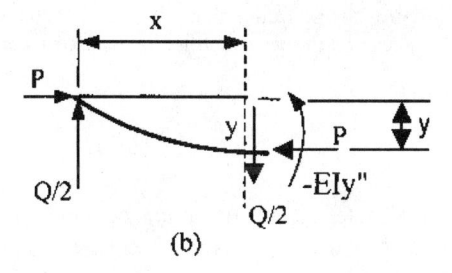

FIGURE 3.28

The boundary conditions are $y = 0$ at $x = 0$, and $dy/dx = 0$ at $x = L/2$. The solution of the differential equation is

$$y = \frac{Q}{2Pk}\left[\frac{\sin kx}{\cos(kL/2)} - kx\right] \tag{3.67}$$

in which

$$k^2 = \frac{P}{EI} \tag{3.68}$$

By assigning $x = L/2$, the maximum deflection δ is found to be

$$\delta = \frac{Q}{2Pk}(\tan u - u) \tag{3.69}$$

where $u = kL/2$

Multiplying and dividing Equation 3.69 by $L^3/24EI$ gives

$$\delta = \frac{QL^3}{48EI}\frac{24EI}{kPL^3}(\tan u - u) \tag{3.70}$$

Substitutions of $P = k^2 EI$ and $u = kL/2$ in Equation 3.70 give

$$\delta = \frac{QL^3}{48EI}\frac{3}{u^3}(\tan u - u) \tag{3.71}$$

This equation can be rewritten as

$$\delta = \delta_0 \frac{3(\tan u - u)}{u^3} \tag{3.72}$$

where $\delta_0 = QL^3/48EI$ is the deflection that would exist if the transverse load were acting by itself. The expression for δ can be further simplified by the expansion of $\tan u$ and substitution of P_{cr},

$$\delta = \delta_0 \frac{1}{1 - (P/P_{cr})} \tag{3.73}$$

This equation indicates that the maximum deflection of the member simultaneously bent by a transverse load Q and an axial force P is equal to δ_0, the maximum deflection that would exist if only Q were acting, multiplied by an amplification factor that depends on the ratio P/P_{cr}. This equation also shows that the deflection of the beam column increases without bound as P/P_{cr} approaches unity. In other words, the resistance of the member to lateral deformation vanishes as the axial load approaches the critical load. For design purpose, the expression for moment M is of greater importance than the deflection δ. It can be shown that the maximum bending moment including the axial compression effect is given as

$$M_{max} = M_0 \frac{1 - (0.18P/P_{cr})}{1 - (P/P_{cr})} \tag{3.74}$$

This equation shows that the effect of axial compression on the bending moment is very similar to the effect that an axial load has on the deflection. The maximum moment that would exist if no axial force were present is magnified by the presence of an axial load.

The equations for other end and load conditions similar to the corresponding expressions for deflection and moment for a concentrated transverse load on a simply supported member can be obtained in a similar way.

3.39 COMBINED STRESSES

In previous sections, we have studied stresses, which were loaded in such a manner that an axial stress (tension or compression), torsional or transverse shearing stress, or a bending stress were acting on a structure one at a time. In this section, we consider a more general state of stresses in which two or more of these stresses act simultaneously on a structure.

The four basic types of loads and the corresponding stress formula may be summarized as follows:

Axial load P Axial stress $$\sigma = \frac{P}{A} \tag{3.75}$$

Flexural (bending) load M Flexural stress $$\sigma_b = \frac{My}{I} \tag{3.76}$$

Torsional shear load T Torsional stress $$\tau = \frac{T\rho}{J} \tag{3.77}$$

Transverse shear load V Transverse shear $$\tau = \frac{VQ}{Ib} \tag{3.78}$$

Possible combinations of these loads include:

1. Axial and flexural
2. Axial and torsional shear
3. Torsional shear and flexural
4. Flexural and transverse shear
5. Axial, flexural, and torsional shear
6. Axial, flexural, torsional shear, and transverse shear.

Since the axial stress P/A and the bending stress My/I both belong to the family of normal stresses (acting normal to the cross section), the resultant stress is equal to the superposition of the two stresses acting separately. The combined axial and flexural stress then, is

$$\sigma = \frac{P}{A} \pm \frac{My}{I} \tag{3.79}$$

As a result of this addition of stresses, the neutral axis shifts away from the centroidal axis of the section. The new location of the neutral axis can be found by setting $\sigma = 0$.

Torsional and transverse shearing stresses, however, are tangential stresses acting along or parallel to area resisting the forces, whereas the axial and bending stresses are perpendicular or normal to the areas on which they act—they cannot be added algebraically. The tool we use to combine the shearing stresses with normal stresses is the Mohr's circle.

3.40 PRINCIPAL STRESSES AND PLANES

As mentioned in Section 3.2, when a prismatic bar is subjected to simple tension or compression, the stresses on a cross section normal to its axis are distributed uniformly and equal to P/A. Now let us consider the stresses on an inclined plane whose outward normal makes an angle θ with respect to the x axis. If the equations of equilibrium are applied to the free-body diagram of Figure 3.29, the resultant force distributed over the inclined plane can be resolved into two components, N and V, which are normal and tangential to the inclined plane respectively.

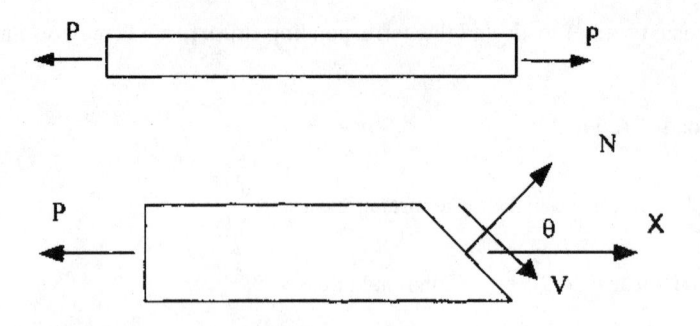

(Stress on inclined plans for a bar in tension)
FIGURE 3.29

These components are $N = P\cos\theta$ and $V = P\sin\theta$

Since the area A' of the inclined section is $A/\cos\theta$, the stresses corresponding to N and V are

$$\sigma_\theta = \frac{N}{A'} = \frac{P\cos\theta}{A/\cos\theta} = \frac{P}{A}\cos^2\theta = \sigma\cos^2\theta = \frac{\sigma}{2}(1 + \cos 2\theta) \quad (3.80)$$

and

$$\tau_\theta = \frac{V}{A'} = \frac{P\sin\theta}{A/\cos\theta} = \frac{P}{A}\sin\theta\cos\theta = \frac{\sigma}{2}\sin 2\theta \quad (3.81)$$

Therefore, once the stress on the cross section normal to the axis of the member through a point is known, the normal and shearing stresses on any specific oblique plane through the same point can be determined from Equations 3.80 and 3.81. The maximum normal stress occurs when $\theta = 0$ and is

$$\sigma_{max} = \frac{P}{A} \quad (3.82)$$

The maximum shear stress occurs when $\theta = \pi/4$ and is

$$\tau_{max} = \frac{P}{2A} \quad (3.83)$$

Note that the normal stress is either maximum or minimum on planes for which the shearing stress is zero.

Consider a more general state of stress in which the normal stresses on an element act in both the x and y directions, known as biaxial stress (Figure 3.30a). The stresses acting on the inclined plane whose outward normal makes an angle θ with the x axis are the normal stress σ_θ and the shear stress τ_θ shown in Figure 3.30b.

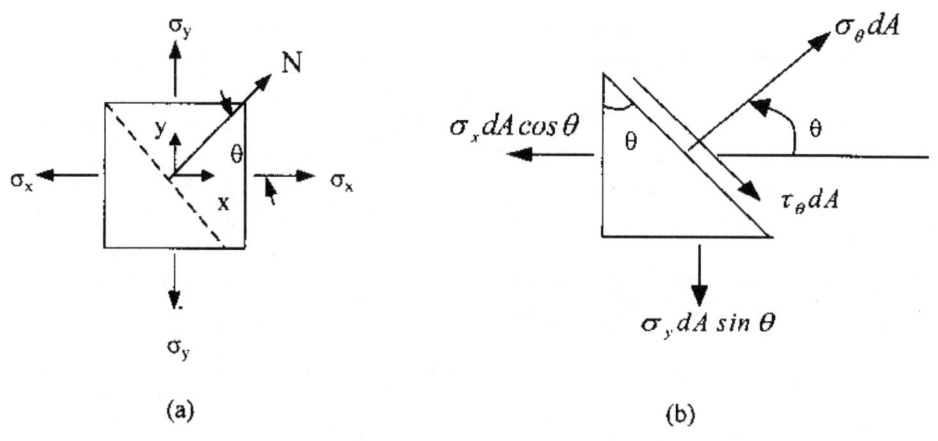

FIGURE 3.30

The force equation of equilibrium in the direction of σ_θ gives

$$\sigma_\theta dA - \sigma_x(dA\cos\theta)\cos\theta - \sigma_y(dA\sin\theta)\sin\theta = 0,$$

from which

$$\sigma_\theta = \sigma_x\cos^2\theta + \sigma_y\sin^2\theta \qquad (3.84)$$

The force equation of equilibrium in the direction of τ_θ gives

$$\tau_\theta dA + \sigma_x(dA\cos\theta)\sin\theta - \sigma_y(dA\sin\theta)\cos\theta = 0$$

or

$$\tau_\theta = (\sigma_x - \sigma_y)\sin\theta\cos\theta \qquad (3.85)$$

Using the trigonometric relations in terms of the double angle, we can rewrite the equations in the following alternate form:

$$\sigma_\theta = \frac{\sigma_x + \sigma_y}{2} + \frac{\sigma_x - \sigma_y}{2}\cos 2\theta \qquad (3.86)$$

$$\tau_\theta = \frac{\sigma_x - \sigma_y}{2}\sin 2\theta \qquad (3.87)$$

Now consider the plane stress situation indicated in Figure 3.31a, in which the stress states can be represented by components that act parallel to a single plane. The components of stress present for plane stress analysis include σ_x, σ_y, and $\tau_{xy} = \tau_{yx}$. The sign convention for stresses is as follows: Normal stresses (indicated by means of a single subscript corresponding to the face on which they act) are positive if they point in the direction of the outward normal. A face takes the name of the axis normal to it. For example, the x face is normal to the x axis. Thus, normal stresses are positive if tensile. A shearing stress is denoted by the symbol τ followed by two subscripts, the first subscript corresponding to the face on which the shearing stress acts and the second indicating the direction in which it acts. Thus τ_{xy} is the shearing stress on the x face acting in the y direction. A positive shearing stress points in the positive direction of the coordinate axis of the second subscript if it acts on a surface with an outward normal in the positive direction. Conversely, if the outward normal of the surface is in the negative direction, then the positive shearing stress points in the negative direction of the coordinate axis of the second subscript.

Figure 3.31b shows the normal and shearing stress components acting on an inclined plane whose outward normal makes an angle θ with the x axis. Applying the equilibrium of forces in the direction of σ_θ gives

$$\sigma_\theta dA - \sigma_x(dA\cos\theta)\cos\theta - \sigma_y(dA\sin\theta)\sin\theta - \tau_{yx}(dA\sin\theta)\cos\theta - \tau_{xy}(dA\cos\theta)\sin\theta = 0$$

from which, since $\tau_{xy} = \tau_{yx}$, we obtain

$$\sigma_\theta = \sigma_x\cos^2\theta + \sigma_y\sin^2\theta + 2\tau_{xy}\sin\theta\cos\theta \tag{3.88}$$

The equilibrium of forces in the direction of τ_θ gives

$$\tau_\theta = (\sigma_x - \sigma_y)\sin\theta\cos\theta + \tau_{xy}(\sin^2\theta - \cos^2\theta) \tag{3.89}$$

Again using the trigonometric relations in terms of the double angle, we can rewrite the equations in the following alternate form:

$$\begin{aligned}\sigma_\theta &= \frac{\sigma_x(1+\cos 2\theta)}{2} + \frac{\sigma_y(1-\cos 2\theta)}{2} + \frac{2\tau_{xy}\sin 2\theta}{2}\\ &= \frac{\sigma_x+\sigma_y}{2} + \frac{\sigma_x-\sigma_y}{2}\cos 2\theta + \tau_{xy}\sin 2\theta\end{aligned} \tag{3.90}$$

(a)

(b)

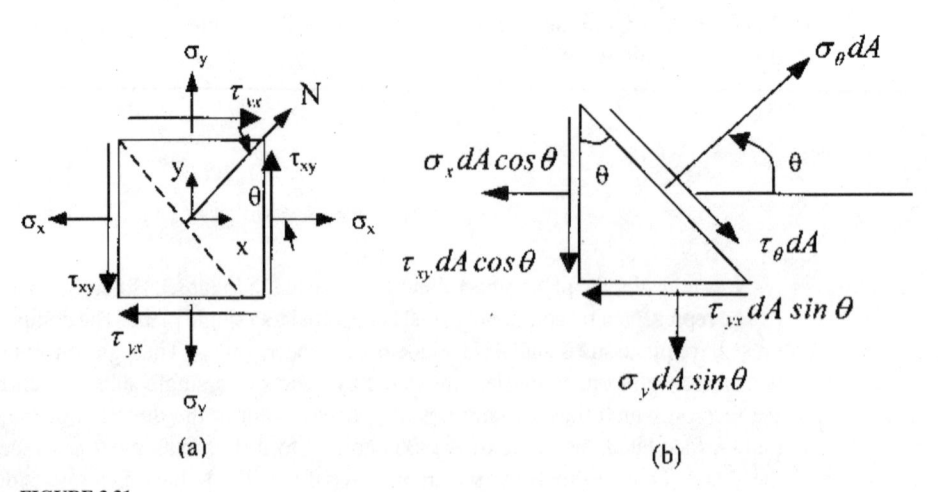

FIGURE 3.31

and

$$\tau_\theta = \frac{\sigma_x - \sigma_y}{2} \sin 2\theta - \tau_{xy} \cos 2\theta \qquad (3.91)$$

The maximum and minimum value of σ_θ and the planes on which they occur can be located at the values of θ for which $d\sigma_\theta/d\theta = 0$. Thus, differentiation of σ_θ of Equation 3.90 with respect to θ gives

$$\tan 2\theta_p = \frac{2\tau_{xy}}{\sigma_x - \sigma_y} \qquad (3.92)$$

The maximum and minimum normal stresses are called the *principal stresses*, and the planes on which they act are called *principal planes*.

Substituting the values of 2θ from Equation 3.92 into Equation 3.90, we obtain

$$\sigma_{1,2} = \frac{\sigma_x + \sigma_y}{2} \pm \sqrt{\left(\frac{\sigma_x - \sigma_y}{2}\right)^2 + \tau_{xy}} \qquad (3.93)$$

where σ_1 and σ_2 denote the algebraically maximum and minimum principal stresses respectively.

If we set $\tau_\theta = 0$ in Equation 3.91 and solve for 2θ; this gives

$$\tan 2\theta = \frac{2\tau_{xy}}{\sigma_x - \sigma_y}$$

which is identical to Equation 3.92. This shows that there are no shear stresses on the principal planes.

Another useful relation of the normal stresses and shear stresses on the orthogonal planes is obtained by replacing θ with $\theta + \pi/2$ in Equations 3.90 and 3.91; this gives

$$\sigma'_\theta = \frac{\sigma_x + \sigma_y}{2} - \frac{\sigma_x - \sigma_y}{2} \cos 2\theta - \tau_{xy} \sin 2\theta \qquad (3.94)$$

and

$$\tau'_\theta = -\frac{\sigma_x - \sigma_y}{2} \sin 2\theta + \tau_{xy} \cos 2\theta \qquad (3.95)$$

Adding these two expressions and Equations 3.90 and 3.91, we find that

$$\sigma_\theta + \sigma'_\theta = \sigma_x + \sigma_y \quad \text{and} \quad \tau'_\theta = -\tau_\theta \qquad (3.96)$$

The result shows that the sum of the normal stresses on any two perpendicular planes is a constant, and the shear stresses on two perpendicular planes are equal in magnitude and opposite in direction.

3.52 CHAPTER 3

The maximum shear stress and the planes on which they act can be determined by taking the derivative $d\tau_\theta/d\theta$ of Equation 3.91 and setting it equal to zero. Thus, differentiating Equation 3.91, we find

$$\tan 2\theta_s = -\frac{\sigma_x - \sigma_y}{2\tau_{xy}} \tag{3.97}$$

Substituting the values of $2\theta_s$ determined from Equation 3.97 into Equation 3.91, we find the following expressions for the maximum shearing stress:

$$\tau_{max} = \pm\sqrt{\left(\frac{\sigma_x - \sigma_y}{2}\right)^2 + \tau_{xy}^2} \tag{3.98}$$

Comparing Equations 3.93 and 3.98, we observed that τ_{max} has the same magnitude as the second term of the expressions for the principal stresses. By subtracting the values of the two principal stresses, we find that τ_{max} also can be expressed as

$$\tau_{max} = \frac{\sigma_1 - \sigma_2}{2} \tag{3.99}$$

A comparison of Equations 3.92 and 3.97 reveals that the two tangents are negative reciprocals. This means that the two angles $2\theta_p$ and $2\theta_s$ differ by 90°. In other words, the planes on which the maximum shearing stresses occur are at 45° with the principal planes.

If we substitute the values of $\sin 2\theta_s$ and $\cos 2\theta_s$ from Equation 3.97 into Equation 3.90, we find

$$\sigma = \tfrac{1}{2}(\sigma_x + \sigma_y) \tag{3.100}$$

Thus on each plane of the maximum shearing stress we have a normal stress of magnitude $\tfrac{1}{2}(\sigma_x + \sigma_y)$.

EXAMPLE 3.6

An element in plane stress is subject to the stresses $\sigma_x = 12{,}000$ psi, $\sigma_y = 15{,}000$ psi, and $\tau_{xy} = -8000$ psi, as shown in Figure 3.32a. Determine the principal stresses and principal planes, the maximum shearing stresses and the directions of the planes on which they occur, and the stresses on an element rotated through an angle of 45°.

Solution

1. In accordance with Equation 3.93, the maximum principal stress is

$$\sigma_1 = \tfrac{1}{2}(\sigma_x + \sigma_y) + \sqrt{\left[\tfrac{1}{2}(\sigma_x - \sigma_y)\right]^2 + (\tau_{xy})^2}$$

$$= \tfrac{1}{2}(12{,}000 + 15{,}000) + \sqrt{\left[\tfrac{1}{2}(12{,}000 - 15{,}000)\right]^2 + (-8000)^2}$$

$$= 13{,}500 + 8140 = 21{,}640 \text{ psi}$$

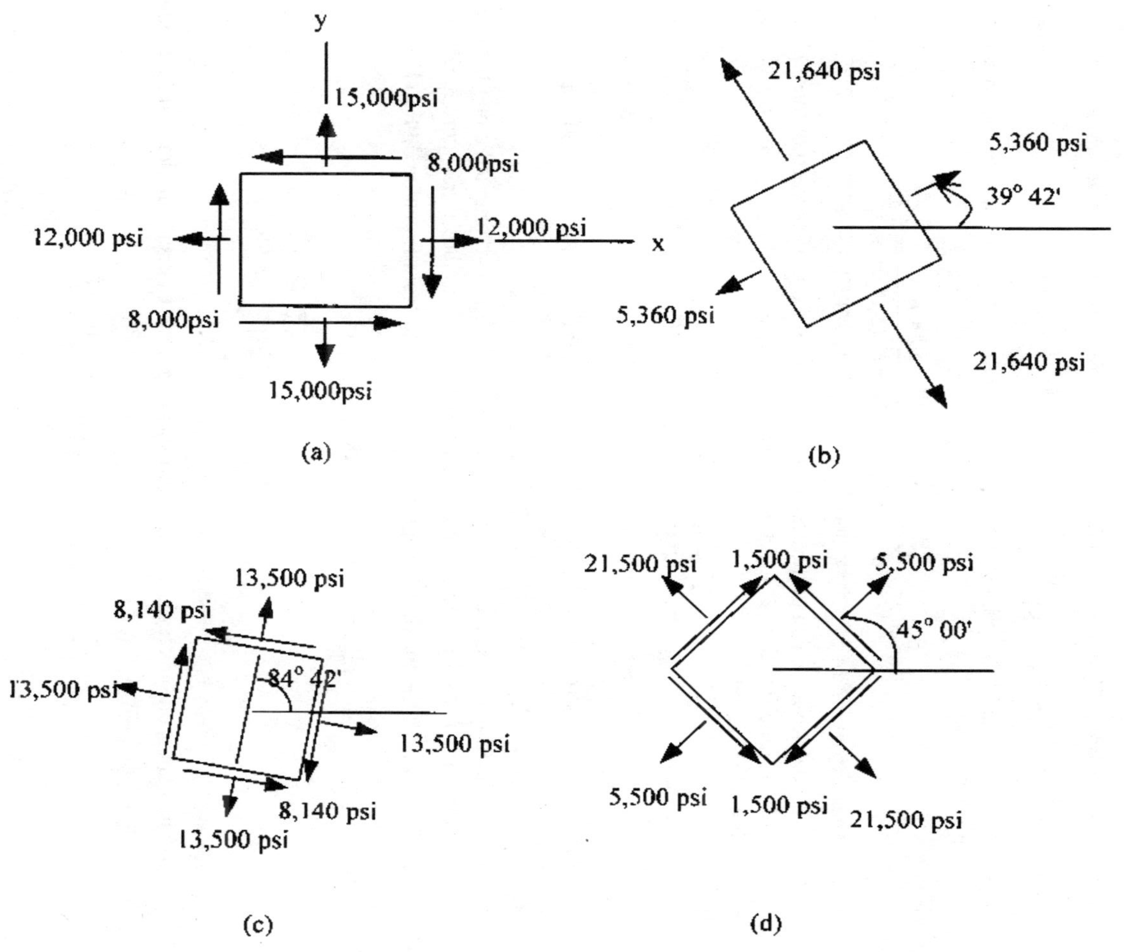

FIGURE 3.32

The minimum principal stress is

$$\sigma_2 = \tfrac{1}{2}(\sigma_x + \sigma_y) - \sqrt{\left[\tfrac{1}{2}(\sigma_x - \sigma_y)\right]^2 + (\tau_{xy})^2}$$

$$= \tfrac{1}{2}(12{,}000 + 15{,}000) - \sqrt{\left[\tfrac{1}{2}(12{,}000 - 15{,}000)\right]^2 + (-8000)^2}$$

$$= 13{,}500 - 8140 = 5360 \text{ psi}$$

The direction of the principal plane on which these principal stresses occur are given by

$$\tan 2\theta_p = \frac{2\tau_{xy}}{\sigma_x - \sigma_y} = \frac{2(-8000)}{12{,}000 - 15{,}000} = 5.33$$

Then $2\theta_p = 79°24'$, $259°24'$ and $\theta_p = 39°42'$, $129°42'$.

To determine which of the above principal stresses occurs on each of these planes, we substitute $\theta = 39°42'$ together with the given values of σ_x, σ_y, and τ_{xy} into Equation 3.90, namely,

$$\sigma_\theta = \frac{\sigma_x + \sigma_y}{2} + \frac{\sigma_x - \sigma_y}{2}\cos 2\theta + \tau_{xy}\sin 2\theta$$

We obtain

$$\sigma = \tfrac{1}{2}(12{,}000 + 15{,}000) + \tfrac{1}{2}(12{,}000 - 15{,}000)\cos 79°24' + (-8000)\sin 79°24'$$

$$= 13{,}500 - 276 - 7863 = 5360 \text{ psi}$$

The shearing stresses on these planes are zero. An element oriented along the principal planes and subject to the above principal stresses is shown in Figure 3.32b.

2. The value of the maximum shearing stress is found from Equation 3.98 to be

$$\tau_{max} = \pm\sqrt{\left[\tfrac{1}{2}(\sigma_x - \sigma_y)\right]^2 + (\tau_{xy})^2}$$

$$= \pm\sqrt{\left[\tfrac{1}{2}(12{,}000 - 15{,}000)\right]^2 + (-8000)^2}$$

$$= \pm 8140 \text{ psi}$$

The planes on which these maximum shearing stresses occur are defined by Equation 3.97, namely,

$$\tan 2\theta_s = -\frac{\sigma_x - \sigma_y}{2\tau_{xy}} = -\frac{12{,}000 - 15{,}000}{2(-8000)} = -0.1875$$

Thus, $2\theta_s = 169°24'$, $349°24'$ and $\theta_s = 84°42'$, $174°42'$. Evidently these planes are located 45° from the planes of maximum and minimum principal stresses.

To determine whether the shearing stresses is positive or negative on the 84°42' plane, we substitute $\theta = 84°42'$ together with the given values of σ_x, σ_y, and τ_{xy} into Equation 3.91, namely

$$\tau_\theta = \frac{\sigma_x - \sigma_y}{2}\sin 2\theta - \tau_{xy}\cos 2\theta$$

We obtain

$$\tau = \tfrac{1}{2}(12{,}000 - 15{,}000)\sin 169°24' - (-8000)\cos 169°24' = -8140 \text{ psi}$$

The magnitude of the normal stress on these planes of maximum shearing stresses are found from Equation 3.100 to be

$$\sigma = \tfrac{1}{2}(12{,}000 + 15{,}000) = 13{,}500 \text{ psi}$$

The orientation of the element for which the shearing stresses are maximum is as shown in Figure 3.32c.

3. The stresses on an element rotated through an angle of 45° can be found from Eqs.3.90 and 3.91. Substituting $\theta = 45°$ into these equations gives

$$\sigma = \tfrac{1}{2}(12{,}000 + 15{,}000) + \tfrac{1}{2}(12{,}000 - 15{,}000)\cos(2 \times 45°)$$
$$+ (-8000)\sin(2 \times 45°) = 5500 \text{ psi}$$

and

$$\tau = \tfrac{1}{2}(12{,}000 - 15{,}000)\sin(2 \times 45°) - (-8000)\cos(2 \times 45°) = -1500 \text{ psi}$$

On the plane $\theta = 135°$, the normal stress is

$$\sigma = \tfrac{1}{2}(12{,}000 + 15{,}000) + \tfrac{1}{2}(12{,}000 - 15{,}000)\cos(2 \times 135°)$$
$$+ (-8000)\sin(2 \times 135°) = 21{,}500 \text{ psi}$$

and

$$\tau = \tfrac{1}{2}(12{,}000 - 15{,}000)\sin(2 \times 135°) - (-8000)\cos(2 \times 135°) = 1500 \text{ psi}.$$

These stresses are shown in Figure 3.32d.

3.41 DETERMINING PRINCIPAL STRESSES USING MOHR'S CIRCLE

In Section 3.41 the equations for finding the principal stresses and the maximum shearing stresses at a point in a stressed member were developed. We can show that Equations 3.90 and 3.91 define a circle by rewriting them as follows:

$$\sigma_\theta - \frac{\sigma_x + \sigma_y}{2} = \frac{\sigma_x - \sigma_y}{2}\cos 2\theta + \tau_{xy}\sin 2\theta$$

and

$$\tau_\theta = \frac{\sigma_x - \sigma_y}{2}\sin 2\theta - \tau_{xy}\cos 2\theta$$

Squaring both sides of these equations and then adding them together, we obtain

$$\left(\sigma_\theta - \frac{\sigma_x + \sigma_y}{2}\right)^2 + \tau_\theta^2 = \left(\frac{\sigma_x - \sigma_y}{2}\right)^2 + \tau_{xy}^2 \qquad (3.101)$$

This is the equation of a circle in terms of the variables σ_θ and τ_θ. The circle is centered on the σ axis at a distance $(\sigma_x + \sigma_y)/2$ from the τ axis, and the radius of the circle is given by the relation

$$R = \sqrt{\left(\frac{\sigma_x - \sigma_y}{2}\right)^2 + \tau_{xy}^2} \qquad (3.102)$$

This graphic interpretation of the stress variation by Equations 3.90 and 3.91 frequently is called *Mohr's circle* in honor of German engineer Otto Mohr, who developed it in 1882.
The procedure for construction of Mohr's circle is shown in Figure 3.33.
The steps are as follows:

1. We begin by establishing the stresses σ_x, σ_y, and τ_{xy} on two mutually perpendicular planes through a known point (if not given). On the σ-τ coordinate axes, plot point A having the coordinates (σ_x, τ_{xy}) and point B having the coordinates $(\sigma_y, -\tau_{xy})$. These points represent the normal and shearing stresses acting on the x face ($\theta = 0$) and y face ($\theta = \pi/2$). In plotting these points, we assumed the normal tensile stress as positive and the normal compressive stress as negative. The convention used for shearing stresses is that shearing stresses that form a clockwise couple are positive.

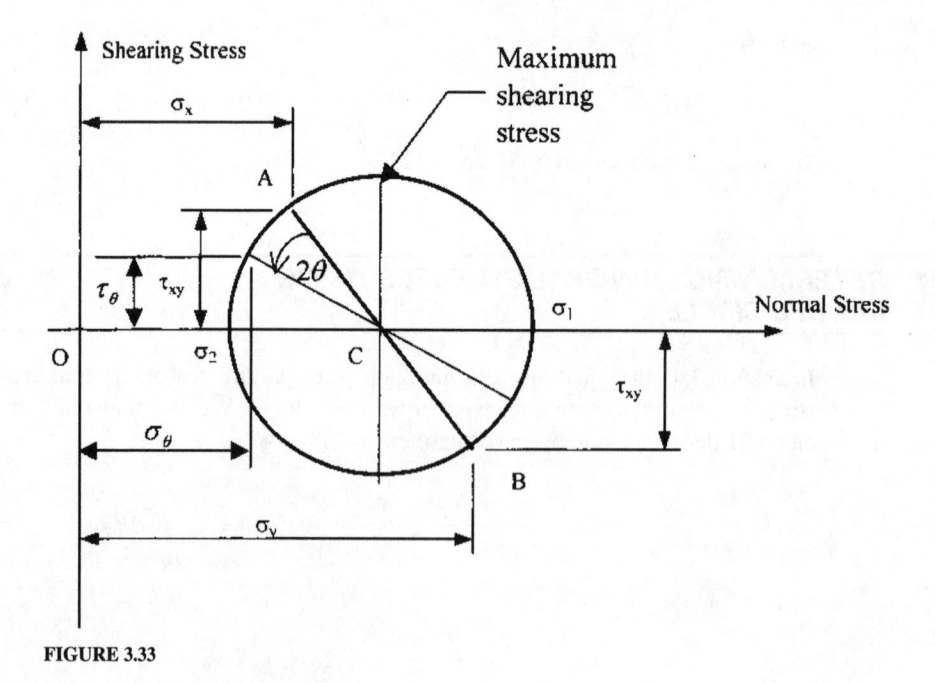

FIGURE 3.33

2. Join the points just plotted by a straight line. The intersection of this line with the σ axis of the coordinate system is the center of the Mohr's circle, and the line is the diameter of Mohr's circle.

3. The two points of intersection of the circle with the σ axis represent the maximum and minimum principal stresses σ_1 and σ_2, respectively.

4. The two points of intersection of the circle with the vertical line through the curve center represent the maximum shearing stresses τ.

Having drawn Mohr's circle, we are able to determine the stresses on any inclined plane at an angle θ with the x axis. It is only necessary to lay off a counterclockwise angle 2θ from the x axis (point A), thereby locating a point on the circle that has coordinates σ_θ, and τ_θ equal to the stresses on the corresponding plane.

EXAMPLE 3.7
An element in plane stress is subject to the stresses $\sigma_x = 12,000$ psi, $\sigma_y = 15,000$ psi, and $\tau_{xy} = -8000$ psi, as shown in Figure 3.32a. Using Mohr's circle, determine the principal stresses and principal planes, the maximum shearing stresses and the directions of the planes on which they occur, and the stresses on an element rotated through an angle of 45°. (Note that this same problem was solved in Example 3.6.)

Solution
Following the procedure outlined in Section 3.41, since the shear stresses on the vertical faces of the given element as shown in Figure 3.32a form a clockwise couple, the shearing stresses are positive, while those on the horizontal faces are negative. Thus the stress condition of $\sigma_x = 12,000$ psi, $\tau_{xy} = 8000$ psi, existing on the vertical faces of the element plots as point A in Figure 3.31. The stress condition of $\sigma_y = 15,000$ psi, and $\tau_{xy} = -8000$ psi, existing on the horizontal faces plots as point B. Line AB is drawn, and the intersection of this line and the σ axis, its midpoint C, is located. A circle of radius $\overline{CA} = \overline{CB}$ with center C is drawn. This is Mohr's circle. The endpoints of the diameter $\overline{AB}$ represent the stress condition existing if it has the original orientation of Figure 3.32a.

1. The principal stresses are represented by points G and H, as shown in Figure 3.34. The principal stress can be determined either by direct measurement from Figure 3.34 or by realizing that the coordinate of C is 13,500 psi = $\frac{1}{2}(15,000 + 12,000)$ psi, and $\overline{CK} = 1500$, and that $\overline{CA} = \sqrt{(1500)^2 + (8000)^2} = 8140$. Thus, the algebraically larger principal stress σ_1 is equal to $13,500 + 8140 = 21,640$ psi (point H in Figure 3.34). Also, the minimum principal stress is $13,500 - 8140 = 5360$ psi (point G in Figure 3.34).

   From the geometry of the circle, the principal plane has an angle $2\theta_P$ given by $\tan 2\theta_P = 8000/1500 = 5.33$, from which $\theta_P = 39°42'$. This value also could be obtained by the measurement of $\angle ACK$ in Mohr's circle. Thus, the angle from the x axis to the principal plane is $39°42'$. The other principal stress acts on the plane for which $\theta = 39°42' + 90°$. Thus, the principal stresses and principal planes have been found from Mohr's circle, and they can be represented on an element, as shown earlier in Figure 3.32b.

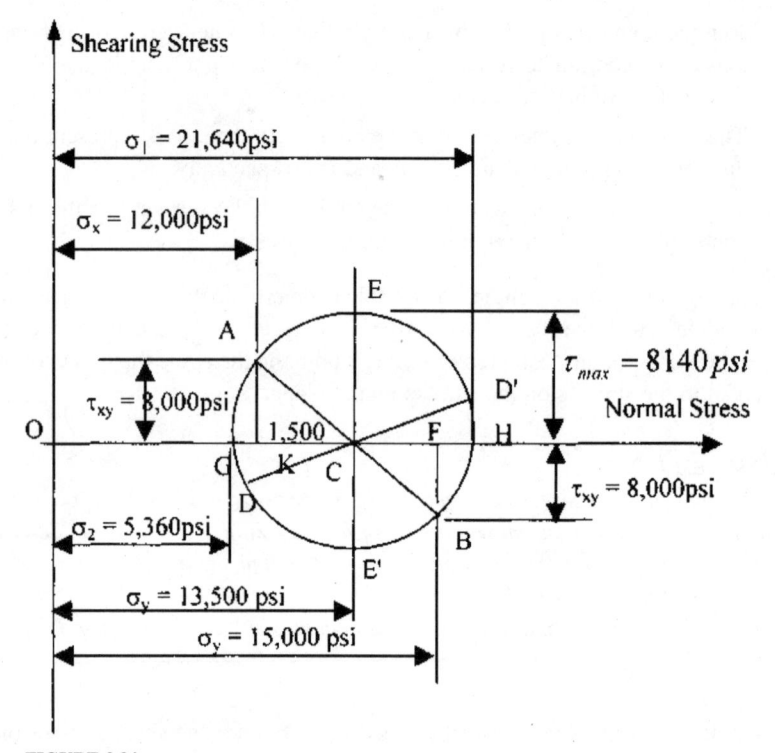

FIGURE 3.34

2. The maximum shear stresses and their planes are represented by points E and E' on Mohr's circle. This radius has been found in (1) to represent 8140 psi. The angle $2\theta_s$ can be found either by direct measurement from the Mohr's circle or by adding 90° to the angle $2\theta_P$, which has already been determined. This leads to $2\theta_s = 169°24'$ and $\theta_s = 84°42'$. The shearing stress represented by point E' is negative; hence, on this plane the shearing stress tends to rotate the element in a counterclockwise direction. Also from the Mohr's circle, the abscissa of point E' is equal to the radius, giving the maximum shear stress of 8140 psi.

3. The stresses on an element rotated through an angle of 45° can be found from the Mohr's circle represented by point D, for which $2\theta = 90°$ and $\theta = 45°$. The angle between line $\overline{CD}$ and the σ_θ axis is 90° minus 79°24', or 10°36'. The cosine of this angle is 0.983, and the normal stress represented by point D is $\sigma_\theta = 13,500 - (8140)(0.983) = 5500$ psi. Also, the shear stress is $\tau_\theta = -(8140)(\sin 10°36') = -1500$ psi. Similarly, the stresses at point D' are $\sigma_\theta = 13,500 + (8140)(0.983) = 21,500$ psi and $\tau_\theta = 1500$ psi. These stresses are shown in Figure 3.32d.

CHAPTER 4
PRINCIPLES OF STATICS

4.1 INTRODUCTION

The design of structural members, connections, and systems requires a thorough understanding of the forces or loads acting on them. We need to learn where the forces or loads come from, how they act on a member, connection, or system, and how these structural components or systems react to the forces or loads. We have gained an understanding of the internal stress and strain caused by outside forces and the reaction or resistance of structural members to external forces and loads in Chapter 3, Strength of Materials. Here, we will discuss the sources and effects of external forces or loads on the structural components and systems.

A set of forces or loads acting on a structure and its components can cause the structure either to remain stationary or be in motion. Statics deals with forces that are in equilibrium or keeping a structure stationary. Dynamics deals with bodies and forces that are in motion. Mechanics is the physical science that deals with both statics and dynamics. Mechanics is one of the oldest physical sciences that has attracted the interest of mathematicians, scientists and engineers over the ages. The earliest recorded studies in this field are those of Archimedes (287–212 BC), about the principles of lever and buoyancy. Stevinus (1564–1642) formulated the laws of vector combination of forces and most of the principles of statics. Galileo (1564–1642) was credited as the first to investigate a dynamic problem in connection with his experiments with falling stones. Hooke (1635–1703) established the relationship between forces and deformation—Hooke's law. Newton (1642–1727) accurately formulated the laws of motion and gravity. Da Vinci, Varignon, D'Alembert, Lagrange, Laplace, Euler, and others also made substantial contributions to the development of mechanics. Modern research and development, and progress in the fields of civil engineering, structural engineering, mechanical engineering, electrical engineering, and electronics depend upon the basic principles of mechanics.

The principles of statics are most important to civil and structural engineers who deal with the equilibrium, strength, and stability of buildings, bridges, and other civil engineering structures. This chapter covers the essential elements of statics applicable to structural design of buildings and bridges.

4.2 BASIC CONCEPTS

Statics considers the equilibrium of bodies under the action and reaction of forces. It is based on physical laws and the mathematics used to represent the physical situations. In analyzing a physical situation, it is important to establish the proper relationship between the physical description and the mathematical expression. Some assumptions or approximations often are necessary to solve an engineering problem rapidly. For example, a force is assumed to act as a point load if the area over which it acts is small compared with pertinent dimensions. Similarly, small distances, angles, and weights may be neglected when compared with relatively large distances, angles, and weight. For small angles, the sine and cosine of the angle may be taken as the value of the angle in radian. The weight of a member may be approximated or neglected during preliminary design. The weight of a cable per foot can be neglected if the tension in the cable is large compared with its total weight. The assumptions or approximations depend on the information desired and the accuracy required. The ability to make the proper assumptions and approximations in solving engineering problems comes through experience. The goal of this chapter, indeed, the book, is to provide opportunity and guidance to develop this ability through illustrative examples and practical problems.

In structural engineering, we deal with a structure at rest and stable under a system of external forces. A force is the action of one body on another. A force tends to move a body in the direction of the application of the force. It results in the push, pull, or rotation of a body. For example, the wind blowing on a roof exerts a force on the structural roof system and the building. A truck driving across a bridge exerts a moving load on the structural system of the bridge deck and other components. The concept of statics helps us calculate the external forces that maintain a body or structure in equilibrium.

A force is a vector, meaning it has magnitude and direction. A force can be decomposed into a set of forces having the same effect on the body. Similarly, a set of forces can be combined into a single force. The composition and decomposition of forces help simplify the solution of certain engineering problems. The characteristics of a force or a set of forces will be discussed in detail in this chapter.

4.3 SCALAR AND VECTOR QUANTITIES

We deal with two kinds of quantities in statics: scalar and vector. Scalar quantities are those with magnitude only. Examples of scalars are volume, density, speed, energy, and mass. Scalars can be combined algebraically. Vector quantities are those with magnitude and direction. Examples of vectors are displacement, force, moment, and velocity.

A vector quantity **V** is represented by a straight line, having the magnitude represented by the length of the line, an arrowhead to indicate the direction and an angle θ to represent the direction with reference to a selected line. The negative of **V** is a vector $-**V**$ directed in the opposite direction. Bold letters in this chapter denote vectors and forces.

Vectors can be added and subtracted in accordance with triangle or parallelogram laws. Two vectors—**V**₁ and **V**₂—are shown in Figure 4.1(a). They can be added head-to-tail to obtain their sum, as shown in Figure 4.1(b) using the triangle law. The same result can be obtained using the parallelogram law shown in Figure 4.1(c)

FIGURE 4.1

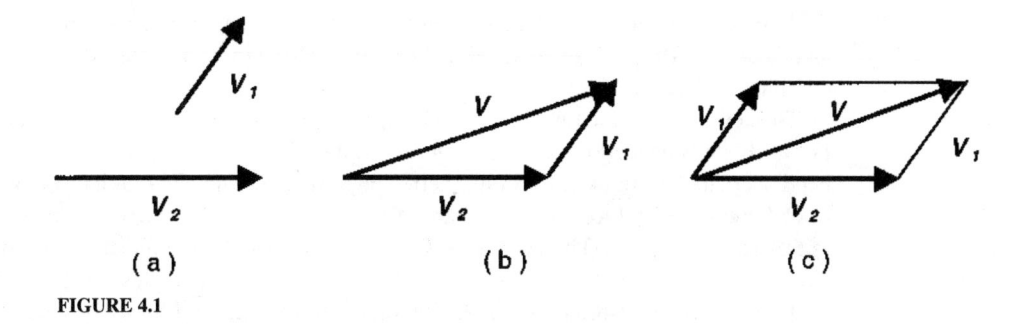

The vector addition can be expressed symbolically

$$V = V_1 + V_2 \qquad (4.1)$$

4.4 NEWTON'S LAWS

Sir Isaac Newton was the first to state correctly the basic laws governing the motion of a particle and to demonstrate their validity. Newton's original formulas may be found in the translation of his Principia (1687) revised by F. Cajori, University of California Press, 1934. Newton's laws are as follows:

> Law I. A particle remains at rest or continues to move in a straight line with a uniform velocity if there is no unbalanced force acting on it.
>
> Law II. The acceleration of a particle is proportional to the resultant force acting on it and is in the direction of this force.
>
> Law III. The forces of action and reaction between contacting bodies are equal in magnitude, opposite in direction, and collinear.

Law I contains the principle of the equilibrium of forces. This is the main topic of concern in our study of statics. Law II forms the basis for the study of dynamics. It will not be of concern to us in this chapter. Law III is basic to our understanding of force. It states that forces always occur in pairs of equal magnitude but opposite direction. This holds for all forces. In analyzing structures under the action of forces, we must be clear as to which pairs of forces are being considered. We will discuss this when we study free-body diagrams.

4.5 SYSTEM OF FORCES

Structural design involves a good understanding of the forces that will act on the members and connections of a structural system. Forces are produced by the loads—generally dead load and live load plus impact—to be supported or resisted by the structure. The behavior of forces acting on the structure and its component must be understood to design properly for the effect of the forces.

A force is a vector quantity. Its effect depends on the magnitude, the direction of action, and the location or point of application. We can resolve a force into components, and we can combine a system of forces into one or more equivalent forces

The action of a force on a body can be separated into two effects, external and internal. For the bolted connection shown in Figure. 4.2, the effects of the external force P to the connection are the reactions of the bolts exerted on the connecting plates as a result of the action of P. Forces external to a body then are of two kinds—applied or active forces and resulting or reactive forces. The effects of P internal to the bolted connection are the internal deformations and forces distributed throughout the connecting plates and bolts of the connection. The relationship between the internal deformations and forces has been studied in Chapter 3, Strength of Materials.

Forces can be concentrated or distributed. In the real world, a force always acts over a finite area. However, for the purpose of engineering analysis, when the dimensions of the area are small compared with other dimensions of the body, the force is normally considered as concentrated at a point. A force may be distributed over a larger area or over the entire length or volume of a member. Considering such a force as concentrated at a point often will not represent the force properly. For computing convenience, however, such a force can be divided into a system of concentrated loads with a practically equivalent effect on the member. The practical applications of concentrated and distributed forces will be dealt with in the design sections.

4.6 COMPOSITION AND RESOLUTION OF FORCES

A system of forces can be combined to produce a resultant force or simply a resultant, which has the same external effect on a body as the system of forces. The resultant of a system of forces is obtained using the triangular or parallelogram law of addition of vectors. For example, two concurrent forces F_1 and F_2 may be added vectorially to obtain the resultant R shown in Figure 4.3(a). If F_1 and F_2 are in the same plane, but acting at two locations, they can

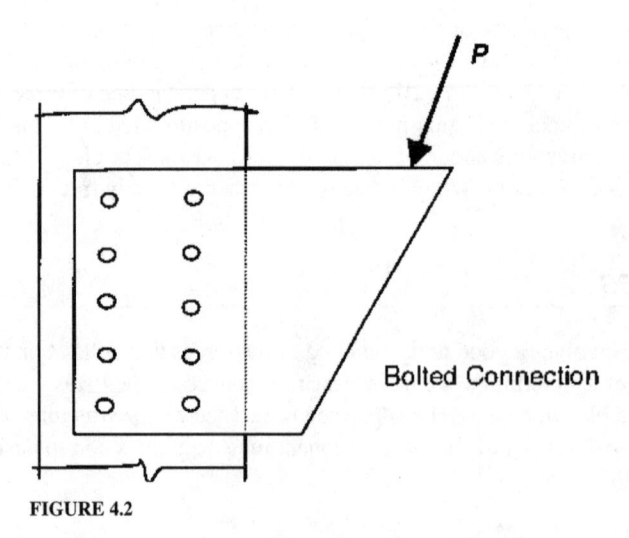

FIGURE 4.2

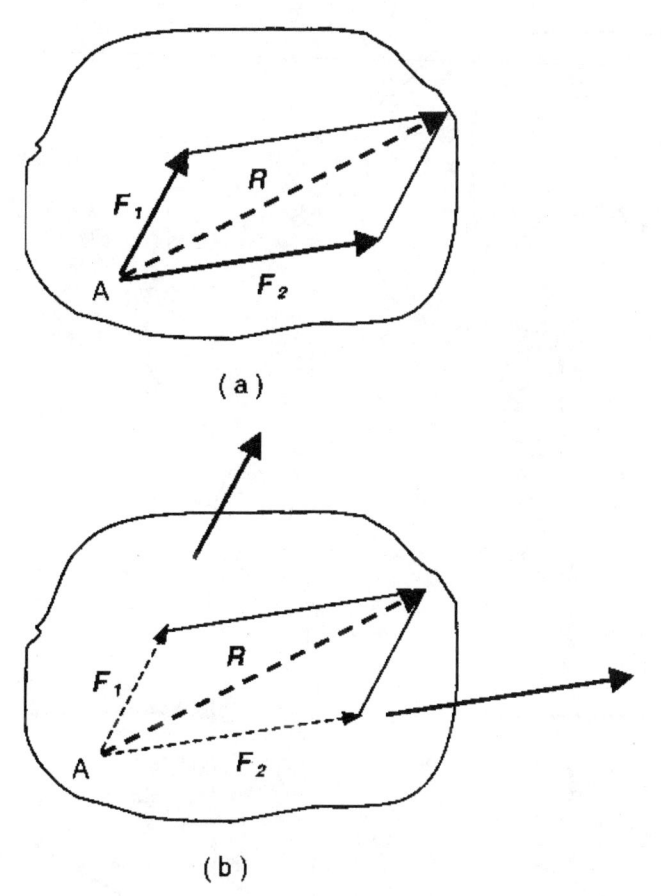

FIGURE 4.3

be moved along their lines of action and the resultant R completed at the point of intersection shown in Figure 4.3(b). The process of adding forces of a system to obtain a resultant is termed composition of forces. The resultant produces identical external effect as the system of forces.

Two parallel forces, F_1 and F_2, can be combined graphically, as shown in Figure 4.4. The force F is selected arbitrarily to facilitate the composition process. The pair of Fs is equal, opposite, and collinear, which produces no external effect on the body. F can be of any convenient magnitude. Adding F_1 and F, and combining with the sum F_2 and F, produce the resultant R. This method can be used to graphically find the resultant of nearly parallel forces.

By the process of resolution, force R can be resolved into two or more components having the same external effect as force R. In Figure 4.5a, R is resolved in two components F_1 and F_2. In engineering practice, it is often convenient to resolve the force R into two mutually perpendicular components—R_x and R_y. R is said to have rectangular components R_x and R_y as shown in Figure 4.5b.

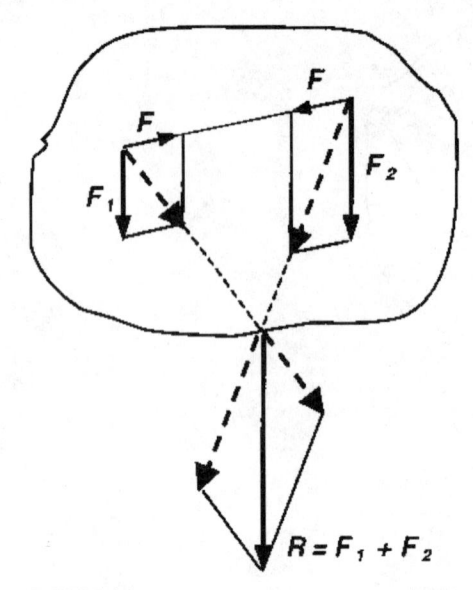

FIGURE 4.4

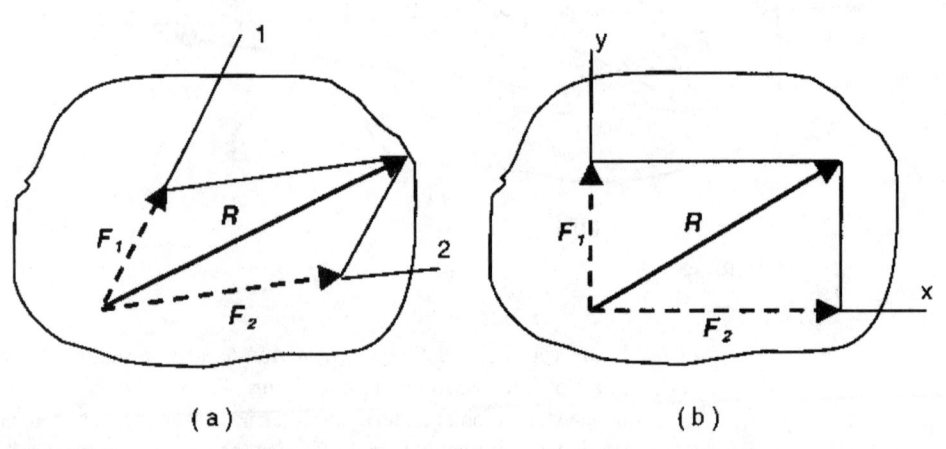

FIGURE 4.5

(a)   (b)

From Figure 4.5(b) we can establish the following relationships

$$R_x = R \cos \theta \qquad R_y = R \sin \theta$$

$$R = \sqrt{R_x^2 + R_y^2} \tag{4.2}$$

$$\theta = \tan^{-1} \frac{R_y}{R_x} \tag{4.3}$$

Many problems in engineering are three-dimensional. It is convenient to consider three mutually perpendicular components of a force as shown in Figure 4.6. Force F is resolved

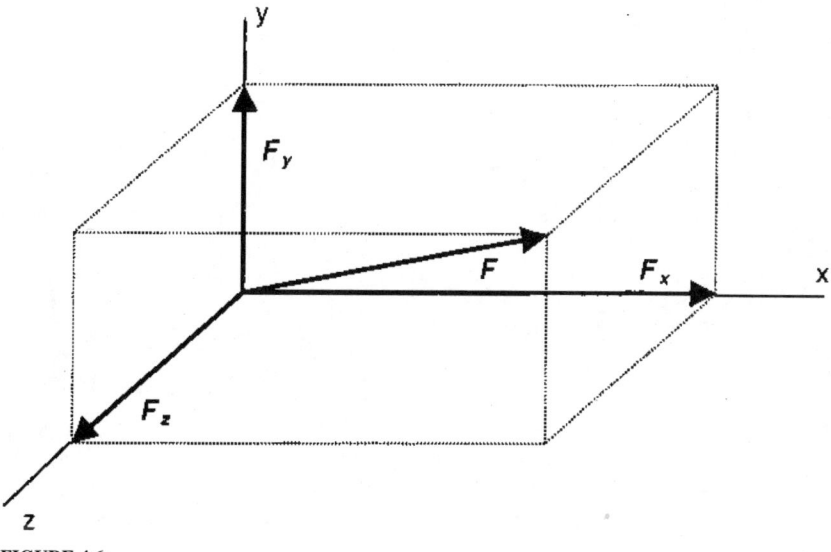

FIGURE 4.6

into rectangular components F_x, F_y, and F_z. The angles made by F with the x, y, and z axes are designed by θ_x, θ_y, and θ_z, respectively.

From Figure 4.6 we can establish the following relationships

$$F_x = F \cos \theta_x$$
$$F_y = F \cos \theta_y \quad \text{and} \quad F = \sqrt{F_x^2 + F_y^2 + F_z^2} \qquad (4.4)$$
$$F_z = F \cos \theta_z$$

The choice of the orientation of the axes is at the discretion of the engineer.

4.7 MOMENT AND COUPLE

Moment is the tendency of a force to rotate the body about a certain point or axis. In Figure 4.7, force F acts in a given plane about point O. The perpendicular distance to force F is d. Then, the moment of force F about O is given by

$$M_o = F \times d \qquad (4.5)$$

The moment, in effect, is rotating the body about an axis perpendicular to the plane and passing through point O.

Moment is a vector quantity. The laws for composition and resolution of forces also apply to moments. The vector direction is along the axis about which the moment is taken. The right-hand rule is used universally to establish the direction of the moment. For example, the direction of the moment of R about the axis O-O in Figure 4.8 is obtained by curling the fingers in the direction of the tendency to rotate the body. Then the thumb points in the positive direction of the moment.

4.8 CHAPTER 4

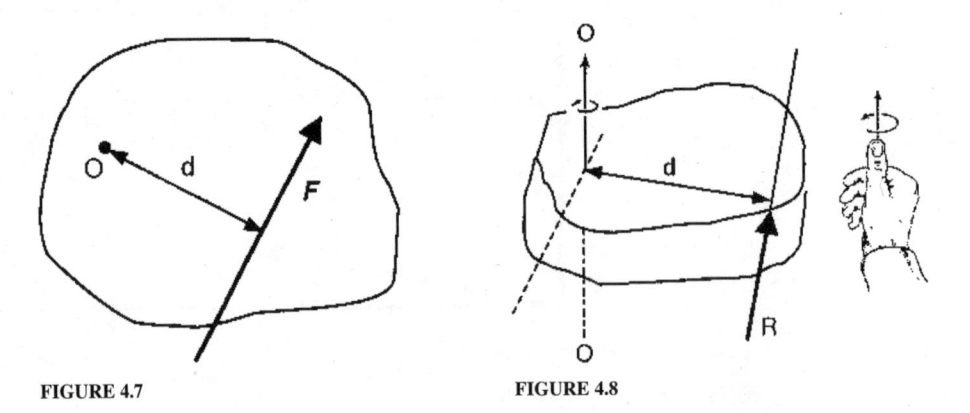

FIGURE 4.7

FIGURE 4.8

Two forces, F_1 and F_2, that are equal in magnitude, parallel but not collinear, and opposite in direction form a couple. A couple tends to rotate but not translate a body. The moment of a couple is equal to the magnitude of one of the forces multiplied by the distance between the two forces, as shown in Figure 4.9.

4.8 VARIGNON'S THEOREM

Varignon's Theorem states that the moment of a force about any point is equal to the sum of the moments of its components about the same point. To prove this theorem, let us consider force R and its components P and Q acting at point A as shown in Figure 4.10.

Any point O is selected as the moment center. The vectors of forces P, Q, and R are projected to a line normal to line AO. Moment arms p, q, and r are the distances of forces P, Q, and R from point O, respectively. The angles of the forces to line AO are designated as α, β, and γ. By geometry, the projections of the parallelogram with P and Q as sides are equal, that is, ac = bd. From line ad, it is evident that

$$Ad = ab + bd = ab + ac$$

Again by geometry, we have

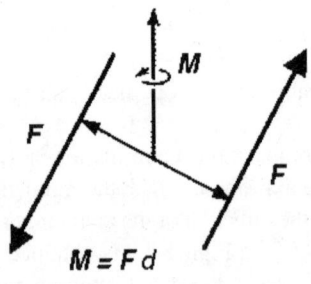

$$R \sin \gamma = P \sin \alpha + Q \sin \beta \tag{4.6}$$

FIGURE 4.9

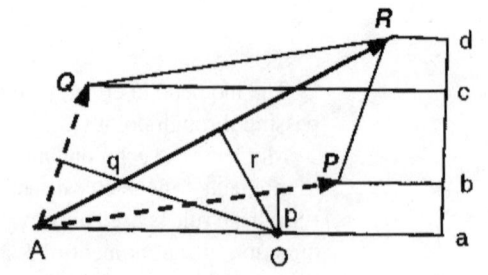

FIGURE 4.10

Also,

$$\sin \gamma = \frac{r}{OA}, \quad \sin \alpha = \frac{p}{OA}, \quad \sin \beta = \frac{q}{OA} \tag{4.7}$$

Substituting Equation 4.7 in Equation 4.6, we have

$$R_r = P_p + Q_q \tag{4.8}$$

which proves Varignon's Theorem for the case of a force with two components. Varignon's Theorem can be proved in a similar way to apply to three or more components, since the rules of vector composition and resolution can always bring the number of components to two. Varignon's Theorem can be used to simplify the computations in finding moment of a force about a point.

EXAMPLE 4.1
Find the moment of force P acting at point B about point O as shown in Figure 4.11 using Varignon's Theorem.

Solution
Resolving the load P into vertical and horizontal components V and H respectively, we have

$$V = \tfrac{3}{5}P \text{ and } H = \tfrac{4}{5}P$$

The moment about O is then

$$Mo = \tfrac{3}{5}P \times 15 + \tfrac{4}{5}P \times 5 = 13P$$

Varignon's Theorem makes the calculation straightforward. Otherwise, we have to first find the shortest distance from point O to the force P and then compute the moment.

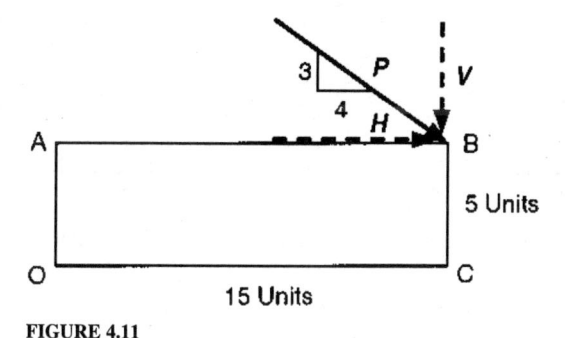

FIGURE 4.11

4.9 STATIC FRICTION

In Figure 4.12, vertical load P acts on body A, which transmits the load to body B. Body A is subjected to horizontal force H, which tends to move body A relative to body B. However, force F is parallel and tangential to the contact surface that resists the action of the horizontal force H. This resisting force F is termed static friction. If body A remains at rest, we have

$$Q = P \quad \text{and} \quad F = H \tag{4.9}$$

Assume that H increases gradually until relative movement between body A and body B is impending, then F has attained its maximum value. For two given materials, the maximum value of F is directly proportional to the normal force P. The constant of proportionality is called the coefficient of static friction and is denoted by μ. Then

$$F_{max} = \mu P \tag{4.10}$$

4.10 EQUILIBRIUM

4.10.1 Equilibrium in Two Dimensions

When a body is in equilibrium under a set of coplanar forces, that is forces acting in the same plane, let the x-y plane be the plane of action of the forces,

$$R = \sqrt{(\Sigma F_x)^2 + (\Sigma F_y)^2} = 0$$

$$M = M_z = 0$$

The above equations may be written as

$$\Sigma F_x = 0$$
$$\Sigma F_y = 0 \tag{4.11}$$
$$\Sigma M_O = 0$$

Equation 4.11 is used most commonly to solve structural engineering problems that can be represented in a plane.

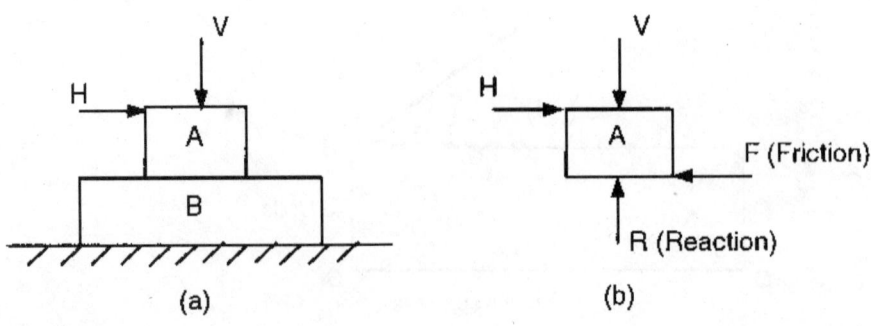

FIGURE 4.12

4.10.2 Equilibrium in Three Dimensions

When solving a general three-dimensional engineering problem, the vector equations of equilibrium, (4.11), can be written in scalar form as

$$\Sigma F_x = 0 \quad \Sigma M_x = 0$$
$$\Sigma F_y = 0 \quad \Sigma M_y = 0 \quad (4.12)$$
$$\Sigma F_z = 0 \quad \Sigma M_z = 0$$

4.11 FREE-BODY DIAGRAM

A free-body diagram is a method used to isolate a body of interest from all contacting and attached bodies and replacing them by forces acting on the body. A free-body diagram, when properly constructed, will account accurately and completely for all external, and internal or resisting forces. The free-body diagram is a basic step in the solution of engineering problems. For difficult and complex problems, it is necessary to use one or more free-body diagrams to represent the external forces and internal resistances accurately and completely to arrive at correct mathematical solutions.

An understanding of the free-body diagram method is essential to understanding statics and the solution to engineering problems. The best way to gain proficiency in constructing accurate and complete free-body diagrams is to study examples of free-body diagrams and to practice drawing free-body diagrams for structural systems with various loading conditions.

Figure 4.13 shows some examples of contact or connection forces that will help draw free-body diagrams. In Figure 4.13a, the force exerted by a flexible cable is always a tension in the direction of the cable. In Figure 4.13b, the force is compressive and is normal to the contacting surfaces for roller and rocker bearings. Friction normally can be neglected. In Figure 4.13c, a pin connection is capable of supporting force in any direction in a plane normal to the pin axis. Friction normally can be neglected. In Figure 4.13d, a built-in, or fixed-end, support can be represented by reactions and moment.

Next we will study some examples of free-body diagrams and develop a general procedure for drawing them systematically and properly. The examples are shown in Figure 4.14. You are encouraged to practice drawing free-body diagrams of structural systems to gain proficiency and accuracy in the construction of such diagrams.

The procedure for drawing a free-body diagram consists of the following steps:

1. Examine the types of connection and contact, and determine the reactions.
2. Draw a sketch of the system.
3. Add the external forces.
4. Add the reactions and moments in the directions you assume.

The assumption will be based on your experience in visualizing the flow of forces. If your assumed directions are correct, the computational values will be positive. If the numerical answer is negative, the resulting force, reaction, or moment is in the opposite direction of assumed directions.

High Tension,
Cable Weight Negligible

Free-body

Cable Weight Only

Free-body

(a)

Roller Support

Rocker Support

Free-Body

(b)

Hinge Support

Free-Body

(c)

Fixed End

Cantilever Beam

Free-body

(d)

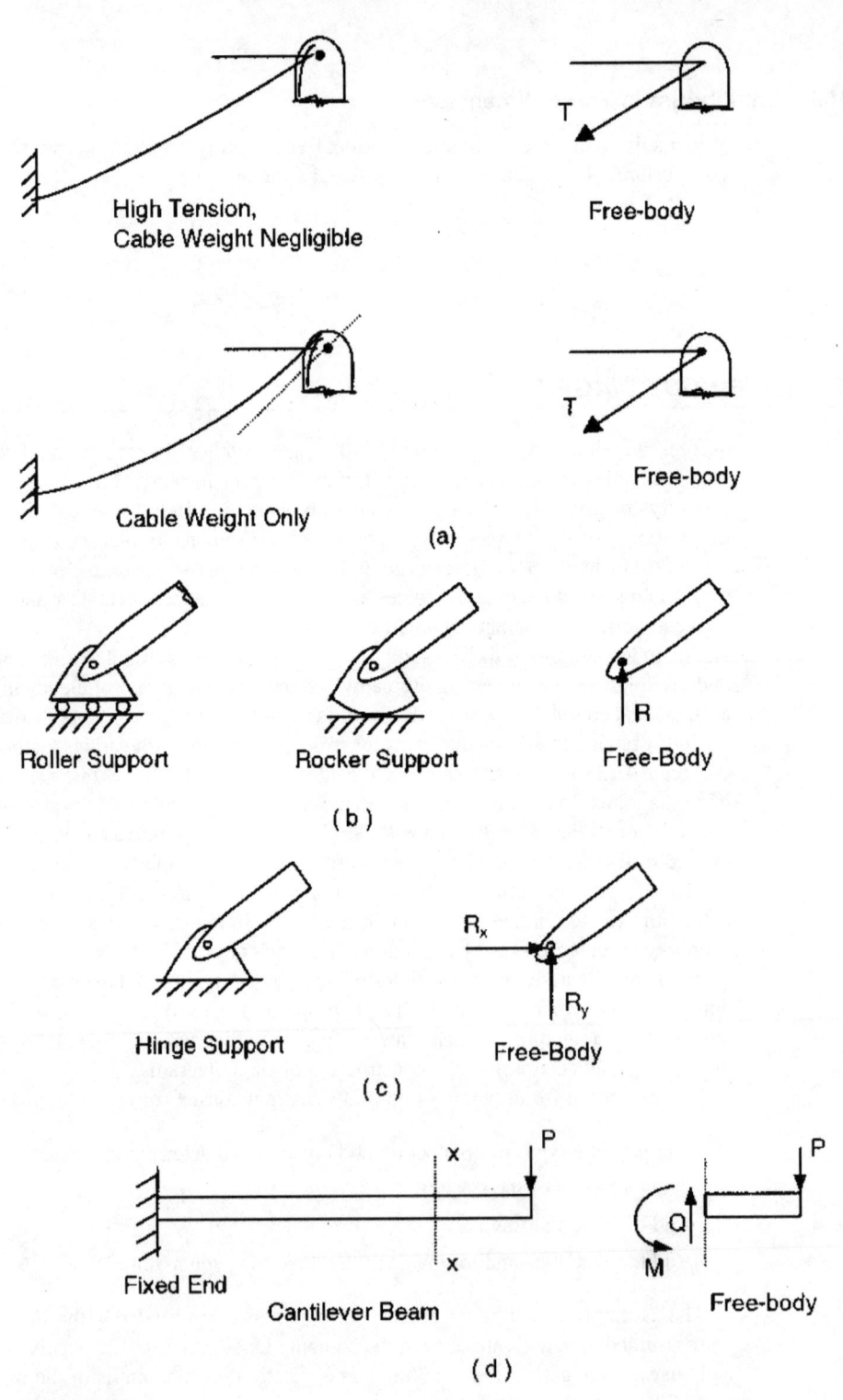

FIGURE 4.13

4.12

FIGURE 4.14

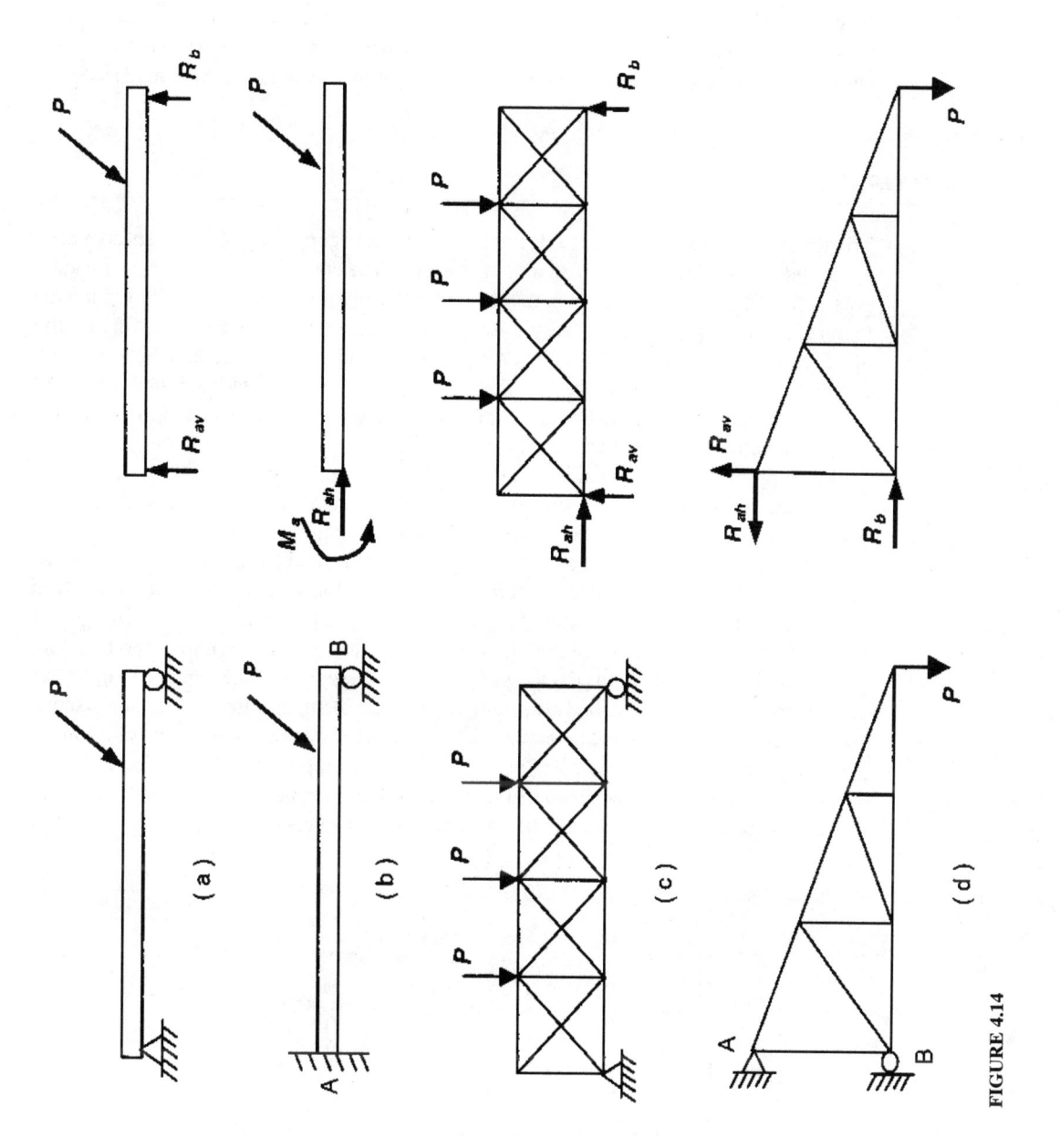

4.12 STRUCTURES

Now, we will investigate the forces internal to any engineering structure. It is necessary to understand the internal forces in the members of trusses and cable-suspended structures to ensure that the members and the connected system will withstand external loads safely.

4.13 TRUSSES

Trusses are a special type of framed structures. A truss is composed of members connected at the ends of the members to form a rigid structure. The connections can be constructed of pins, welds, rivets, and bolts. When the pins are relatively frictionless, the members are free to rotate about the pinned joint. The pinned joints often are called hinged joints. On the other hand, a welded connection offers total resistance to rotation. A welded connection usually is called a rigid connection. Joints that offer some, but not total, resistance to rotation are semi-rigid connections. Riveted and bolted connections are good examples of semirigid connections. In practice, it is appropriate to assume pin-jointed connection for these connections. Additionally, it is assumed that all external forces are applied at the pin connections.

All members of a truss are assumed to be two-force members with negligible weight. A two-force member is in equilibrium under the action of only two forces. These forces are applied at the ends of the member, and are equal, opposite, and collinear. The member may be in tension (T) or compression (C). A member is in tension when the force acts away from the joint, and in compression when the force acts towards the joint.

In the analysis of trusses, the directions of forces usually are shown in a free-body diagram of a joint or a portion of the truss. The correct directions of the forces might be evident by inspection of the free-body diagram. If it is not evident, the directions can be assumed and corrected after analysis. The assumption is correct if the computed value of a force is positive. A negative value indicates that the direction is opposite to that assumed.

Trusses are used for bridges, roofs of industrial and residential buildings, derricks, cranes, etc. Terms commonly used for identifying members and joints of trusses are shown in Figure 4.15.

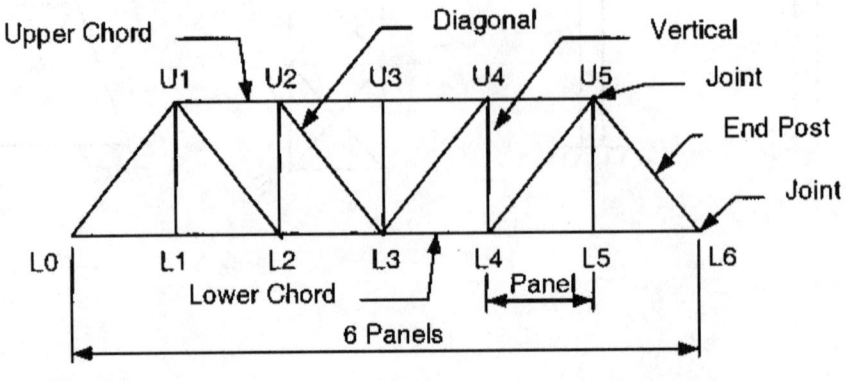

FIGURE 4.15

The basic element of a truss is a triangle. Three members connected by pins at their ends form a rigid frame. The term rigid is used to mean stable and noncollapsible, and also to mean small deformations in the members because of induced internal stresses and strains.

Structures that are built from basic triangles are known as simple or statically determinate trusses. In this chapter, we are dealing mainly of statically determinate structures. When more members are used than needed to form basic triangles or to prevent collapse, the truss is termed statically indeterminate. The additional or extra members that are not necessary to maintain equilibrium are called redundant members. Statically indeterminate structures cannot be analyzed by the equations of equilibrium alone. Other conditions will have to be used to solve such structures.

Figure 4.16 shows some trusses commonly used for roof supports and bridges.

4.14 DETERMINACY

Statically determinate and stable trusses are basic triangles. A rigid truss is formed by using three members to connect the three joints in the form of a triangle as shown in Figure 4.17a and then using two members to connect each additional joint to form another triangle to the one already constructed, shown in Figure 4.17b. We can continue adding additional basic triangles to form simple or statically determinate trusses as shown in Figure 4.17c and Figure 4.17d. Thus, to form a rigid simple truss of n joints, it is necessary to use the three members of the first triangle plus two additional members for each of the remaining joints $(n - 3)$. If b denotes the total number of members required, then

$$b = 3 + 2(n - 3) = 2n - 3 \qquad (4.13)$$

This is the minimum number of members that can be used to form a rigid simple or statically determinate truss. To use more is unnecessary and to use fewer results in an unstable or collapsible truss.

We can conclude that

1. If there are three independent reactions and $(2n - 3)$ members, the truss is stable and statically determinate with respect to reactions and member stresses.

2. If there are more than three independent reactions, the truss is statically indeterminate with respect to its reactions.

3. If there are three independent reactions and more than $(2n - 3)$ members, the truss is indeterminate with respect to member stresses.

4. If there are more than three independent reactions and more than $(2n - 3)$ members, the truss is indeterminate with respected to both reactions and member stresses.

Next let us discuss determinacy and stability more generally. Suppose a truss has R independent reactions, b bars and n joints. At each joint, we have two independent equations of static equilibrium, which are

$$\Sigma F_x = 0 \quad \text{and} \quad \Sigma F_y = 0$$

A total of $2n$ independent equations would be obtained for the n joints. However, there are R unknown reactions and b member stresses, giving a total of $(R + b)$ unknowns. The

FIGURE 4.16

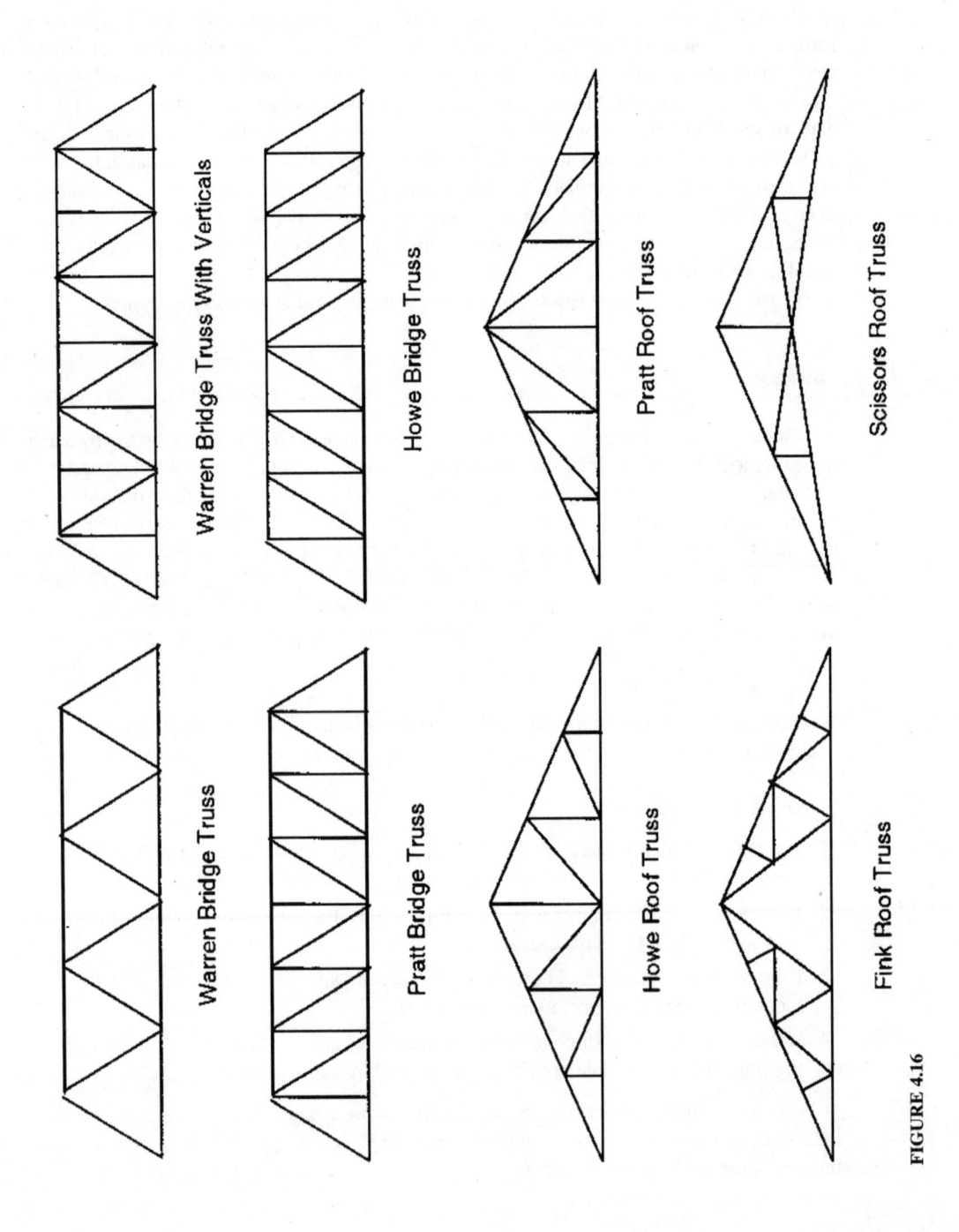

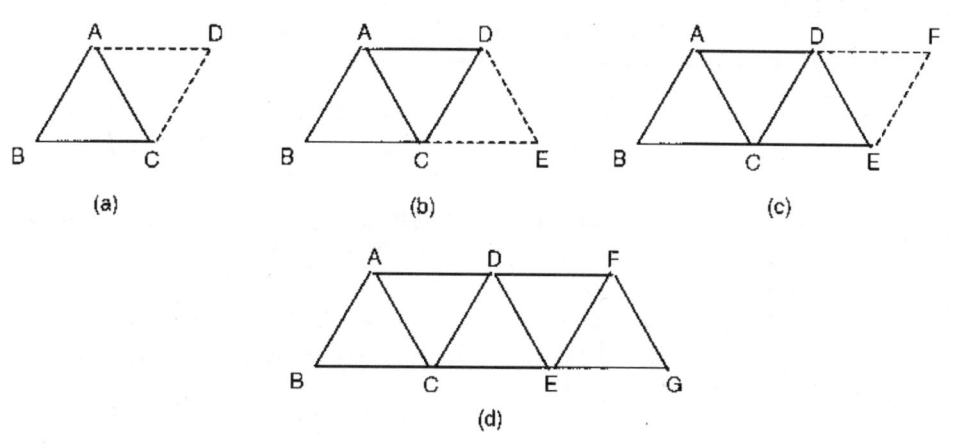

FIGURE 4.17

$2n$ equations must be satisfied simultaneously by the $(R + b)$ unknowns. By comparing the number of unknowns with the number of independent equations, we can determine whether a truss is unstable, statically determinate, or indeterminate.

We may generalized the conditions as follows:

1. If $R + b$ is less than $2n$, there are not enough unknowns available to satisfy the $2n$ equations simultaneously. The truss is said to be statically unstable.
2. If $R + b$ is equal to $2n$, the unknowns can then be obtained from the $2n$ simultaneous equations. The truss is said to be statically determinate.
3. If $R + b$ is greater than $2n$, there are too many unknowns to be determined from the $2n$ equations alone. The truss is said to be statically indeterminate.

The above criteria establish the degree of indeterminacy and stability with respect to reactions and member stresses.

4.15 INFLUENCE LINES FOR TRUSSES

When designing a member of a truss, it is necessary to find the greatest stress in the member due to live load. The stress produced in a given member by the live load varies with the position of the load on the truss. There is a position of loading on the truss that causes the maximum live load stress in a particular member of the truss. The member of the truss and the type of stress involved may be the reaction at a support, the tension or compression of a member, or the load at a joint.

To illustrate the concept and significance of influence lines, let us consider a couple of cases of a simple truss as shown in Figure 4.18.

A vertical unit load is applied at L_0, L_1, L_2, and so forth. The reaction R_0 at L_0 is computed and plotted as the unit load is moved across the truss. When the unit load is at L_0, the reaction $R_0 = 1$. When the unit load is at L_1, $R_0 = \frac{5}{6} \times 1 = \frac{5}{6}$. When the unit load is at L_2, $R_0 = \frac{4}{6} \times 1 = \frac{2}{3}$ and so forth until the unit load is at L_6, when the reaction at L_0 is zero.

We can draw some conclusions from examining the influence line for the reaction R_0 of the truss shown in Figure 4.18:

1. The ordinate at any point on the curve is equal to the reaction at L_0 due to a unit load applied at that point. Since a unit load is used to generate the curve, the ordinates can be used to determine the influence of higher magnitude loads acting on the truss.
2. As the unit load moves from L_0 to L_6, the reaction R_0 decreases linearly. The maximum value of R_0 occurs when the load is at L_0 and the value is zero when the load is at L_6.
3. The ordinates of the curve are all positive. This means that any load applied between L_0 and L_6 will have a positive influence in the reaction; that is, it increases the reaction at L_0.
4. If the truss is to be loaded with a uniform live load, the entire span should be loaded with the uniform load to obtain the maximum reaction R_0 at L_0.

This simple example shows an influence line is useful to visually and quickly guide the designers in loading a structure to obtain maximum effect on the reactions and forces in the structure.

4.16 METHOD OF JOINTS

One method of determining the member forces in a truss after the external reactions are known is to consider the equilibrium of the joints successively. For plane trusses, there are two equations of equilibrium. Therefore, we start at a joint where the number of unknown forces is also two. We can solve for the two unknowns with the two simultaneous equations of

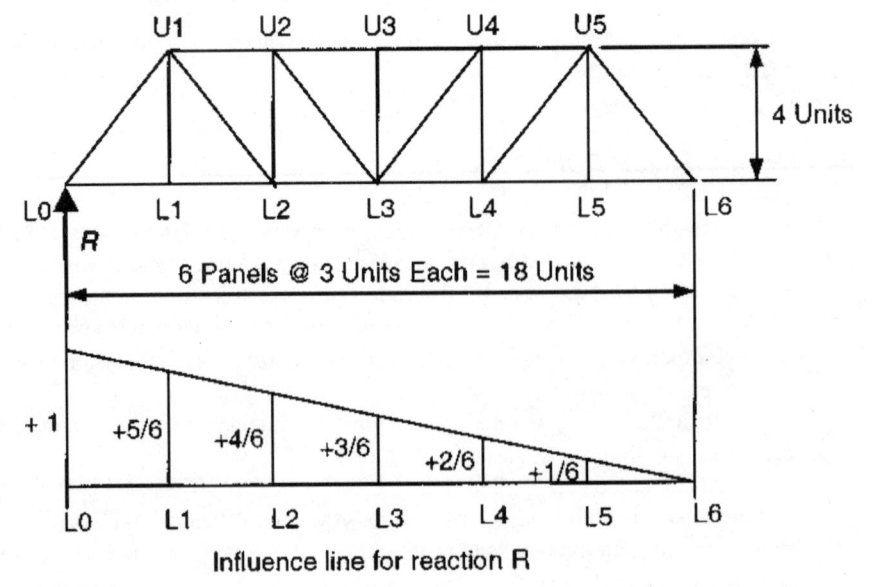

FIGURE 4.18

equilibrium, and then move on to the next joint where the number of unknowns is two. We continue this way until the member forces are found for the whole truss. This is the method of joints. This method is very useful when all or most of the member forces are necessary for the design or analysis.

EXAMPLE 4.2

Determine the member forces in the truss shown in Figure 4.19 by the method of joints.

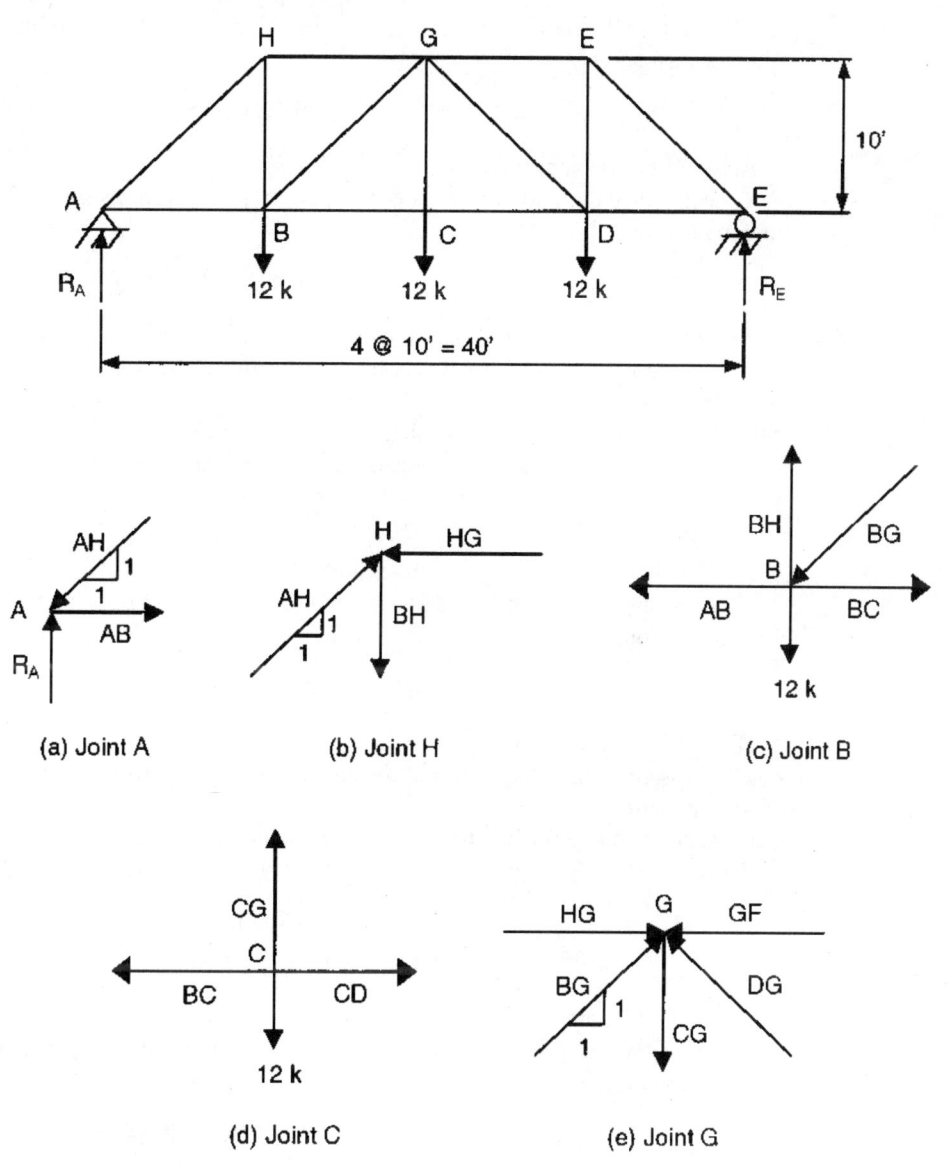

FIGURE 4.19

4.20 CHAPTER 4

Solution
Step 1: Find the reactions at Joints A and E.
By symmetry
$$R_A = R_E = 18 \text{ k}$$

Step 2: Start with Joint A, where the reaction R_A is known and there are only two unknowns.
Draw a free-body diagram at joint A and apply the equilibrium equations. (Note that the directions of forces acting at the joint can be evident by inspection of the free-body diagram or assumed.)

$$\Sigma F_y = 0, \quad -0.707\text{AH} + 18 = 0$$

∴ AH = 25.5 k (Compression)
(The assumed direction of AH is correct; i.e., acting toward the joint. Therefore AH is in compression.)

$$\Sigma F_x = 0, \quad -0.707\text{AH} + \text{AB} = 0$$

∴ AB = 0.707 (25.5) = 18 k (Tension)
(The assumed direction of AB is correct; i.e., acting away from the joint. Therefore AB is in tension.)

Step 3: Next, go to joint H—now there are only two unknowns.
Draw a free-body diagram at Joint H and apply the equilibrium equations.

$$\Sigma F_x = 0, \quad 0.707\text{AH} - \text{HG} = 0$$

∴ HG = 0.707 (25.5) = 18 k (Compression)
(The assumed direction of HG is correct; i.e., acting toward the joint. Therefore HG is in compression.)

$$\Sigma F_y = 0, \quad 0.707\text{AH} - \text{BH} = 0$$

∴ BH = 0.707 (25.5) = 18 k (Tension)
(The assumed direction of BH is correct; i.e., acting toward the joint. Therefore BH is in compression.)

Step 4: Next go to joint B, which has only two unknowns.
Draw a free-body diagram at joint B and apply the equilibrium equations

$$\Sigma F_y = 0, \quad \text{BH} - 0.707 \text{ BG} - 12 = 0$$
$$(18.0) - 0.707 \text{ BG} - 12 = 0$$

∴ BG = 8.5 k (Compression)
(The assumed direction of BG is correct; i.e., acting toward the joint. Therefore BG is in compression.)

$$\Sigma F_x = 0, \quad \text{AB} + 0.707 \text{ BG} + \text{BC} = 0$$
$$(-18) - 0.707(8.5) + \text{BC} = 0 \therefore \text{BC} = 24 \text{ k}$$

(The assumed direction of BC is correct; i.e., acting away from the joint. Therefore BC is in tension.)

Step 5: Next go to joint C, which now has only two unknowns.
Draw a free-body diagram at Joint C and apply the equilibrium equations

$$\Sigma F_x = 0, \quad CD = BC = 24 \text{ k (Tension)}$$
$$\Sigma F_y = 0, \quad CG = 12 \text{ k (Tension)}$$

Step 6: Next go to joint G, which now has only two unknowns
Draw a free-body diagram at Joint G and apply the equilibrium equations.
By symmetry

$$GF = HG = 18 \text{ k (Compression)}$$
$$DG = BG = 8.5 \text{ k (Compression)}$$

Check $\quad \Sigma F_y = 0$,

$$-CG + 0.707BG + 0.707DG = -(12) + 0.707(8.5) + 0.707(8.5) = 0 \quad \text{OK}$$

(The joint is in equilibrium.)
By symmetry, the rest of the members of the truss can be found.

4.17 METHOD OF SECTIONS

In the method of joints, only two of the three equilibrium equations are used to analyze the truss. In the method of sections, the third equilibrium equation ($\Sigma M = 0$) is used to advantage by considering a section of the truss as a free-body in equilibrium under the action of a nonconcurrent system of forces. The method of sections has the advantage of quickly finding the force in any member of interest. It is not necessary to go joint-by-joint to reach the member under consideration. The application of the method of sections is illustrated by the following example.

EXAMPLE 4.3
Determine the force in member DK by the Method of Sections of the truss shown in Figure 4.20.

Solution
Step 1: Find the reactions R_A and R_G at Joints A and G.

$$\Sigma M_G = 0, \quad R_A(60) - 3(50) - 3(40) - 3(20) = 0$$

$\therefore R_A = 5.5 \text{ k}$

$$\Sigma F_y = 0, \quad R_A + R_G - 3 - 3 - 3 = 0$$
$$5.5 + R_G - 3 - 3 - 3 = 0$$

$\therefore R_G = 3.5 \text{ k}$

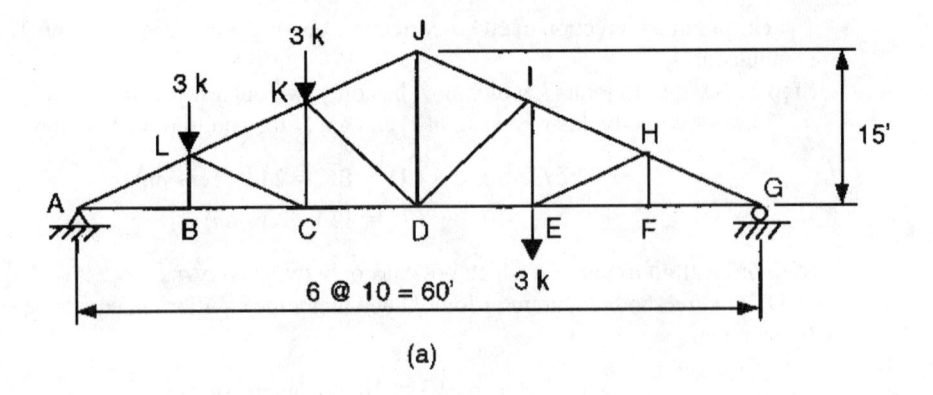

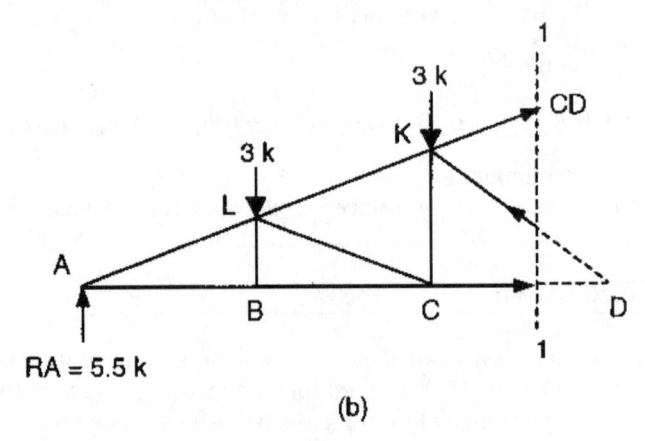

FIGURE 4.20

Step 2: Draw a free-body diagram to the left of Section 1.

(Section 1 is taken such that it cuts across DK, and two other members CD and JK, which intersect at A. When moment is summed about A, the effects of these two forces vanish, leaving the force for member DK. The force in member DK is assumed to be in compression, acting in the direction shown in Figure 4.20.)

Step 3: Calculate the force in member DK.

The horizontal and vertical components of DK near joint D are

$$DK_h = DK_v = 0.707 \, DK$$
$$\Sigma M_A = 0, \quad -0.707 \, DK(30) + 3(20) + 3(10) = 0$$

$\therefore DK = 4.24 \, k$

(The positive sign of DK indicates that the assumed direction of DK is correct, i.e. the member force is in compression. With respect to joint D, the compressive force acts toward the joint. With respect to member DK, the force acts upward and to the left of the section.)

Hence, DK = 4.24 k (Compression)

4.18 METHOD OF SUPERPOSITION

Method of superposition is used frequently in structural design and analysis. It is a process in which the effects of individual loads can be computed separately and then added algebraically to obtain the total effects, such as stresses, shears, moments, deflections, and slopes. The method of superposition is valid when the members of the structure remain below the elastic limit and the geometric changes of the structure are relatively small.

4.19 FLEXIBLE CABLES

Flexible cables are used in suspension bridges, traffic-signal supports, cable-car systems, telephone lines, guys for radio and antenna towers, and many other applications. In the design of cable-suspended structures, it is necessary to establish the relations between the tension in the cables, the span, the sag, and the length of the cables. The relations can be determined by applying the equations of equilibrium. The cable is assumed to be perfectly flexible, so the bending moment at any point on the cable is zero, and the force in the cable is in the direction of the cable.

The cables may support a series of concentrated loads, or loads distributed over the length of the cable as shown in Figure 4.21. The weight of the cable may be neglected in cases where the support loads are much heavier than the weight of the cable. In other cases the weight of the cable may not be neglected. In any case, the equations of equilibrium can be applied in the same manner.

Next, let us consider a finite length of the loaded cable at the lowest point as shown in Figure 4.21(c). Let T_o be the horizontal tension of the cable at the origin of the coordinate system, and T be the tension in the cable at a point with coordinates x and y. The resultant of the applied load is shown as R. In this example, the applied load is a distributed vertical

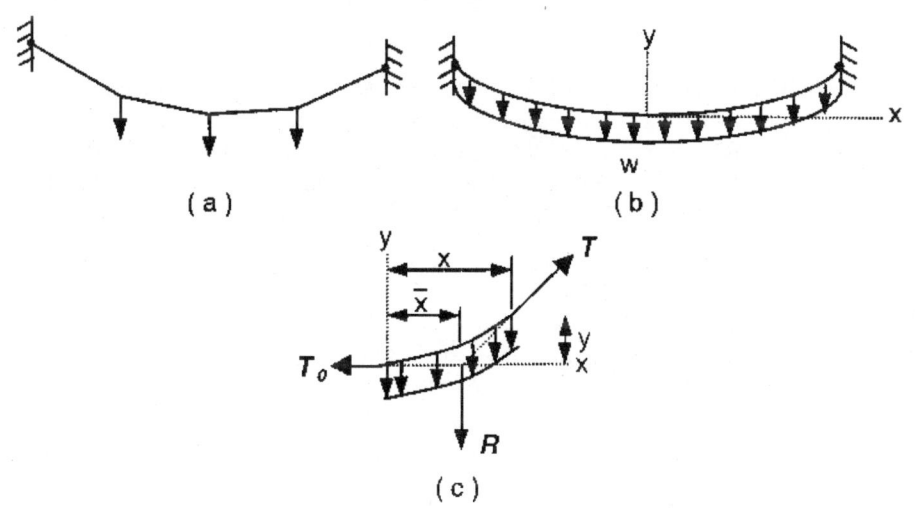

FIGURE 4.21

load. If the cable supports concentrated loads, the resultant R will be the sum of the concentrated loads within the finite length selected.

In this case, ω is the distributed vertical load per horizontal length of cable. Hence

$$dR = wdx \quad \text{and} \quad R = \int dR = \int wdx \tag{4.14}$$

The position of R can be determined by the principle of moments, so that

$$R\bar{x} = \int xdR, \quad \bar{x} = \frac{\int xdr}{R} \tag{4.15}$$

Applying the two force equations of equilibrium to the free-body diagram, we have

$$\left[\sum F_y = 0\right] T \sin\theta = R \tag{4.16}$$

$$\left[\sum F_x = 0\right] T \cos\theta = T_0 \tag{4.17}$$

Dividing Equation 4.16 by Equation 4.17, we have

$$\tan\theta = \frac{R}{T_0} \quad \text{or} \quad \frac{dy}{dx} = \frac{R}{T_0} \tag{4.18}$$

Equation 4.18 is the differential equation of the cable.

Eliminating θ in Equations 4.16 and 4.17 by squaring and adding, we have

$$(T \sin\theta)^2 + (T \cos\theta)^2 = R^2 + T_0^2 \quad \text{or} \quad T^2 = R^2 + T_0^2 \tag{4.19}$$

Equations 4.18 and 4.19 are applicable to any cable that supports vertical loads, distributed or concentrated.

When the weight of the cable is negligible compared to the applied loads, and the applied load is uniformly distributed along the horizontal, the shape of the cable may be assumed to be parabolic. This is the case for a suspension bridge. When the cable hangs under the action of its own weight, as in a transmission line, the cable takes the shape of a catenary.

We will consider the case of a parabola and a catenary in the next two sections.

4.20 PARABOLIC CABLES

Let us analyze the case of parabolic cables in a suspension bridge as shown in Figure 4.22.

Let the uniform load supported by each cable be w pounds per unit of horizontal length.

The resultant R of the vertical loading for a section of the horizontal length x is given by $R = \omega x$ acting at a distance $x/2$ from the origin. Equation 4.18 then becomes

$$\frac{dy}{dx} = \frac{wx}{T_0} \tag{4.20}$$

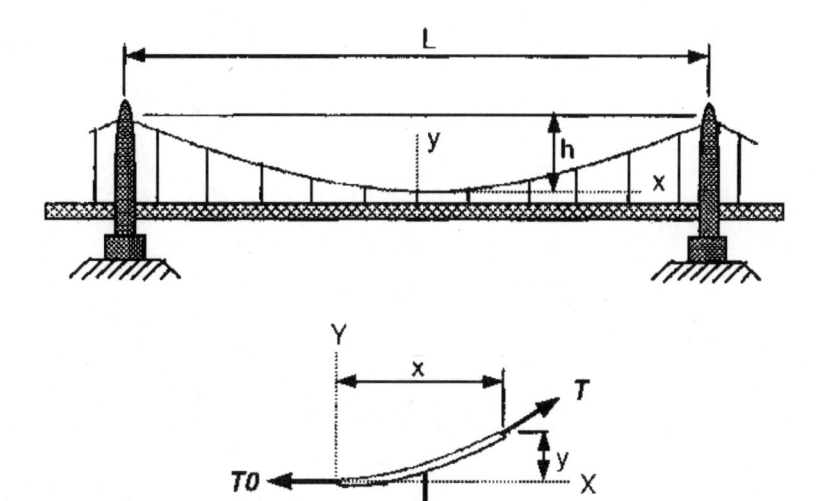

FIGURE 4.22

By direct integration, we have

$$\int_0^y dy = \frac{w}{T_0} \int_0^x x\, dx \quad \text{or} \quad y = \frac{wx^2}{2T_0} \qquad (4.21)$$

This is the equation to the parabolic shape of a cable in terms of w and T_o. Substituting the boundary conditions $x = L/2$ and $y = h$, we have

$$T_0 = \frac{wL^2}{8h} \quad \text{and} \quad y = \frac{4hx^2}{L^2} \qquad (4.22)$$

From Equation 4.19, we have

$$T = \sqrt{T_0^2 + w^2 x^2}$$

Substituting the value of T_0 from Equation 4.22 and simplifying, we have

$$T = w\sqrt{x^2 + \frac{L^4}{64h^2}} \qquad (4.23)$$

The maximum tension occurs at $x = L/2$ and is

$$T_{max} = \frac{wL}{2}\sqrt{1 + \frac{L^2}{16h^2}} \qquad (4.24)$$

The length S of the cable is obtained from the relationship

$$dS = \sqrt{(dx^2 + dy^2)}$$

Therefore

$$\frac{S}{2} = \int_0^{L/2} \sqrt{1 + \left(\frac{dy}{dx}\right)^2}\, dx = \int_0^{L/2} \left[1 + \left(\frac{wx}{T_0}\right)^2\right]^{1/2} dx \quad (4.25)$$

A convergent series of the following form will be used to simplify integration

$$(1 + X)^n = 1 + nX + \frac{n(n-1)}{1.2} X^2 + \frac{n(n-1)(n-2)}{1.2.3} X^3 + \cdots \quad (4.26)$$

Here we have

$$X = \left(\frac{wx}{T_0}\right)^2 \quad \text{and} \quad n = \frac{1}{2}$$

Using the first three terms from Equation 4.26 in Equation 4.25, we have

$$\frac{S}{2} = \int_0^{L/2} \left[1 + \frac{1}{2}\frac{w^2 x^2}{T_0^2} + \frac{\frac{1}{2}\left(\frac{1}{2}-1\right)}{1.2}\frac{w^4 x^4}{T_0^4} + \cdots\right]$$

or

$$S = 2\int_0^{L/2} \left[1 + \frac{w^2}{2T_0^2} x^2 - \frac{w^4}{8T_0^4} x^4 + \cdots\right]$$

Integrating term by term, we have

$$S = 2\left[x + \frac{w^2}{2T_0^2}\frac{x^3}{3} - \frac{w^4}{8T_0^4}\frac{x^5}{3} + \cdots\right]_0^{L/2}$$

$$= 2\left[\frac{L}{2} + \frac{w^2}{2T_0^2}\frac{L^3}{3.8} - \frac{w^4}{8T_0^4}\frac{L^5}{5.32} + \cdots\right]$$

$$= L\left(1 + \frac{w^2 L^2}{24 T_0^2} - \frac{w^4 L^4}{640 T_0^4} + \cdots\right) \quad (4.27)$$

In the above equation, the length of the cable S is expressed in terms of the span length L, the horizontal unit weight of the cable w, and the horizontal cable tension T_0. The cable

$$S = L\left(1 + \left[\frac{8h}{L^2}\right]^2 \frac{L^2}{24} - \left[\frac{8h}{L^2}\right]^4 \frac{L^4}{640} + \cdots\right)$$

length S can be expressed in terms of span length L and sag h by substituting $w/T_0 = 8h/L^2$ from Equation 4.22.

Simplifying we have

$$S = L\left(1 + \frac{8}{3}\left[\frac{h}{L}\right]^2 - \frac{32}{5}\left[\frac{h}{L}\right]^4 + \cdots\right) \qquad (4.28)$$

The series converges for all values of $h/L \leq \frac{1}{4}$. In most practical cases sag h is much smaller than $L/4$. Therefore the first three terms of Equation 4.28 give very close approximation of the cable length S.

4.21 CATENARY CABLES

Next let us consider a cable of span L suspended at points A and B in the same horizontal plane and hanging under its own weight μ per unit length as shown in Figure 4.23a. The free-body diagram of a segment of the cable length s is shown in Figure 4.23b.

The resultant R of the segment is $R = \mu$. Equation 4.18 becomes

$$\frac{dy}{dx} = \frac{\mu}{T_0} s$$

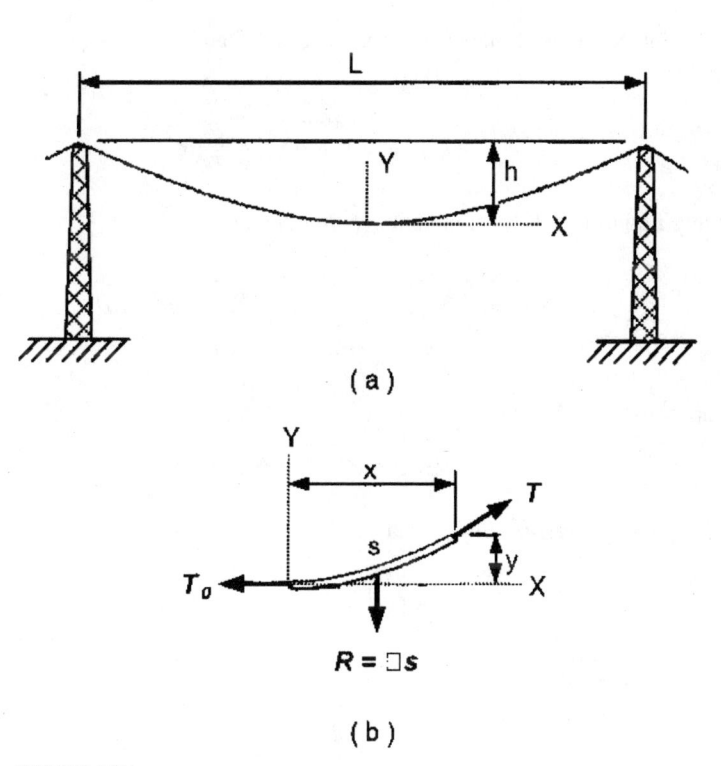

FIGURE 4.23

Differentiating with respect to x, we have

$$\frac{d^2y}{dx^2} = \frac{\mu}{T_0}\frac{ds}{dx} \qquad (4.29)$$

Substituting $(ds)^2 = (dx)^2 + (dy)^2$ in Equation 4.29, we have

$$\frac{d^2y}{dx^2} = \frac{\mu}{T_0}\sqrt{1 + \left(\frac{dy}{dx}\right)^2} \qquad (4.30)$$

Equation 4.30 represents the catenary of the suspended cable. To solve this equation, let us substitute $p = dy/dx$ or $dp/dx = d^2y/dx^2$ in Equation 4.30

$$\frac{dp}{dx} = \frac{\mu}{T_0}\sqrt{1 + p^2} \quad \text{or} \quad \frac{dp}{\sqrt{1+p^2}} = \frac{\mu}{T_0}dx \qquad (4.31)$$

Integrating, we have

$$\log\left(p + \sqrt{1+p^2}\right) = \frac{\mu}{T_0}x + C$$

where C is the constant of integrating and is zero, because $dy/dx = p = 0$ when $x = 0$.
Therefore

$$\log\left(p + \sqrt{1+p^2}\right) = \frac{\mu}{T_0}x \qquad (4.32)$$

Expressing Equation 4.32 in exponential form, we have

$$\left(p + \sqrt{1+p^2}\right) = e^{\mu x/T_0} \quad \text{or} \quad \sqrt{1+p^2} = e^{\mu x/T_0} - p$$
$$1 + p^2 = e^{2\mu x/T_0} - 2pe^{\mu x/T_0} + p^2$$

Squaring to clear the radical, we have

$$2pe^{\mu x/T_0} = e^{2\mu x/T_0} - 1$$

Simplifying and transposing, we have

$$p = \frac{e^{\mu x/T_0} - e^{-\mu x/T_0}}{2}$$

Therefore, substituting

$$p = \frac{dy}{dx} = \frac{\mu s}{T_0}$$

we have

$$\frac{\mu}{T_0}s = \frac{e^{\mu x/T_0} - e^{-\mu x/T_0}}{2}$$

Solving for s, we have

$$s = \frac{T_0}{\mu}\frac{e^{\mu x/T_0} - e^{-\mu x/T_0}}{2} \qquad (4.33)$$

Equation 4.33 gives the cable length s in exponential form, which can be expressed in hyperbolic form as follows

$$s = \frac{T_0}{\mu} \sinh \frac{\mu x}{T_0} \qquad (4.34)$$

Substituting $\mu s/T_0 = dy/dx$ in Equation 4.34 and simplifying, we have

$$\frac{dy}{dx} = \sinh \frac{\mu x}{T_0}$$

Integrating, we have

$$y = \frac{T_0}{\mu} \cosh \frac{\mu x}{T_0} + C$$

where C is the constant of integration. From the boundary condition that $y = 0$ when $x = 0$, we have $C = -T_0/\mu$.
Therefore

$$y = \frac{T_0}{\mu}\left(\cosh \frac{\mu x}{T_0} - 1\right) \qquad (4.35)$$

This is the catenary equation to the cable shape hanging under its own weight.
The cable tension T can be found from Equation 4.19, which gives

$$T^2 = \mu^2 s^2 + T_0^2$$

Substituting the value of s from Equation 4.34, we have

$$T^2 = T_0^2 \sinh^2 \frac{\mu x}{T_0} + T_0^2$$

or

$$T^2 = T_0^2\left(1 + \sinh^2 \frac{\mu x}{T_0}\right) = T_0^2 \cosh^2 \frac{\mu x}{T_0}$$

Therefore the cable tension is given by

$$T = T_0 \cosh \frac{\mu x}{T_0} \qquad (4.36)$$

From Equation 4.35, we have

$$T_0 \cosh \frac{\mu x}{T_0} = T_0 + \mu y$$

Substituting this in Equation 4.36, we have

$$T = T_0 + \mu y \qquad (4.37)$$

The cable tension T is given in terms of y. The maximum tension T_{max} is at $y = h$. Hence

$$T_{max} = T_0 + \mu h \qquad (4.38)$$

Equations 4.34 through 4.38 are sufficient to solve most problems dealing with catenary. For small sag-to-span ratios, the catenary problems can be solved simply by the equations for the parabolic cable. For problems in which the suspension points are not at the same level, the parabolic and catenary cables can be solved by applying the equations to each part of the cable on either side of the lowest point.

CHAPTER 5
INTRODUCTION TO DESIGN AND ANALYSIS

5.1 INTRODUCTION

A structural designer generally is concerned with two related problems: design and analysis. In a design problem, the external loads and the overall dimensions and shapes of the structures are known. The task of the structural engineer is to select, arrange, and proportion the members and connections of the structures so that they will safely support the external loads and meet the dimensional, architectural, and functional requirements of the structures. The owner may engage an architect to help him with the dimensions and shapes of the structures and then engage an engineer to help him select the materials, size, shape, and connections of the structural members. In an analysis problem, the materials, size, shape, and connections of the structures are known. The task of the structural engineer is to analyze or determine whether the given design is satisfactory with regards to the design criteria, codes, and other special requirements established by the owner. The elastic analysis and plastic analysis methods may be used where applicable.

Design and analysis are interrelated. An analysis verifies the stresses, deflections, dynamic response, and so forth caused by the loads imposed on the structures. An analysis checks the adequacy and efficiency of a design, and shows where the design can be resized or refined to improve the proportioning of members and optimize cost. The design then can be revised accordingly. Depending on the extent of the revisions, another analysis might be needed to check the revised design. The design-analysis process continues until the structural designer determines the optimum design has been reached.

5.2 RESPONSIBILITIES OF THE STRUCTURAL DESIGNER

Assuming responsibility for the safety and comfort of the structure's users, the structural designer must understand and be proficient in the engineering principles presented in Chapters 1-4, and other special knowledge and experience. The structural designer must select structural systems that can safely support the design loads, are in harmony with the environment, can be practically erected, and are cost-effective. The structural designer must consider the following key elements in the design.

5.2.1 Safety

An engineered structure must safely support the design loads to which it is subjected. The structure must have service life consistent with the design requirements, and it should not have excessive deflections and vibrations that would cause discomfort or frighten its occupants or users. Failure of structures is inadmissible.

Past failures provide invaluable lessons to the structural designer. Failures are seldom due to the selection of members of inadequate size or strength; instead, they usually occur because of insufficient attention to the details of connections, excessive deflections, vibrations, erection problems, buckling of compression elements, shear and fatigue of members, foundation settlements, and inadequate support width for beams. Benjamin Franklin observed that "a wise man learns more from failures than from success."

Engineering specifications and building codes developed by various organizations are available to help the structural designer assure that good and modern engineering practices are followed. Logically written specifications and codes are very helpful to designers. There are far fewer structural failures in regions that have good structural specifications and strictly enforced codes.

5.2.2 Aesthetics

Aesthetics is an important structural aspect of design. An ugly structure is as bad as failure. It causes discord in the community and distaste to many people who might wish that the structure never had been built or could be demolished.

A structure should be in harmony with the surroundings with respect to size, shape, and color. The members of a structure should be well-proportioned and balanced to be in harmony among the members of the structure, nature, and neighboring structures.

The main aim of aesthetics is to design a structure of inherent beauty. It is our nature to enjoy good music, soft lights, and beautiful things. We each can express opinions about aesthetics. For major structures, however, it is best to consult an architect who is trained and has experience in the aesthetics of structures. The architect can integrate the views of owners, engineers, and the community, and transform them into a beautiful structure that is in harmony with the environment as well as the pride of the community.

5.2.3 Constructibility

Constructibility is another very important aspect of design. A structural designer must ensure that the structure can be built without problems in the field, checking availability of materials, workforce, fabrication facilities, transportation modes and routes, and erection equipment and techniques. Constructibility is using knowledge of construction, inspection, maintenance, and operations to ensure that the structure can be built with no big field problems, easily inspected for condition evaluation, accessible for maintenance or replacement of wearing parts, and operated efficiently throughout the expected service life of the structure.

For major and complex structures, it is important for the structural designer to work closely with the fabricators, contractors, and maintenance and operation personnel to ensure that all constructibility issues are addressed adequately during design and preparation of construction plans and specifications of the project. Some unusual projects might need inspection and maintenance guidelines for personnel who will be responsible for maintenance. The structural designer can play an important role in the preparation of inspection, maintenance, and operations manuals so key elements of the designs receive proper attention throughout the service life of the structures.

5.2.4 Economy

Cost-effective designs result from taking advantage of the engineering properties of construction materials, the proper proportioning of members and framing, and the careful selection of connections to ensure long service life with minimal maintenance and operation problems. The structural designer will need to work closely with local suppliers, fabricators, and contractors to share knowledge on cost-saving ideas for short- and long-term performance of the structure.

The lowest initial construction cost is not necessarily the lowest cost in the long term. The cost to the owner of a structure is the sum of all costs involved in initial construction, operating, maintaining, repairing, rehabilitating, disposing, and other incidental costs to the owner, users, and third parties. The structural designer plays a very influential role in the initial construction cost and subsequent operation, maintenance, repair, and rehabilitation costs. During preliminary design, the structural designer should perform life-cycle cost analysis to compare short- and long-term cost-effectiveness of corrosion protection systems, alternative construction materials, and types of structural systems to arrive at an optimum system that has the lowest total cost over the service life of the structure.

5.3 DESIGN METHODS

Designers over the years have attempted to achieve structural safety by ensuring that the effects of the applied loads on the structures will not exceed the resistance or capacity of the structural members and connections, with a certain margin of safety. It may be expressed as

$$\text{Load effects } Q \leq \text{Resistance } R \tag{5.1}$$

This inequality is the basis for the design methods that have been continuously improved over many decades through research and development in materials, design procedures, analytical techniques, computer applications, erection and fabrication methods, structural performance, and construction experience. The evolution of these methods will be discussed in the following subsections.

5.3.1 Allowable Stress Design (ASD)

This method has been in use since before the 1900s. It is based on the assumption that the material behaves elastically, i.e., follows Hooke's law, under service load combinations. Certain fractions of the yield stress F_y or the ultimate strength F_u are used as allowable stresses for design purposes. For example, the allowable stresses for steel in building design is $0.60F_y$ or $0.50F_u$ and the allowable stresses for concrete is $0.4f_c'$, where f_c' is the 28-day compressive strength of concrete. In other words, ASD is a method of proportioning structural members so computed stresses produced in the members by the allowable stress load combinations do not exceed the specified allowable stress. The ASD method is used generally in the design of wood structures. This method is also called the working stress design (WSD) method.

The fractions usually are expressed in terms of factors of safety FS given by

$$FS = \frac{F_y}{0.60F_y} = 1.67 \text{ against yielding}$$

and

$$FS = \frac{F_u}{0.50F_u} = 2.0 \text{ against fracture}$$

The inequality Equation 5.1 for ASD can be expressed as

$$\text{Load effects (Stresses)} \leq \frac{\text{Yield Stress}}{FS} \quad \text{or} \quad \frac{\text{Ultimate Strength}}{FS} \quad (5.2)$$

The factor of safety FS is based on experience and judgment and not on scientific approach. It is applied to the resistance side of the inequality equation and does not account for the variability of loads. The ASD method does not recognize that most construction materials can resist stresses beyond their yield stresses. Tests have shown that strength of a structural member is a better measure of resistance than allowable stress. Since the early 1970s, the structural design profession has been moving to design methods that use strength as a fundamental measure of resistance and recognize variability in the loads.

5.3.2 Strength Design Method

Strength design is a method of proportioning structural members so the computed forces produced in the members by the factored load combinations do not exceed the member strength. The strength-design method is used for concrete and masonry structures. This method also is referred to as the load factor design (LFD) method.

In the strength-design method, the structure is analyzed by the methods of elastic analysis using the factored loads. Member strength is based on ultimate strength times a strength-reduction factor. For example, the member strength of reinforced concrete is based on the concrete reaching crushing stress of $0.85f_c'$ and steel reaching yield stress F_y. The computed member strength is then multiplied by the resistance factor ϕ to account for the uncertainty involved in the prediction of the material strength.

In the strength design or load factor design method, the inequality Equation 5.1 can be expressed as

$$\text{Factored Load effects} \leq \text{Strength Reduction Factor} \times \text{Ultimate Strength} \quad (5.3)$$

5.3.3 Load and Resistance Factor Design (LRFD)

Load and resistance factor design (LRFD) is a method of proportioning structural members using load and resistance factors so no applicable limit state is reached when the structure is subjected to all appropriate load combinations. LRFD is used in the design of steel in this book. The term *limit state* is used to describe the condition at which a structure or component ceases to perform its intended function.

LRFD uses different factors for each load and for the resistance. The load and resistance factors are based on extensive research, trial designs, and experience. The different factors account for the degree of uncertainty in the different types of loads and load combinations, and the variability in the materials. By using different load and resistance factors, we are able to attain a more uniform level of safety and reliability in the structures.

Usually, there are two categories of limit states—serviceability and strength—for building structures. Serviceability limit states refer to the performance of structures under normal service loads, and are concerned with the uses and occupancy of the structures, such as excessive deflection, vibrations, cracking, and deterioration. For instance, a high-rise building must be designed so that the lateral deflection or drift is not excessive during normal storms. We want to make sure the occupants will not be frightened or seasick because of deflections of the building. Strength limit states are based on the safety or load-carrying capacity of structures, such as ultimate strength, buckling, fatigue, fracture, and overturning. For instance, a high-rise building must be designed to resist the ultimate load occurring during major storms that may only hit every 50 or 100 years. The building might suffer some minor damage and the occupants might experience some discomfort.

In LRFD, each member and connection must satisfy the following equation for each limit state:

$$\sum \gamma_i Q_i \leq \phi R_n \quad (5.4)$$

where γ = Load factor
Q = Load effects
ϕ = Resistance factor
R_n = Nominal resistance

In the AASHTO LRFD Bridge Design Specifications, there are five limit states—service limit state, fatigue and fracture limit states, strength limit state, and extreme event limit state.

1. The service limit state is taken as restrictions on stress, deformation, and crack width under regular service conditions.
2. The fatigue limit state is taken as restrictions on stress range as a result of a single design truck occurring at the number of expected stress range cycles.

3. The fracture limit state is taken as a set of material toughness requirements of the AASHTO Material Specifications.
4. The strength limit state is taken to ensure that strength and stability, both local and global, are provided to resist the specified statistically significant load combinations a bridge is expected to experience in its design life.
5. The extreme event limit state is taken to ensure the structural survival of a bridge during a major earthquake or flood, or when hit by a vessel, vehicle, or ice floe.

The LRFD equation in bridge design is similar to that of Equation 5.4 for building, except an additional factor η is added to the load side of the equation to recognize ductility, redundancy, and operational importance. The AASHTO LRFD equation is:

$$\sum \eta_i \gamma_i Q_i \leq \phi R_n \tag{5.5}$$

A probability-based limit states design method has been used in Canada and European countries for many years. AISC introduced LRFD method in the 1986 Edition of the Specifications for Structural Steel Buildings, based on a LRFD research project conducted at Washington University in St. Louis, Missouri, from 1969 through 1976 under the direction of T. V. Galambos and M. K. Ravindra. AASHTO adopted the LRFD method in 1993, and the first edition of the AASHTO LRFD Bridge Design Specifications was published in 1994. The AASHTO LRFD was based on NCHRP Project 12-33, directed by J. M. Kulicki, of Majeski and Masters, with the support of more than 70 experts in bridge engineering and thousands of hours of review and trial designs from states, industry, consultants, academia, and special agencies. Since the first editions of LRFD, AISC and AASHTO have introduced new editions to incorporate technological advances and improved engineering practices. Structural-design engineers are moving gradually to LRFD from the traditional Allowable Stress Design (ASD) or Working Stress Design (WSD) and from the Ultimate Strength Design (USD) or Load Factor Design (LFD) methods.

LRFD is based on new and old developments in structural engineering, sound principles, and logical approaches to ensure constructibility, safety, serviceability, inspectability, economies, and aesthetics. LRFD incorporates the best of allowable or working stress design, strength design, load-factor design, and plastic design, which are already familiar to designers. Because of the many advantages of LRFD, and because of the fact that many colleges and universities are teaching LRFD in design classes, the structural-design profession is moving rapidly to adopt the LRFD method.

5.4 DESIGN LOADS

The most important task for a structural designer is to determine the loads that will be imposed on a structure during its service life. Not all loads have the same effect on the structural elements, and not all loads act on the structure at the same time. The next task is to determine the load factors and combinations of loads that are most likely to occur at the same time. These are not simple tasks. To assist the designers, codes and specifications have been prepared by knowledgeable and experienced professionals and adopted by agencies for selecting the proper loads and load combinations for the design of buildings and bridges. The most commonly used specifications are:

1. The Uniform Building Code (UBC) for buildings
2. The American Institute of Steel Construction (AISC) Manual of Steel Construction, Load and Resistance Factor Design for Steel Structures
3. The American Association of Highways and Transportation Officials (AASHTO) Specifications for Highway Bridges
4. The American Railway Engineering Association (AREA) Specifications for Railroad Bridges

For special cases, project-specific and/or site-specific studies might need to be carried out to determine the most appropriate loads and load combinations for the design to assure trouble-free, short- and long-term performance. The American Society of Civil Engineers (ASCE) has developed special guidelines that can be referenced by the structural designers to determine the proper design parameters. Examples are: Minimum Design Loads for Buildings and Other Structures, ASCE Standard 7, and Wind Forces on Structures, Task Committee on Wind Forces, Committee on Loads and Stresses, Structural Division, ASCE.

Some examples on loads and load combinations, in accordance with the AISC LRFD Specifications are:

$$1.4D$$
$$1.2D + 1.6L + 0.5(L_r \text{ or } S \text{ or } R)$$
$$1.2D + 1.6(L_r \text{ or } S \text{ or } R) + (0.5L \text{ or } 0.8W)$$
$$1.2D + 1.3W + 0.5L + 0.5(L_r \text{ or } S \text{ or } R)$$
$$1.2D \pm 1.0E + 0.5L + 0.2S$$
$$0.9D \pm (1.3W \text{ or } 1.0E)$$

The numerals in the above expressions are the load factors, and the alphas take on definitions as follows:

where D = Dead load due to the weight of the structural elements and the permanent features on the structure
L = Live load due to occupancy and movable equipment (reduced as permitted by the governing code)
L_r = Roof live load
W = Wind load
S = Snow load
E = Earthquake load
R = Nominal load due to initial rainwater or ice exclusive of the ponding contribution

The magnitudes of the loads can be taken from the specifications referenced above or from the governing building code.

5.5 DESIGN SPECIFICATIONS AND CODES

States, counties, and cities publish or adopt building codes to ensure public safety. These codes are actually laws or ordinances of the governmental agencies. They control the construction of various types of structures and specify design loads, design stresses, material quality, and other good practices. The structural designer performing engineering work for an agency must follow the codes published or adopted by the agency.

Several engineering organizations publish recommended practices for regional or national use. Among these organizations are the ICBO (International Conference of Building Officials), the ACI (American Concrete Institute), the AISC (American Institute of Steel Construction), ASCE (American Society of Civil Engineers), and AASHTO (American Association of State and Transportation Officials). Most, if not all, municipal and state building codes have adopted the Uniform Building Code published by ICBO. All state highway and transportation departments have adopted the AASHTO Specifications in their entirety or with modifications. These specifications and codes represent the collective knowledge and experience of the professional membership of these organizations, which are dedicated to public safety and advancement in engineering practice. The structural designer is encouraged to use these specifications and codes diligently, while exercising good engineering judgment to accomplish excellence in the constructed projects.

The 1997 Uniform Building Code, Volume 2: Structural Engineering Design Provisions, on which the PE examination is based, will be used extensively, if not exclusively, in the subsequent chapters on design.

CHAPTER 6
CONCRETE DESIGN

6.1 INTRODUCTION

Concrete is a mixture of various sizes of aggregates blended with cement and water. The semi-fluid mixture is placed in forms to produce desired structural members. After hardening, these artificial stone-like structural members are relatively strong in resisting compressive loads, yet weak in tension. Steel reinforcement is added and positioned to compensate for the lack of tensile strength and to provide ductility. The resulting material is versatile, economical, strong, and durable, and hence one of the most widely used construction materials today.

We will discuss some general requirements of constructing concrete structural members, as well as physical properties of concrete. Then, we'll proceed with some basic principles for the analysis of reinforced concrete members subjected to flexural, axial, and shear forces—either alone or with some combinations of two or more forces simultaneously. Examples are provided to clarify the applications of the principles. Two frequently encountered practical concrete design problems—cantilever retaining wall and isolated footing—are used to illustrate how simple design principles for reinforced concrete structures are applied.

The ACI 318-95 Code (Building Code Requirements for Structural Concrete and Commentary) will be used in the discussions because they have been adopted/referenced by all three model building codes: UBC (Uniform Building Codes), NBC (National Building Codes), and SBC (Standard Building Codes). Other codes, such as AASHTO (American Association of State Highway and Transportation Officials) and AREA (American Railway Engineering Association), are similar.

The ACI Code bases the design on the strength, or *ultimate strength* of the concrete structural member. In this method of design, the margin of safety is provided by two sets of factors:

- Load Factors: To account for uncertainties in the estimation of load effects at the structural *element* level—including the effects of overload and inaccuracies/assumptions in structural analysis (typically elastic analysis) of the structural system—the specified service loads are multiplied by load factors (usually greater than 1). Different load factors are assigned to different categories of loads and load combination. In the most common case, which involves only dead and live loads, ACI specifies $1.4D + 1.7L$. Other load factors for various load combinations are given in ACI 9.2. In AASHTO and AREA codes, because of greater uncertainties associated with the live load and more severe consequence of any structural failure, the load factors specified are generally larger than those shown in ACI.

- Strength Reduction Factors: To provide for uncertainties in the estimation of member capacity, including the effects of material understrength, future deterioration of section, and inaccuracies in strength theories, the nominal member strength is multiplied by a strength reduction factor ϕ (smaller than 1) to obtain the *design strength*.

Note that in concrete design, this is often referred to as *strength design method*, not *load and resistant factor design*, as used in steel design.

Structural concrete refers to all concrete used for structural purposes, including plain concrete, (conventionally) reinforced concrete, and prestressed concrete. In this chapter, the emphasis will be on reinforced concrete design, with a short discussion of prestressed concrete design at the end.

Concrete design in a general sense involves the "guessing" of a member section, and then iteratively adjusting it to improve safety and economy based on the analysis. Many tables, charts (e.g., *CRSI Handbook*), and computer programs (e.g., PCA column program) have been developed to facilitate this process. Detailed illustrations of using these design aids are beyond the scope of this brief review.

It is recommended that a copy of the ACI code be used with this chapter. The ACI code commentary provides additional information on the equations cited, and the notations used here are identical to those in the code.

6.2 MECHANICAL PROPERTIES OF CONCRETE

6.2.1 Compressive Strength

The concrete strength of a structural member can be reliably estimated using a simple cylindrical specimen (typically 6 in. × 12 in.), prepared and tested according to ASTM standards. Because concrete strength increases with time at an ever-decreasing rate (as long as moisture is maintained for the hydration of cement), the specimen is tested at 28 days to determine the standard value of compressive strength, f'_c. Concrete used for structural purposes may have f'_c varying from 2500 to 7000 psi—usually between 3000 and 5000 psi, although in recent years for some high-performance concrete, f'_c as high as 10,000 psi is not unusual.

6.2.2 Tensile Strength

Concrete tensile strength is much lower than f'_c—typically one order of magnitude lower. Studies indicate that tensile strength is related more to the square root of f'_c than to f'_c.

Tensile flexural stress generally is neglected in reinforced concrete design.

6.2.3 Stress-Strain Relationship

While the compressive stress-strain relations of the constituents of the concrete—namely, the aggregate and the cement paste—are linear, the compressive stress-strain curve for concrete is distinctly non-linear.

The general shape of stress-strain curve is primarily a function of concrete strength and consists of a rising curve from zero to a maximum at a compressive strain between 0.0015 and 0.002 followed by a descending curve to an ultimate strain (crushing of concrete) from 0.003 to higher than 0.008. ACI sets the maximum usable strain for design at 0.003.

6.2.4 Modulus of Elasticity

The slope of the stress-strain curve is the *modulus of elasticity*. Since the curve for concrete is nonlinear, there are several ways to approximate the modulus of elasticity: the initial tangent modulus, the actual tangent modulus, and the secant modulus, which is most commonly used.

ACI code suggests a formula for estimating the secant modulus of elasticity based on concrete strength and unity weight:

$$E_c = 33\omega^{1.5}\sqrt{f'_c}$$

where E_c = modulus of elasticity of concrete, psi
ω = unit weight of concrete, pcf
f'_c = strength of concrete, psi

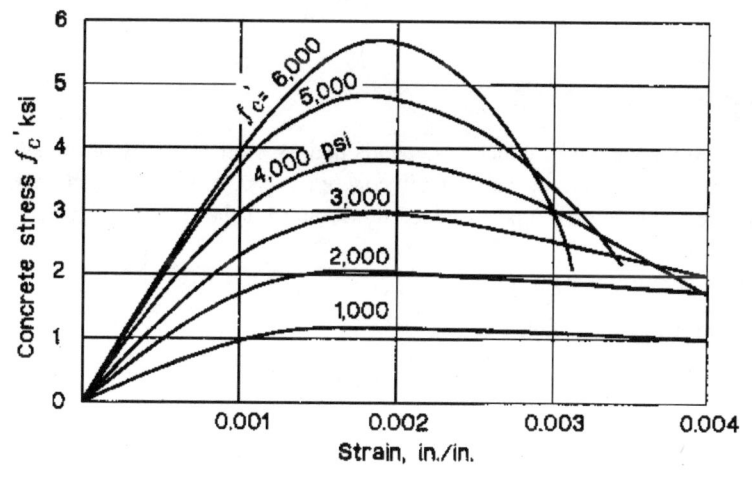

FIGURE 6.1

For normal concrete,

$\omega = 145$ pcf, and
$E_c = 57{,}000\sqrt{f'_c}$

6.2.5 Creep

If concrete is held under sustained stress, its strain will continue to increase with time (at an ever-decreasing rate). This time-dependent phenomenon is called *creep*. The creep strain will eventually reach a level several times the instantaneous strain. A reduced secant modulus for the sustained stress is used to account for the increased strain. The modulus is often taken as $\frac{1}{2}$ to $\frac{1}{5}$ of the normal E_c (e.g., $\frac{1}{3}$ is used for superimposed dead load for composite steel bridge design).

6.2.6 Shrinkage

Shrinkage strain is associated with water evaporation from hardened concrete. Among other factors, it is controlled by the water content of the concrete during mixing, ambient environmental conditions during curing, and constraints of the concrete structural members after hardening.

6.2.7 Thermal Coefficient

The thermal coefficient of concrete is about $5.5 \times 10^{-6}/°F$ or $10 \times 10^{-6}/°C$. It is close to the thermal coefficient of steel, and that makes the composite of the two materials practical. ACI 7.12 specifies some minimum reinforcement in the structural concrete to control cracks resulting from the shrinkage and temperature movements/restraints.

6.2.8 Unit Weight

Although the weight of concrete depends mainly on the density of the aggregates, which varies somewhat, normal-weight plain concrete is typically taken as 145 pcf. For reinforced concrete, an additional 5 pcf is added to account for the reinforcing steel.

6.3 REINFORCEMENT

6.3.1 Grades

Reinforcement used in concrete structure typically consists of deformed steel bars (which have ribbed projections to provide better bond between steel and concrete), although plain reinforcing bars are allowed for spirals. The most commonly used bars today are Grade 60 (ASTM-615), with a yield strength of 60,000 psi. When welding of reinforcing bars is required, ASTM 706 should be specified.

6.3.2 Sizes

Reinforcing bars referred to in the contract documents vary in size from #3 to #18. For bars up to and including #8, the number of the bar coincides with the bar diameter in eighths of an inch. The #9, #10 and #11 bars have diameters that provide areas equal to 1 in. × 1 in. square bars, $1\frac{1}{8}$ in. × $1\frac{1}{8}$ in. square bars, and $1\frac{1}{4}$ in. × $1\frac{1}{4}$ in. square bars, respectively. Similarly, the #14 and #18 bars correspond to $1\frac{1}{2}$ in. × $1\frac{1}{2}$ in. and 2 in. × 2 in. bars, respectively.

6.3.3 Development Length

The development length is the length of anchorage required so that the steel stress can *develop* from zero at one end to the full strength f_y at the other. Experimentally, it was found that the development length l_d depends on:

Concrete strength f'_c

Yield point of steel f_y

Size of bar d_b

Different development lengths for deformed bars in tension and in compression are specified in ACI 12.2 and 12.3, respectively. Often, design aids (e.g., CRSI charts) in the form of tables/charts are prepared to ease the lengthy calculations for various bars with different modifying factors.

The effect of a standard hook at the end of a bar in tension is reflected in a reduced development length l_{dh}, which is measured from the critical section to the outside end of the hook. Hook shall not be used to develop bars in compression.

6.3.4 Splice

Three methods are used to splice reinforcing bars: lap splices, mechanical splices, and welded splices. Lap splice is the most common among the three, but are not allowed for bars larger than #11. The contract plans should clearly show the location and lengths of lap splices.

6.3.5 Lengths

Reinforcing bars normally are stocked in lengths of 60 feet. Where possible, specify lengths 60 feet and less for bar sizes #8 to #18. Because of placement and transportation considerations, the overall length of bar size #3 should be limited to 30 feet, and bar sizes #4 and #5 to 40 feet.

6.3.6 Concrete Protection (Cover) of Reinforcement

Concrete cover should be provided as protection of reinforcement against weather and other effects. When minimum cover is prescribed for a class of structural member, it is measured from the concrete surface to the outermost surface of the reinforcing steel being protected, and to the edges of stirrups, ties, or spirals in the case of transverse reinforcement enclosed main bars. ACI 7.7 specifies the minimum concrete cover for non-prestressed cast-in-place concrete as follows:

	Minimum cover, in.
Concrete cast against and permanently exposed to earth	3
Concrete exposed to earth or weather:	
No. 6 through No. 18 bars	2
No. 5 bar, W31 or D31 wire, and smaller	$1\frac{1}{2}$
Concrete not exposed to weather or in contact with ground:	
Slabs, walls, joists:	
No. 14 and No. 18 bars	$1\frac{1}{2}$
No. 11 bar and smaller	$\frac{3}{4}$
Beams, columns:	
Primary reinforcement, ties, stirrups, and spirals	$1\frac{1}{2}$
Shells, folded plate members:	
No. 6 bars and larger	$\frac{3}{4}$
No. 5 bar, W31 or D31 wire, and smaller	$\frac{1}{2}$

6.4 CONCRETE QUALITY, PROPORTIONING, PLACING, AND CURING

6.4.1 Types of Cement

Five types of portland cements are commonly used in concrete:

- Normal, Type I: Used in ordinary construction
- Modified, Type II: C_3A content is lower than Type I, and hence, lower heat of hydration; moderate sulfate resistance
- High early strength, Type III: high in C_3S and C_3A; used when high early strength is required; early form removal and fast construction
- Low heat, Type IV: C_3A is low and C_2S is high; used in mass concrete such as dams and high footings where low heat of hydration is required; concrete develops strength slowly
- Sulfate resisting, Type V: very low C_3A content; used in concrete in a severe sulfate exposure; relatively low heat generation.

6.4.2 Aggregates, Water, Admixture

The nominal maximum aggregate size is limited to the following:

- One-fifth the narrowest dimension between sides of forms
- One-third the depth of the slab
- Three-quarters the minimum clear spacing between reinforcing bars.

The limitation may be waived if the engineer judges that the workability and method of consolidation are such that the concrete can be placed without honeycomb or voids.

Water used in reinforced concrete should not contain deleterious amount of chloride ion. The main concern over a high chloride content in mixing water is the possible effect of chloride ions on the corrosion of embedded reinforcing steel.

Admixtures are used to improve workability, hardening, and strength characteristics in concrete. Types include air-entraining agents, accelerators, retarders, water reducers, gas-forming agents, fly ash, pozzolans, ground granulated blast-furnace slag, and silica fume.

6.4.3 Proportioning

Concrete mix proportioning is specified either through *prescription specifications* or *performance specifications* to ensure minimum strength requirements and certain workability and durability considerations.

In prescription specifications, the purchasers (or their representative engineers/architects) provide the concrete producers the necessary prescription for mixing/proportioning the concrete and assume the responsibility for obtaining the desired concrete quality.

On the other hand, in performance specifications the concrete producer is responsible for the concrete mix design, and the purchasers provide only minimum strength (in terms of f'_c) and workability criteria (in terms of slump range and maximum aggregate size).

For selecting proportions for normal weight concrete, ACI 211.1-91 (in Part 1 of ACI Manual of Concrete Practice, 1999: ACI Committee 211, "Standard Practice for Selecting Proportions for Normal, Heavyweight, and Mass Concrete") provides two methods: the *estimated weight* and the *absolute volume* method. The methods describe procedures for an initial approximation of proportions, followed by trial batches in the laboratory or field, and adjusted, as necessary, to produce the desired characteristics of concrete. The procedures involve the following steps:

1. Choice of slump
2. Choice of maximum size of aggregate
3. Estimation of mixing water and air content
4. Selection of water-cement or water-cementitious materials ratio
5. Calculation of cement content
6. Estimation of coarse aggregate content
7. Estimation of fine aggregate content (either the *estimated weight* method or the *absolute volume* method can be used to determine the remaining ingredient of the

batch after the quantities of other ingredients have been determined through Steps 1 to 6)

8. Adjustments for aggregate moisture
9. Trial batch adjustments.

6.4.4 Placing and Curing

Placing concrete in the forms consists of depositing and consolidating. The basic rule in depositing concrete is that the concrete should be as close as possible to its final location and should be as nearly vertical as possible when dropped into place. To avoid honeycomb, voids, rock pockets, and sand streaks, the concrete should be consolidated with high-frequency vibrators immediately after it is deposited. The vibrator should be operated in a nearly vertical position. Application of vibrators should be spaced uniformly and close enough to ensure that the concrete is consolidated to the maximum practical density.

Curing ensures that the freshly placed concrete maintains the required temperature and moisture for the necessary time period and is free from impact, loading, and other mechanical disturbances until the desired strength and hardness of concrete is reached. Maintaining moisture is of particular importance because it keeps the water available for hydration, which strongly affects the compressive strength of the concrete. It has been found that prolonged moist curing leads to higher concrete compressive strength.

6.5 DESIGN FOR FLEXURAL (PURE BENDING) LOADING

6.5.1 Assumptions

Flexural strength of reinforced concrete beams can be analyzed using the following assumptions:

- Plane section remains plane (i.e., linear strain distribution)

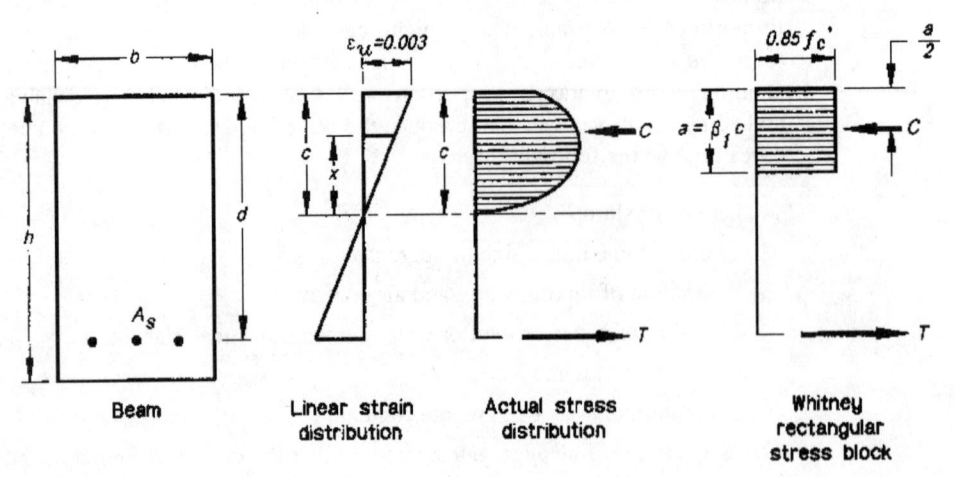

FIGURE 6.2

- Concrete crushes when strain reaches 0.003
- Stresses in concrete and reinforcement based on stress-strain curves
- Strain in reinforcement equals strain in concrete at same level (i.e. no slippage at steel-concrete interface)
- Tensile strength of concrete is negligible
- Concrete stress distribution can be simplified as Whitney's stress block (i.e. the distribution is rectangular with a value of $0.85f_c'$).

6.5.2 Rectangular Singly Reinforced Beam

Based on the listed assumptions, the *neutral axis* of the section under pure bending can be determined, and the *nominal bending strength* of the section can, in turn, be calculated. Using a relatively simple and common case of rectangular section with tensile reinforcement for illustration, here are the steps:

Step 1. From the linear strain assumption and simple geometry:

$$\epsilon_c = \epsilon_u \left(\frac{x}{c}\right)$$

In addition, concrete ultimate strain:

$$\epsilon_u = 0.003$$

Step 2. Concrete stress:

f_c varies with ϵ_c in compression
$f_c = 0$ in tension
Steel stress:
$f_s = (E_s)(\epsilon_s)$ if $\epsilon_s < \epsilon_y$
$f_s = f_y$ if $\epsilon_s > \epsilon_y$

Step 3. Compressive force:

$$C = (0.85f_c')(\text{area of stress block}) = (0.85f_c')(a)(b)$$

where $a = \beta_1 c$
c is the distance from the neutral axis to the extreme compressive fiber
β_1 is a function of f_c':

β_1	f_c'
0.85	≤ 4 ksi
0.80	5 ksi
0.75	6 ksi
0.70	7 ksi
0.65	≥ 8 ksi

Step 4. Tensile force:

$T = (A_s)(f_s)$

$C = T$ Because it is pure bending, total external force equals zero.

$$\therefore c = \frac{A_s f_s}{b\beta_1(0.85f_c')}$$

$$M_n = C\left(d - \frac{a}{2}\right) = T\left(d - \frac{a}{2}\right)$$

Note in Step 2, the steel strain may or may not exceed yield point when the section reaches its ultimate stage (i.e. $\epsilon_u = 0.003$). Therefore, for the same *concrete* section, the neutral axis and the nominal bending strength of the section depend on the amount of reinforcement provided.

When a relatively large amount of tensile steel is provided, the section reaches its ultimate stage and the steel does not yield (i.e., $f_s < f_y$). The section will fail in a sudden and brittle way by crushing of the concrete. The section is called *over-reinforced* section. The neutral axis of the section can be determined through iteration process (because f_s in the equation for calculating c is unknown initially) or by solving a quadratic equation. Since this is an undesirable type of section, the calculations are rather academic and have little practical value.

One special case is where a specific amount of steel reinforcement is provided such that the yielding of steel and crushing of concrete occur simultaneously. This is called *balanced section*. For the balanced section, the strain diagram at ultimate stage is shown as:

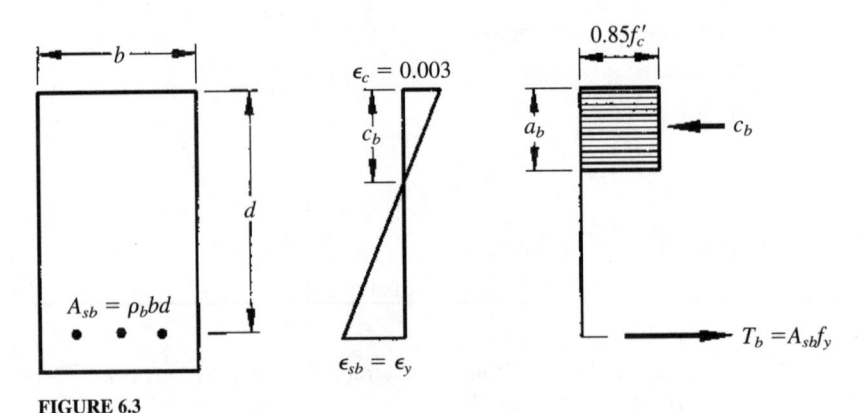

FIGURE 6.3

From simple geometry:

$$\epsilon_{sb} = \frac{0.003(d - c_b)}{c_b} = \epsilon_y$$

Neutral axis:

$$c_b = \frac{0.003}{0.003 + \epsilon_y}d = \frac{87,000}{87,000 + f_y}d$$

Balanced reinforcement:

$$A_{sb} = \frac{C_b}{f_y} = \frac{0.85 f'_c \beta_1 c_b b}{f_y}$$

When the amount of steel is less than the balanced reinforcement, the section will fail initially by yielding of steel. This is called *under-reinforced* section. Resistance will slightly increase due to rising of the neutral axis, accompanied by substantial increase of curvature. Finally, the section fails when the concrete crushes. Since this type of failure mode is gradual and ductile, allowing time for users to escape or to apply remedy measures, it is a desirable design.

To provide an additional safety margin, ACI requires that reinforcement should not exceed 0.75 of that amount which creates the balanced condition, i.e.:

$$A_s < 0.75 A_{sb}$$

EXAMPLE 6.1
Find the design strength ϕM_n of the rectangular section where $b = 12$ in., $d = 30$ in., $A_s = 6$ in^2, $f'_c = 4000$ psi, and $f_y = 60,000$ psi as shown in Figure 6.4.

Solution
First guess the section is under-reinforced (to be verified later)

$$T = A_s f_y = 6 \times 60 = 360 \text{ k} = C$$

$$a = \frac{C}{(0.85 f'_c b)} = \frac{360}{(0.85 \times 4 \times 12)} = 8.82 \text{ in.}$$

$$\therefore c = \frac{a}{\beta_1} = \frac{8.82}{0.85} = 10.38 \text{ in.}$$

$$\epsilon_s = \frac{0.003(d-c)}{c} = \frac{0.003(30 - 10.38)}{10.38} = 0.00567$$

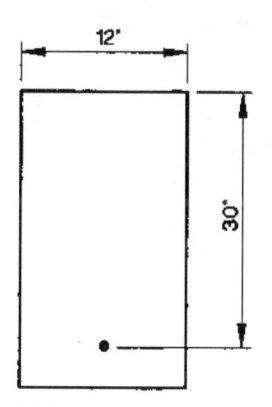

FIGURE 6.4

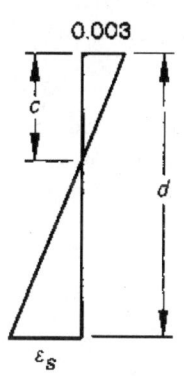

FIGURE 6.5

$$\epsilon_y = \frac{f_y}{E_s} = \frac{60}{29,000} = 0.00207$$

$$\epsilon_s > \epsilon_y$$

∴ Section is under-reinforced, initial guess is correct.

$$M_n = T\left(d - \frac{a}{2}\right) = 360\left(30 - \frac{8.82}{2}\right) = 9212 \text{ k-in.}$$

Design strength = $\phi M_n = 0.9 \times 9212 = 8291$ k-in.

Alternate solution
(This equation can be readily derived from the equation in Step 4 in the text for an under-reinforced case)

$$\rho = \frac{A_s}{bd} = \frac{6}{(12 \times 30)} = 0.0167 < \rho_b \quad \text{(see below)}$$

$$\omega = \frac{\rho f_y}{f'_c} = 0.0167 \times \frac{60}{4} = 0.25$$

$$M_n = f'_c bd^2 \omega(1 - 0.59\omega)$$
$$= 4(12)(30)^2(0.25)(1 - 0.59 \times 0.25)$$
$$= 9207 \text{ k-in.}$$

Design strength = $\phi M_n = 0.9 \times 9207 = 8286$ k-in.

6.5.3 Rectangular Doubly Reinforced Beam

A section containing reinforcement steel both in tension and in compression is referred to as *doubly reinforced*. The analysis of such a section is similar to the case of singly reinforced section discussed above except there is one additional set of parameters (d', A'_s, ϵ'_s, and f'_s) relating to the compressive reinforcement.

EXAMPLE 6.2
Find the design strength ϕM_n of the rectangular section where $b = 12$ in., $d = 16$ in., $d' = 2.5$ in., $A_s = 6$ in^2, $A'_s = 3$ in^2, $f'_c = 4000$ psi, and $f_y = 60,000$ psi, as shown in Figure 6.6. Does this section satisfy ACI 10.3.3?

Solution
First find the balanced amount of tensile reinforcement A_{sb}:
Note if the section is without compressive reinforcement:

$$\rho_b = \frac{(0.85)\beta_1 f'_c (87,000)/f_y}{(87,000 + f_y)} = \frac{0.85 \times 0.85 \times 4 \times 87,000/60}{(87,000 + 60,000)} = 0.0285$$

$$\rho = \frac{A_s}{bd} = \frac{6}{(12 \times 16)}$$
$$= 0.03125 > 0.75\rho_b = 0.0214$$

The section is over-reinforced, if considered as a singly reinforced section.

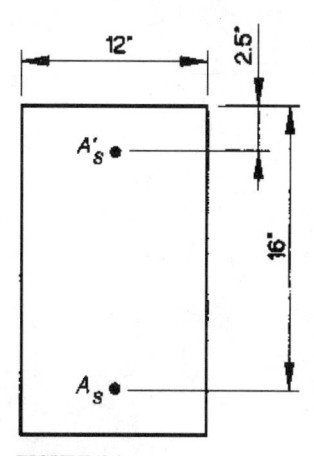

FIGURE 6.6

However, the compressive reinforcement has been added to counteract the extra tensile reinforcement. If the section is balanced, i.e. concrete crushes at the same time tensile reinforcement yields, the distance from top fiber to neutral axis is

$$c_b = \frac{(d)(87,000)}{(87,000 + f_y)} = \frac{16 \times 87,000}{(87,000 + 60,000)} = 9.47 \text{ in.}$$

Compressive force attributed to concrete, if section is balanced. See Figure 6.7.

$$C_{cb} = 0.85 f'_c b \beta_1 c_b$$
$$= 0.85 \times 4 \times 12 \times 0.85 \times 9.47$$
$$= 328.4 \text{ k}$$

$$\epsilon'_{sb} = \frac{0.003(c_b - d')}{c_b} = \frac{0.003(9.47 - 2.5)}{9.47} = 0.00221 > \epsilon_y = \frac{60}{29,000} = 0.00207$$

$$\therefore f'_{sb} = f_y = 60 \text{ ksi}$$

$$C_{sb} = A'_s(f_y - 0.85 f'_c) = 3(60 - 0.85 \times 4)$$
$$= 169.8 \text{ k}$$

$$A_{sb} = \frac{(C_{cb} + C_{sb})}{f_y} = (328.4 + 169.8)/60 = 8.30 \text{ in}^2$$

$$A_{s(max)} = \frac{(0.75 C_{cb} + C_{sb})}{f_y} = \frac{(0.75 \times 328.4 + 169.8)}{60} = 6.94 \text{ in}^2$$

$$A_s = 6 < A_{s(max)} \quad \text{OK.}$$

Check ACI 10.3.3:
Compressive reinforcement yields compressive force, attributed to steel reinforcement, if section is balanced.

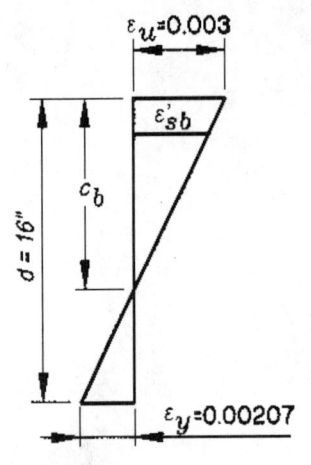

FIGURE 6.7

First, $T = A_s f_y = 6 \times 60 = 360$ k

Guess A'_s yields (to be verified later),

$C_s = 169.8$ k

$C_c = 360 - 169.8 = 190.2$ k

$$\therefore a = \frac{190.2}{(0.85 \times 4 \times 12)} = 4.66 \text{ in.}$$

$$c = \frac{4.66}{0.85} = 5.48 \text{ in.}$$

$$\epsilon'_s = \frac{0.003(c - d')}{c} = \frac{0.003(5.48 - 2.5)}{5.48}$$

$$= 0.0016 < \epsilon_y = 0.00207$$

The initial guess was incorrect.

$f'_s = E_s \epsilon'_s = 29{,}000 \times 0.0016 = 46.4$ ksi

$$M_n = T\left(d - \frac{a}{2}\right) + C_s\left(\frac{a}{2} - d'\right) = 360\left(16 - \frac{4.66}{2}\right) + 46.4 \times 3\left(\frac{4.66}{2} - 2.5\right)$$

$= 4897$ k-in.

$\phi M_n = 0.9 \times 4897 = 4408$ k-in.

Note that unlike the example shown, the doubly reinforced section is often under-reinforced if considered as a singly reinforced section (discounting the compressive reinforcement in the section). Therefore, the nominal bending strength of the section can be conservatively calculated by ignoring the compressive reinforcement (as if it does not exist). The compressive reinforcement often is only provided for shrinkage and temperature (see ACI 7.12 for shrinkage and temperature reinforcement requirements).

6.5.4 Check Crack Width Limitation

ACI Equation 10-5 provides a check on crack width disguised in the form of a stress value z, in kips/in.

$$z = f_s \sqrt[3]{d_c A}$$

where f_s = Steel stress under service load, may be taken as $0.6 f_y$
d_c = Concrete cover thickness to the center of the lowest steel bar
A = Average tensile area per longitudinal bar. The total concrete tensile area may be taken as $2 b_w(h - d)$. If mixed bar sizes are used, the tributary area for each bar is taken as proportional to its area, and the design is controlled by the largest bar.

The z value should not exceed 175 kips/in. for interior exposure and 145 kips/in. for exterior exposure.

EXAMPLE 6.3
Find the stress value z for the exterior beam with width $b_w = 11$ in., height $h = 24$ in., depth $d = 21.5$ in., $f_y = 60$ ksi, and there are two #9 and one #8 reinforcement in the bottom layer of steel as shown in Figure 6.8.

Solution

$d_c = 24 - 21.5 = 2.5$ in.

$2(h - d) = 2(24 - 21.5) = 5$ in.

$A = (b_w)(2)(h - d)/[A_s/\text{bar area}_{(\text{largest bar})}]$
$= (11)(5)/[(2 \times 1.0 + 0.79)/1]$
$= \dfrac{55}{2.79} = 19.7$ in^2

$z = f_s(d_c A)^{1/3}$ (ACI 10.5)
$= 36 (2.5 \times 19.7)^{1/3}$
$= 132$ kip/in < 145 OK.

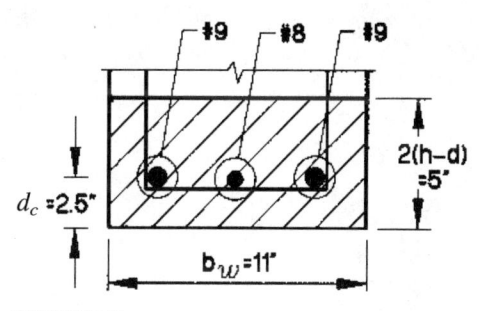

FIGURE 6.8

6.5.5 Detailing

To avoid sudden failure, ACI specifies a minimum amount of tensile reinforcement when required by analysis:

$$A_{s(min)} = \frac{3\sqrt{f'_c}\,b_w d}{f_y} \geq \frac{200 b_w d}{f_y}$$

This requirement can be waived if at every section the area of tensile reinforcement provided is at least one-third greater than that required by analysis.

6.6 DESIGN FOR AXIAL AND FLEXURAL LOADING

6.6.1 (Pure) Axial Loading

In order to analyze a reinforced concrete member under combined axial and flexural loading, an *axially loaded* member, similar to the *pure bending* case, is necessary as a reference point for discussion. The *axially loaded column* is a highly idealized member which rarely, if ever, exists in reality. Because of construction inaccuracies, end restraints, and load applications, some bending is almost always present. The axially loaded column is defined to correspond to a condition of uniformly distributed compressive strain. The column reaches its ultimate strength when the strain is 0.003.

Because of the possible long-term effects due to sustained loading, as well as size effect, the uniform stress in concrete is less than f'_c. Empirically, $f_c = 0.85 f'_c$.

The steel will reach its yield stress if the strain exceeds 0.003, therefore, $f_s = f_y$.

The strength of axially loaded reinforced concrete column P_{n0} consists of two components: one is the contribution from concrete and the other from steel:

$$P_{n0} = C_c + \Sigma C_s = A_g(0.85 f'_c) + A_{st}(f_y - 0.85 f'_c)$$

where A_g = Gross area of section
 A_{st} = Total steel area.

The point of application of the force P_{n0} is called the *plastic centroid*. For an unsymmetrical column, the plastic centroid does not coincide with the geometrical center.

EXAMPLE 6.4

Find the strength P_{n0} and the plastic centroid of the axially loaded reinforced concrete column with $b = 15$ in., $h = 18$ in., $f'_c = 6000$ psi, $f_y = 60,000$ psi, and all the bars are #11 as shown in Figure 6.9.

Solution
1. Find each force component:

 $C_c = 15 \times 18 \times 0.85 \times 6 = 1377$ k
 $C_{s1} = 2 \times 1.56 \times (60 - 0.85 \times 6) = 171.3$ k
 $C_{s2} = 4 \times 1.56 \times (60 - 0.85 \times 6) = 342.6$ k
 $P_{n0} = C_c + \Sigma C_s = 1377 + 171.3 + 342.6 = 1891$ k

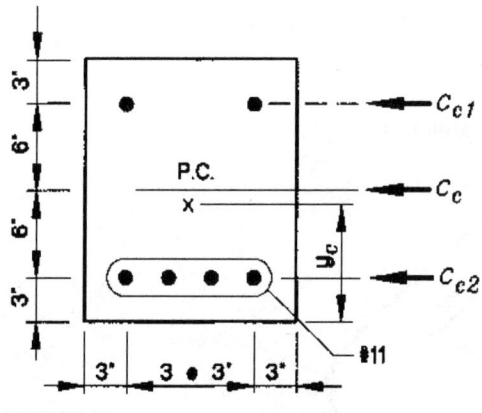

FIGURE 6.9

2. Summing moment about AB:

$(P_{n0})(y_c) = 1377 \times 9 + 171.3 \times 15 + 342.6 \times 3 = 15{,}990.3$ k-in.

$\therefore y_c = 15{,}990.3/1891 = 8.46$ in.

6.6.2 Combined Axial and Flexural Loading

The strength (capacity) of a given section to resist combined action of axial and flexural loading can be shown by a *P-M* diagram where:

Ordinate: P_n = Normal compressive load

Abscissa: M_n = Moment about plastic centroid axis.

Every point on the diagram represents one combination of loading that caused the section to fail. In other words, the *P-M* curve depicts how the section capacity (strength) decreases/increases due to the interaction of the axial force and the bending moment. The construction of the *P-M* curve for a specific section is similar to the analysis of a doubly reinforced section shown in the previous section, except:

- The summation of all internal forces does not equal zero. (It equals the external axial load.)
- The neutral axis is no longer fixed by the section alone. (It depends on the loading condition—the ratio of the applied bending moment to the applied axial force—as well.)

Previous discussions demonstrated how to calculate the P_{n0} (pure axial load) and M_{n0} (pure bending) where the curve intersects the ordinate and the abscissa axes respectively.

Connecting a point on the curve with the origin, the slope of the line is the reciprocal of the eccentricity e (i.e., the ratio of the applied bending moment to the applied axial force).

The interaction curve is separated by the *balanced point B* (M_{nb}, P_{nb}) into two parts: tension control region below and compression control region above.

In the tension control region, M_n increases as P_n increases, i.e., existence of axial load increases the moment capacity. This is because under compression force, when the section

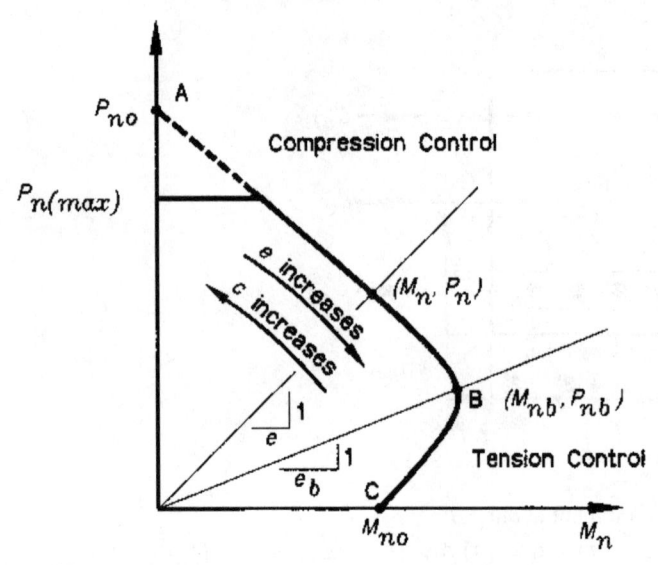

FIGURE 6.10

is subjected to predominantly bending moment, more concrete becomes effective in resisting stress.

In the compression control region, the diagram is almost linear, and M_n decreases as P_n increases, i.e., existence of axial load reduces the moment capacity. This is because when the axial load becomes predominant for the same section, in order to increase the axial load capacity, more concrete in the compression side must be engaged. This results in rapid decrease of the moment arm of the compressive force C_c, and, in turn, the moment capacity.

The following example shows how to construct two points on a *P-M* curve for a section. (One is the balanced point and the other is an arbitrary point)

EXAMPLE 6.5
Find

1. P_{nb}, e_b and M_{nb}, and
2. For $c = 9$ in., corresponding P_n and M_n of the section with $b = 12$ in., $h = 14$ in., $d = 11.5$ in., $d' = 2.5$ in., four #11 bars, $A_s = A'_s = 3.12$ in^2, $f'_c = 6000$ psi, and $f_y = 60{,}000$ psi, as shown in Figure 6.12.

Solution

1. $c_b = \dfrac{d(0.003)}{(0.003 + \epsilon_y)} = \dfrac{(11.5)(0.003)}{(0.003 + 60/29{,}000)} = 6.81$ in.

   $\epsilon'_s = \dfrac{(0.003)(c_b - d')}{c_b} = \dfrac{0.003(6.81 - 2.5)}{6.81} = 0.0019 < 0.00207 = \epsilon_y$

   $\therefore$ Compressive steel does not yield.

   $f'_s = E_s \epsilon'_s = 29{,}000 \times 0.0019 = 55.1$ ksi

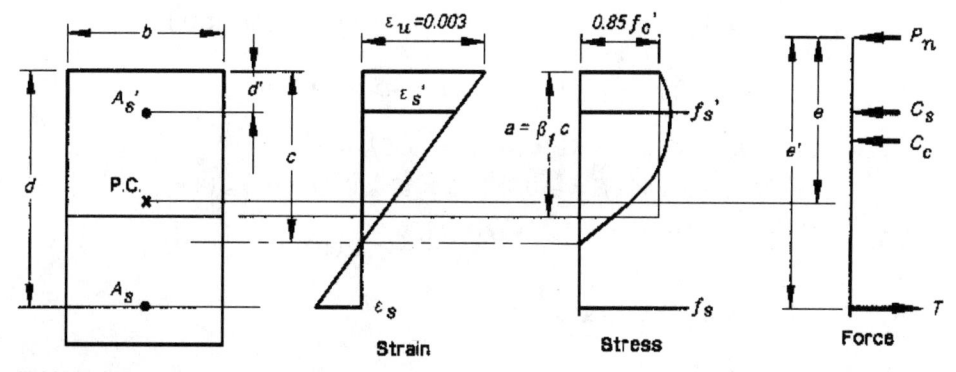

FIGURE 6.11

Compressive force due to compressive steel:

$$C_{sb} = A'_s(f'_s - 0.85f'_c) = 3.12(55.1 - 0.85 \times 6) = 156 \text{ k}$$

Compressive force due to concrete:

$$C_{cb} = 0.85f'_c(b)(\beta_1)(c) = 0.85 \times 6 \times 12 \times 0.75 \times 6.81 = 313 \text{ k}$$
$$T = A_s f_y = 3.12 \times 60 = 187 \text{ k}$$
$$P_{nb} = C_{sb} + C_{cb} - T = 156 + 313 - 187 = 282 \text{ k}$$

Moment with respect to the centroid of tensile steel:

$$P_{nb}e'_b = C_{sb}(d - d') + C_{cb}\left(d - \frac{a}{2}\right)$$

$$= 156(11.5 - 2.5) + 313\left(11.5 - 0.75 \times \frac{6.81}{2}\right)$$

$$= 1404 + 2800 = 4204 \text{ k-in.}$$

$$e'_b = \frac{4204}{282} = 14.91 \text{ in.}$$

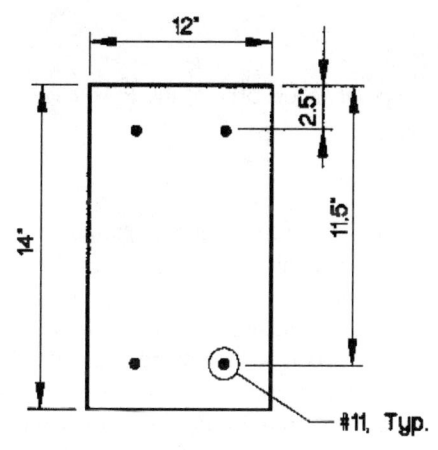

FIGURE 6.12

Plastic centroid is at center of column (due to symmetry):

$$\therefore e_b = e'_b - \frac{(d-d')}{2} = 14.91 - 4.5 = 10.41 \text{ in.}$$

$$M_{nb} = P_{nb}e_b = 282 \times 10.41 = 2935 \text{ k-in.}$$

2. $C_c = 0.85 f'_c b \beta_1 c = 0.85 \times 6 \times 12 \times 0.75 \times 9 = 413 \text{ k}$

$$\epsilon'_s = \frac{0.003(c-d')}{c} = \frac{0.003 \times 6.5}{9} = 0.00217 > \epsilon_y$$

Compressive steel yields:

$$\therefore f'_s = f_y = 60 \text{ ksi}$$
$$C_s = A'_s(f_y - 0.85f'_c) = 3.12(60 - 5.1) = 171 \text{ k}$$

$$\epsilon_s = \frac{0.003(d-c)}{c} = \frac{0.003 \times 2.5}{9} = 0.000867 < \epsilon_y$$

Compression controls, i.e., concrete crushing instead of steel yielding:

$$\therefore f_s = E_s \epsilon_s = 29{,}000 \times 0.000867 = 24.17 \text{ ksi} < 60 \text{ ksi}$$
$$T = A_s f_s = 3.12 \times 24.17 = 75.4 \text{ k}$$
$$P_n = C_c + C_s - T = 413 + 171 - 75.4 = 509 \text{ k}$$

$$P_n e' = C_c \left(d - \frac{a}{2}\right) + C_s(d - d')$$

$$= 413\left(11.5 - 0.75 \times \frac{9}{2}\right) + 171(11.5 - 2.5)$$

$$= 3356 + 1542 = 4898 \text{ k-in.}$$

$$e' = \frac{4898}{509} = 9.62 \text{ in.}$$

$$e = e' - \frac{(d-d')}{2} = 9.62 - 4.5 = 5.12 \text{ in.}$$

$$M_n = P_n e = 509 \times 5.12 = 2606 \text{ k-in.}$$

Now with M_n, P_n, we can plot one point (2606, 509) at the *P-M* curve.

6.6.3 Detailing

Area of longitudinal reinforcement for non-composite columns should be from 1 percent to 8 percent of the gross area of the section. The minimum requirement is to ensure that the incidental bending moment (whether existing by calculation or not) can be carried by the column, and to reduce the effects of creep and shrinkage of the concrete under sustained compressive stresses. The maximum requirement can be considered as a practical limit in terms of economy and for placing.

Two types of lateral reinforcement are used in reinforced concrete columns: reinforcement in the shape of closely pitched spirals (spiral columns) and reinforcement in the form of individual rectangular loops (tied columns). To ensure that the spiral steel yields before the core concrete crushes abruptly, ACI specifies a minimum ratio of spiral:

$$\rho_{smin} = \frac{0.45 f'_c}{f_y}\left(\frac{A_g}{A_c} - 1\right)$$

Recognizing the brittle failure mode (less warning and more dangerous) of tied columns, ACI specifies a lower ϕ factor than that used for spiral columns, i.e.,

$\phi = 0.75$ for spiral columns

$\phi = 0.70$ for tied columns.

In addition, ACI 10.3.5, ϕP_n is further restricted to a fraction of ϕP_{n0} to reflect the likely presence of accidental eccentricity.

$\phi P_{n(max)} = 0.85 \phi P_{n0}$ for spiral columns

$\phi P_{n(max)} = 0.80 \phi P_{n0}$ for tied columns.

This limitation is shown in the P-M diagram as the cut-off portion near the top of the curve where the member is subject to predominantly compressive load. These coefficients (0.85 and 0.80) should not be confused with the ϕ factors.

6.6.4 Long Columns

When a concrete member under combined axial force and bending moment becomes relatively long and slender, the internal moment due to the applied axial load is magnified by the deflection caused by the applied moment. ACI 10.11 describes an approximate design procedure that uses the moment magnifier concept to account for slenderness effect. The slenderness effect can be ignored if:

$$\frac{kl_u}{r} \leq 34 - 12\left(\frac{M_1}{M_2}\right) \text{ for members in laterally braced frame}$$

$\dfrac{kl_u}{r} < 22$ for members in laterally unbraced frame.

6.7 DESIGN FOR SHEAR

6.7.1 Shear Strength (Contribution of Concrete and of Reinforcement)

Shear reinforcement in the form of vertical stirrups, inclined stirrups, and bent-up longitudinal bars have been used in concrete beams to increase shear capacity. Hence, shear strength of reinforced concrete consists of two main components: one from the concrete and the other from the reinforcement.

$$V_n = V_c + V_s$$

The concrete component V_c is determined empirically as

$$V_c = \left(1.9\sqrt{f'_c} + 2500\rho_w \frac{V_u d}{M_u}\right) b_w d \leq 3.5\sqrt{f'_c}\, b_w d$$

and

$$\frac{V_u d}{M_u} \leq 1$$

or, alternatively,

$$V_c = 2\sqrt{f'_c}\, b_w d$$

The steel reinforcement component V_s can be computed as

$$V_s = \frac{A_v f_y d(\sin\alpha + \cos\alpha)}{s}$$

when shear reinforcement perpendicular to axis of member is used, $\alpha = 90°$

$$V_s = \frac{A_v f_y d}{s}$$

To prevent the shear-compression failure of concrete before yielding of web reinforcement, ACI 11.5.6.8 sets a limit to the amount of web reinforcement

$$V_s \leq 8\sqrt{f'_c}\, b_w d$$

and ACI 11.5.2 to the yield strength of shear reinforcement

$$f_y \leq 60{,}000 \text{ psi.}$$

EXAMPLE 6.6
Find the shear reinforcement required for a simply supported beam with span = 20 ft, $w_{DL} = 1.45$ k/ft, $w_{LL} = 3.5$ k/ft, $A_s = 6.06$ in², $b = 16$ in., $d = 22$ in., $f'_c = 3$ ksi, and $f_y = 60$ ksi, as shown in Figure 6.13.

Solution
1. Compute the factored load:

   $w_u = 1.4 w_{DL} + 1.7 w_{LL}$
   $\quad = 1.4(1.45) + 1.7(3.5)$
   $\quad = 8$ kips/ft

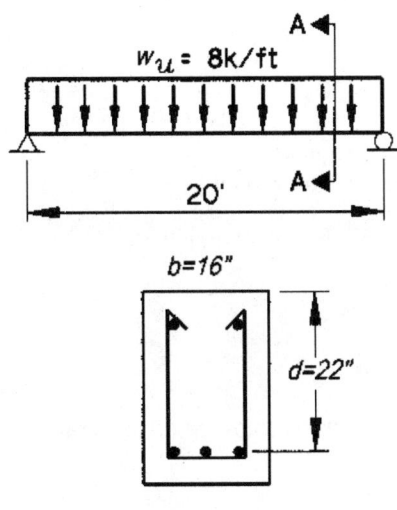

FIGURE 6.13

2. Compute shear and moment on the cross section a distance $d = 22$ in. from the support:

$$V_u = 80 - 8\left(\frac{22}{12}\right) = 65.3 \text{ kips}$$

$$M_u = (80 + 65.3)\frac{\left(\frac{22}{12}\right)}{2} = 133.2 \text{ k-ft}$$

3. Compute the nominal shear strength of the concrete:

$$V_c = \left(1.9\sqrt{f_c'} + 2500\rho\frac{V_u d}{M_u}\right)b_w d \leq 3.5\sqrt{f_c'}\, b_w d$$

$$\rho = \frac{A_s}{(b_w d)} = \frac{6.06}{(16 \times 22)} = 0.0172$$

$$\frac{V_u d}{M_u} = \frac{(65.3 \times 22)}{(133.2 \times 12)} = 0.9$$

$$V_c = \left(1.9\sqrt{3000} + 2500 \times 0.0172 \times 0.9\right)(16)(22)$$

$$= 50254 \text{ lb} = 50.25 \text{ k} \leq 3.5\sqrt{f_c'}\, b_w d = 3.5\sqrt{3000}\,(16)(22)$$

$$= 67{,}479 \text{ lb} = 67.5 \text{ k} \quad \text{OK.}$$

We could also use $V_c = 2\sqrt{f_c'}\, b_w d$ here.

Since $V_u > \frac{1}{2}\phi V_c$, stirrups are required.

4. Compute stirrup spacing:

$$V_s = \frac{V_u}{\phi} - V_c = \frac{65.3}{0.85} - 50.25 = 26.57 \text{ k}$$

$$s = A_v f_y \frac{d}{V_s} = 0.22 \times 60 \times \frac{22}{26.57}$$

$$= 10.93 \text{ in.} \quad \text{Use 10 in.}$$

Here we use stirrup spacing 10 in. throughout.
Can we use a larger spacing for section closer to midspan?

5. Determine the location where the stirrups are not required:

$$V_c = 2\sqrt{f_c'}\, b_w d$$

$$= \left(2\sqrt{3000}\right)(16)(22)$$

$$= 38{,}560 \text{ lb} = 38.56 \text{ k}$$

Beyond the location where $V_u \leq \frac{1}{2}\phi V_c$ the stirrups are no longer needed.

$$V_u = 0.5 \times 0.85 \times 38.56 = 16.4 \text{ k}$$

$$\frac{16.4 \text{ k}}{8 \text{ k/ft}} = 2.05 \text{ ft or approximately 2 ft}$$

Stirrups are not required *within 2 ft of the midspan* section.

6.7.2 Shear Friction

Sometimes concrete members are designed to resist direct shear. Examples of this type of shear transfer include corbel, base of shear wall, precast beam near bearing, and girder stops (shear keys). The design rule presented in ACI code assumes that a crack exists along the shear plane and reinforcement is provided across the crack. The shear strength is computed as

$$V_n = A_{vf} f_y \mu \leq 0.2 f_c' A_c \text{ or } 800\, A_c$$

where μ is the coefficient of friction and taken as:

1.4 λ for concrete placed monolithically

1.0 λ for concrete placed against hardened concrete with the surface intentionally roughened to an amplitude of approximately $\frac{1}{4}$ in.

0.6 λ for concrete placed against hardened concrete not intentionally roughened

0.7 λ for concrete anchored to as-rolled structural steel by headed studs or rebars

where $\lambda = 1.0$ for normal-weight concrete, 0.85 for *sand-lightweight* concrete, and 0.75 for *all-light-weight* concrete.

EXAMPLE 6.7
Find the steel reinforcement required at the support region of a precast concrete beam of normal-weight concrete and $f_y = 60$ ksi, as shown in Figure 6.14. The factored beam reactions are (1) vertical force = 86 k and (2) horizontal force = 20 k.

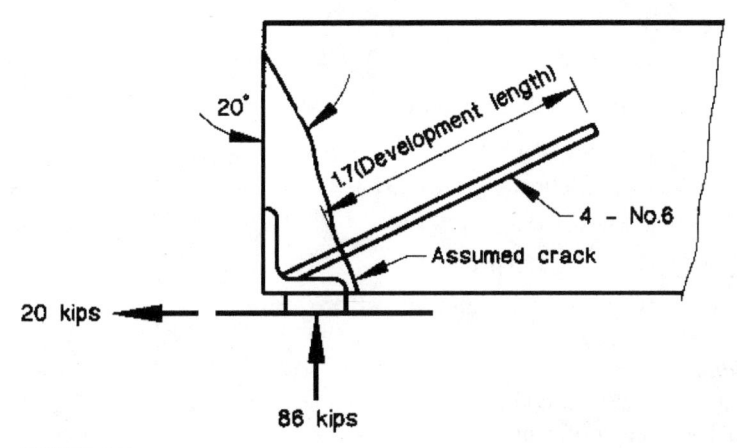

FIGURE 6.14

Solution
Assume the cracked plane (say, 20° from the vertical plane) to provide reinforcement across the assumed crack plane for possible sliding. Compute the area of steel required.

$$V_n = A_{vf} f_y \mu$$

$$\frac{V_u}{\phi} = A_{vf} f_y \mu$$

$$\therefore A_{vf} = \frac{V_u}{\phi f_y \mu} = \frac{86}{(0.85 \times 60 \times 1.4)} = 1.2 \text{ in}^2$$

The tensile force needs to be transferred across the crack by reinforcement

$$\therefore A_n = \frac{N}{(\phi f_y)} = \frac{20}{(0.85 \times 60)} = 0.39 \text{ in}^2$$

Total steel across the crack = $1.2 + 0.39 = 1.59 \text{ in}^2$

Use four #6 bars across the assumed crack. These bars need to be anchored on both sides of the crack.

6.8 DESIGN OF WALLS

Design of cantilever concrete retaining walls can follow the simple steps below:

1. Choose tentative proportion.
2. Determine all forces acting above the bottom of the base.
3. Check overturning, bearing, and sliding.
4. Structural design of the stem and of the toe and heel of the base.

EXAMPLE 6.8
Design the cantilevered retaining wall shown in Figure 6.15 with the following given data:

Wall: Top of the wall is 16 ft above the final level of earth at the toe
Concrete: $f'_c = 3000$ psi and $f_y = 60,000$ psi
Soil: α (internal friction angle) $= 35°$
q (allowable bearing stress) $= 5$ ksf
γ_{soil} (unit weight) $= 120$ pcf
Friction between soil and wall base $= 0.45$

Solution
1. Choose tentative proportion:
   a. Total height of wall
      The base of the wall should be below the frost penetration depth. Approximately 3 ft to 4 ft in northern United States.
      $\therefore$ 16 ft + 4 ft = 20 ft
   b. Thickness of footing
      Approximately 7 percent to 10 percent of the total height, use 2 ft
   c. Base length
      Assume the resultant force of soil pressure passes through the intersection of the front face of wall and bottom of footing

   $$K_A = \frac{1 - \sin\alpha}{1 + \sin\alpha} = 0.271$$

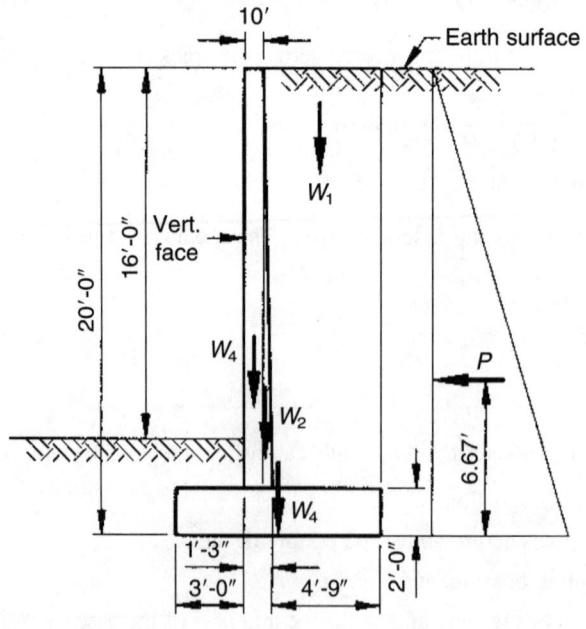

FIGURE 6.15

$P = \left(\frac{1}{2}\right)(0.271)(0.120)(20)^2 = 6.5$ k
$W = (0.120)(20)(x) = 2.4x$
$\Sigma M = 0$
$W\left(\dfrac{x}{2}\right) = P(6.67)$
$1.2x^2 = 43.35$
$x = 6$ ft
$(6)(1.5) = 9$ ft

d. Stem thickness
   The base thickness of the stem is estimated as 12 percent to 16 percent of the base width. In addition, the front face of the wall has a batter of $\frac{1}{4}$ in./ft to offset the deflection.
   $(9)(12)(14\%) = 15.12$ in.
   10 in. $+ (18)(0.25) = 14.5$ in.
   $\therefore$ use 15 in.

2. Determine all forces acting above the bottom of the base:

		Arm	Moment
$W_1 = (0.120)(18)(6 - 0.83)$	$= 11.2$	2.58	28.9
$W_2 = (0.03)(18)(0.42)(0.5)$	$= 0.11$	5.03	0.55
$W_3 = (0.15)(18)(0.83)$	$= 2.24$	5.58	12.5
$W_4 = (0.15)(2)(9)$	$= 2.7$	4.50	12.2
Total		16.25	54.1

   Resultant from heel = 3.33 ft

3. Check overturning, bearing and sliding:
   a. Factor of safety against overturning
      Resisting moment $= (16.25)(9 - 3.33) = 92.14$ k-ft
      Overturning moment $= P\left(\dfrac{1}{3}\right)(20) = (6.5)\left(\dfrac{1}{3}\right)(20) = 43.33$ k-ft
      F.S. against overturning $= \dfrac{92.14}{43.33} = 2.13 > 2.0$ OK.
   b. Check bearing
      $R = 16.25$ k
      $x = \dfrac{(54.1 + 43.33)}{16.25} = 6$ ft
      The resultant is within middle $\frac{1}{3}$ (in fact right at the edge of the middle $\frac{1}{3}$)
      $\therefore$ Effective base length = Full length
      $R = \left(\dfrac{1}{2}\right)(p_{max})(\text{effective base length})$
      $16.25 = \left(\dfrac{1}{2}\right)(p_{max})(9)$
      $\therefore p_{max} = 3.6$ ksf $<$ allowable bearing, 5 ksf   OK.
   c. Factor of safety against sliding
      Force causing sliding $= P = 6.5$ k

Friction force = μR = (0.45)(16.25) = 7.31 k
Factor of safety = 7.31/6.5 = 1.13 < 1.5   NG.
Use a key with depth and width $\simeq \frac{2}{3}$ of footing depth
i.e. key = 1 ft 6 in. × 1 ft 6 in.

4. Structural design of the stem and of the toe and heel of the base:

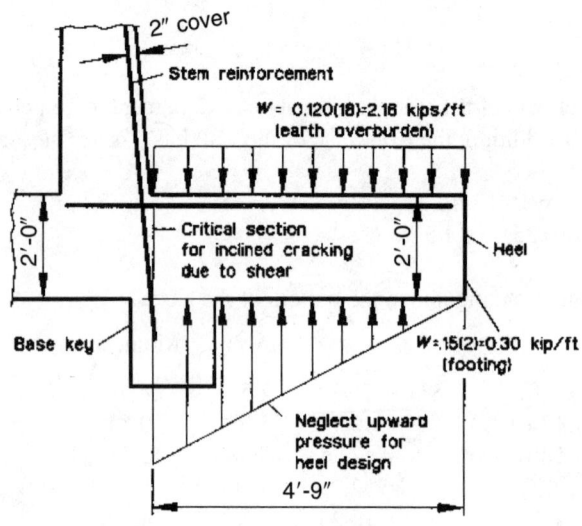

FIGURE 6.16

a. Design of heel cantilever
$W_u = 1.4 W_{DL}$
    $= 1.4(2.16 + 0.3) = 3.44$ k/ft
$M_u = \left(\frac{1}{2}\right)(3.44)(6 - 1.25)^2$
    $= 38.81$ k-ft
$V_u = (3.44)(4.75) = 16.34$ k
$V_c = \phi 2\left(\sqrt{f_c'}\right)bd$
$= \dfrac{(0.85)(2)\left(\sqrt{3000}\right)(12)(21.5)}{1000}$
$= 24$ k > 16.34   OK.

No shear reinforcement required.

$A_{s(min)} = \dfrac{\left(3\sqrt{f_c'}\, bd\right)}{f_y} \geq \dfrac{200 b_w d}{f_y}$

$\therefore A_{s(min)} = 0.86$ in^2

Use #6 @ 6 in. o.c. ($A_s = 0.88$ in^2/ft)

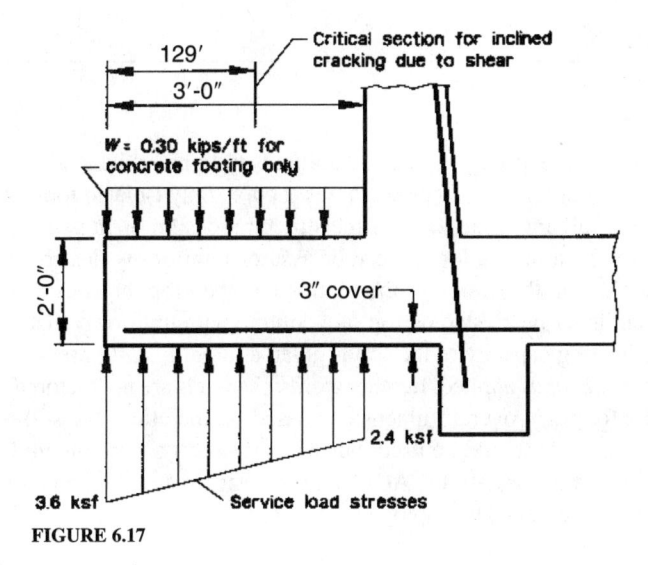

```
           129'
          3'-0''         ┌─ Critical section for inclined
                            cracking due to shear
    W = 0.30 kips/ft for
    concrete footing only

                    3" cover

                    2.4 ksf
    3.6 ksf      Service load stresses
```

FIGURE 6.17

b. Design of toe cantilever

$$V_u = 1.7\left[\frac{(3.6 + 2.4)}{2} - 0.3\right](3 - 1.71)$$

$$= 5.92 \text{ k}$$

$\phi V_c > V_u$ OK.

$$M_u = 1.7\left[\left(\tfrac{1}{2}\right)(3.6)(3)^2\left(\tfrac{2}{3}\right) + \left(\tfrac{1}{2}\right)(2.4)(3)^2\left(\tfrac{1}{3}\right) - \left(\tfrac{1}{2}\right)(0.3)(3)^2\right] = 22.2 \text{ k-ft}$$

Again use #6 @ 6 in. o.c.

c. Reinforcement at wall base

$$M_u = \frac{1.7(0.0325)y^3}{6} = \frac{1.7(0.0325)(18)^3}{6} = 53.7 \text{ k}$$

$$\frac{M_u}{\phi} = \frac{(53.7)(12)}{0.9} = 716 \text{ k-in}$$

$b = 12$ in., $d = 15 - 2 - 0.5 = 12.5$ in.

Guess $a = 2$ in. (adjust later, if necessary)

$$A_s = \frac{(716)}{\left[(60)\left(12.5 - \tfrac{2}{2}\right)\right]} = 1.04 \text{ in}^2$$

$$a = \frac{(A_s)(f_y)}{[(b)(0.85)(f'_c)]} = 2.04 \text{ in}^2$$

Close enough. Use #7 @ 6 in. o.c.

6.9 DESIGN OF FOOTINGS

6.9.1 Sizing of the Footing

Footings are used at the bottom of walls and columns to transfer the loads from the structures above into the supporting soil below. In this section, only isolated footings supporting a single column or wall are discussed. For the structural design of an isolated footing, the soil pressure on the bottom of a footing can be assumed uniformly distributed. If we turn the footing upside down, the design of the footing can therefore be considered as designing a concrete cantilever beam/slab on top of a column with uniformly distributed loads.

Currently the bearing capacity of the soil is given as an *allowable* stress, which is associated with the *service loads* applied. In other words, the loads are not factored. This means, when we size the footing's overall dimensions based on the allowable soil bearing pressure, the unfactored loads should be used, but when we design the structural components (e.g., rebar size and arrangement), the ACI strength design method with factored loads, as discussed in previous sections, is employed.

6.9.2 Flexure Check

The critical section for bending is taken as a vertical plane extending through the entire width of the footing. Its location relative to the face of column depends on the column construction:

Concrete column, pedestal, or wall:	At face of column
Masonry wall:	Halfway between middle and edge of wall
Column with steel base plate:	Halfway between face of column and edge of steel plate

The reinforcement in long direction should be distributed uniformly across entire width of the footing, but the reinforcement in short direction should be distributed so that more reinforcement is placed near the column. ACI 15.4.4.2 specifies the central band width, which equals the length of the short side of footing, should contain the total reinforcement required in short direction times $2/(\beta + 1)$ where β is the ratio of long to short dimension of the footing (i.e., $\beta > 1$). Reinforcement of small diameters is usually used in order to satisfy spacing and development limitations.

6.9.3 Shear Check (Beam Shear and Punching Shear)

Shear in isolated footing should be investigated for two separated conditions, and the more severe condition controls the design.

First is the *beam action* shear, where the critical section extends across the entire width of footing and is located d distance from the face of the column. The shear strength of the footing is estimated using the same formula as for beam design.

The second shear condition is the *two-way action* or *punching action* shear. The critical section follows the perimeter of the loaded area in the shape of a truncated pyramid, and critical shear stress is dependent on the ratio of column size to depth of footing. The shear strength calculation can be estimated by assuming a *pseudo-critical* section located at $d/2$ from the face of the column at all sides. Thus,

$$V_c = \left(2 + \frac{4}{\beta_c}\right)\sqrt{f'_c}\, b_0 d \leq 4\sqrt{f'_c}\, b_0 d$$

where b_0 = Total perimeter of the critical section at $d/2$ from the face of column
β_c = Ratio of long side to short side of column

6.9.4 Bearing/Dowels

Loads are transferred from column into footing primarily through the bearing of concrete, which is augmented with steel dowels when the bearing is insufficient.

Bearing of column concrete and footing concrete should be checked first. Although the footing concrete strength is typically weaker than that of the column, the presence of material outside of the loaded area provides a magnifying factor

$$\sqrt{A_2/A_1} \leq 2$$

for its bearing strength.

For any dowels to be effective, they must have sufficient embedment length both above and below the interface to develop the desired stress. ACI specifies that the minimum area of the dowels should be 0.5 percent of the gross column area.

EXAMPLE 6.9

Design the footing to support the column with the following information, and as shown in Figure 6.18.

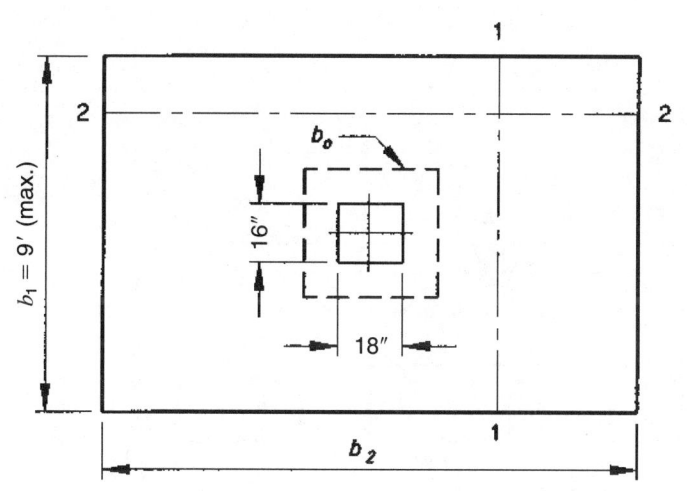

FIGURE 6.18

6.32 CHAPTER 6

Column: 16 in. × 18 in., eight #11 bars, f'_c = 5000 psi, f_y = 60,000 psi
Load at base: P_{DL} = 400 k, P_{LL} = 300 k
Footing: f'_c = 3000 psi, f_y = 60,000 psi
Footing dimension parallel to 16 in. face of column dimension ≤ 9 ft
Soil: Allowable bearing capacity, q = 6000 psf

Solution
1. Sizing of the footing:

   Total load $P = P_{DL} + P_{LL} = 400 + 300 = 700$ k

   Bearing area required $= \dfrac{P}{q} = \dfrac{700}{6} = 116.7$ ft^2

   b_1 is limited to 9 ft; try 9 ft × 13 ft = 117 ft^2

2. Select thickness for shear: (normally controlled by punching shear)

   $P_u = 1.4 P_{DL} + 1.7 P_{LL}$
   $= 1.4(400) + 1.7(300) = 1070$ k

   $q_u = \dfrac{P_u}{(b_1 b_2)} = \dfrac{1070}{(9 \times 13)} = 9.14$ ksf

   Check for punching shear:
   critical section at $\frac{1}{2} d$ from faces of column
   $b_0 = 2[(16 + d) + (18 + d)] = 68 + 4d$

   $V_u = 1070 - 9.14 \left[\dfrac{(16 + d)}{12} \right] \left[\dfrac{(18 + d)}{12} \right]$

   $= (1052 - 2.16d - 0.06347 d^2)$ k

   $\beta_c = \dfrac{18}{16} = 1.125 < 2$

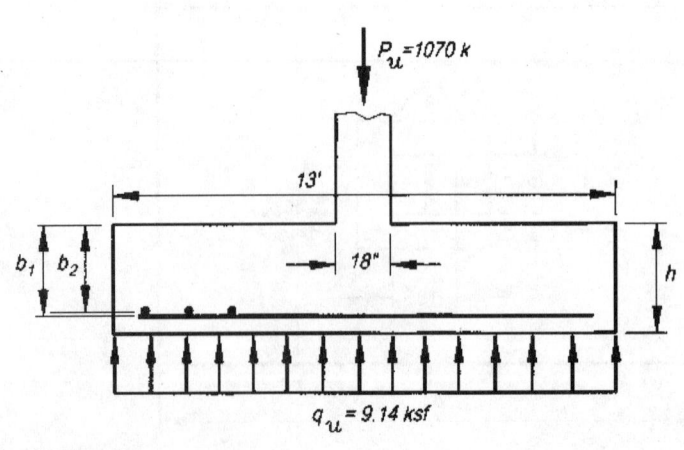

FIGURE 6.19

$$V_n = V_c = 4\sqrt{f'_c}\, b_0 d = 4\sqrt{3000}\,(68 + 4d)d$$
$$= (876\, d^2 + 14{,}900d)\text{ lb}$$

(Stirrups are seldom used in the footing, i.e., $V_s = 0$)
Equate $V_n = V_u$ and solve for d,

$$\frac{0.85(876 d^2 + 14{,}900 d)}{1000} = 1052 - 2.16 d - 0.06347 d^2$$

$$0.808 d^2 + 14.825 d - 1052 = 0$$

∴ Required $d = 28$ in.

Footing depth can be designed by trial-and-error method: first, assume a depth d, calculate V_n and V_u and adjust d as necessary. This way the solution of the quadratic equation can be avoided.

Try 32 in. deep footing ($h = 32$ in.) with min. cover at bottom = 3 in.
$d_1 = 32 - 3 - 0.5 = 28.5$ in.
$d_2 = 27.5$ in.
∴ Provided $d = 28$ in.   OK.

$$V_u = 1070 - 9.14\left[\frac{(16 + 28)}{12}\right]\left[\frac{(18 + 28)}{12}\right] = 1070 - 128.5 = 940\text{ k}$$

$$V_n = 4\sqrt{3000}\,(68 + 4 \times 28)(28) = 1{,}104{,}200\text{ lb} = 1104\text{ k}$$
$\phi V_n = 0.85(1104) = 940$ k
$\phi V_n > V_u$ OK.

Check for one-way action shear (beam) shear:
critical section at d_1 from face of column

$$l_1 = \frac{(13\text{ ft} \times 12 - 18\text{ in.})}{2} - 28.5\text{ in.} = 40.4\text{ in.} = 3.375\text{ ft}$$

$V_u = 9.14 \times 3.375 \times 9 = 277.6$ k

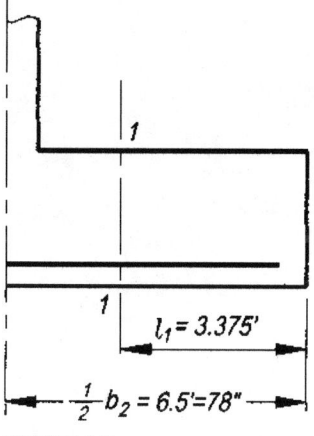

FIGURE 6.20

6.34 CHAPTER 6

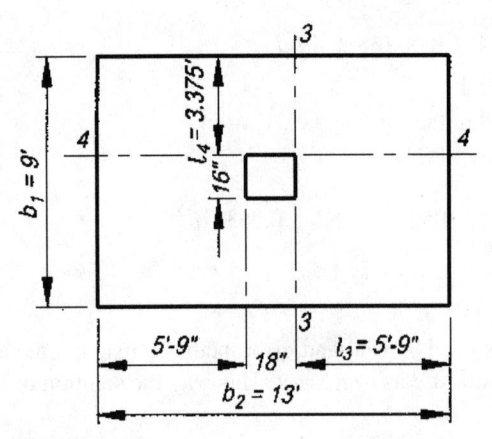

FIGURE 6.21

$\phi V_n = 0.85(2)\sqrt{3000}\,(108)(28.5) = 286.6$ k

$\phi V_n > V_u$ OK.

3. Design for bending: (selection of horizontal bars—bottom mat)
   Selection of long bars (critical section 3-3)

$$l_3 = \frac{\left(13\text{ ft} - \dfrac{18\text{ in.}}{12}\right)}{2} = 5.75\text{ ft}$$

$$M_u = 0.5 q_u b_1 l_3^2$$
$$= 0.5 \times 9.14 \times 9 \times (5.75)^2$$
$$= 1360\text{ k-ft} = 16{,}318\text{ k-in.}$$

Required nominal strength, $M_n = \dfrac{M_u}{\phi}$

$$\frac{M_u}{\phi} = \frac{16318}{0.9} = 18131\text{ k-in.}$$

$b_1 = 9$ ft $= 108$ in., $d_1 = 28.5$ in.

Guess $a = 4$ in.   (adjust later)

$$A_s = \frac{M_n}{\left[f_y\!\left(d - \dfrac{a}{2}\right)\right]} = \frac{18{,}131}{\left[60\!\left(28.5 - \dfrac{4}{2}\right)\right]} = 11.40\text{ in}^2$$

$$a = \frac{A_s f_y}{[b_1(0.85)f_c']}$$
$$= \frac{(11.4)(60)}{[(108)(0.85)(3)]} = (11.4)(0.218) = 2.48\text{ in.}$$

$$A_s = \frac{18{,}131}{\left[60\left(28.5 - \frac{2.48}{2}\right)\right]} = 11.08 \text{ in}^2$$

$a = (11.08)(0.218) = 2.42$ in. Close enough.
Choose 15 #8 @ spacing $A_s = 11.85 \text{ in}^2$
Selection of short bars: (critical section 4-4)

$$l_4 = \frac{\left(9 \text{ ft} - \dfrac{16 \text{ in.}}{12 \text{ in./ft}}\right)}{2} = 3.83 \text{ ft}$$

$$M_u = 0.5 q_u b_2 l_4^2$$
$$= 0.5 \times 9.14 \times 13 \times (3.83)^2 = 873 \text{ k-ft} = 10{,}476 \text{ k-in.}$$

Required nominal strength, $M_n = M_u/\phi$

$$\frac{M_u}{\phi} = \frac{10{,}476}{0.9} = 11{,}640 \text{ k-in.}$$

$b_2 = 13 \text{ ft} = 156 \text{ in.}$, $d_2 = 27.5$ in.

Guess $a = 2$ in. (adjust later)

$$A_s = \frac{M_n}{\left[f_y\left(d - \dfrac{a}{2}\right)\right]} = \frac{11{,}640}{\left[60\left(27.5 - \dfrac{2}{2}\right)\right]} = 7.32 \text{ in}^2$$

$$a = \frac{A_s f_y}{[b_2(0.85)f_c']}$$
$$= \frac{(7.32)(60)}{[(156)(0.85)(3)]} = (7.32)(0.151) = 1.14 \text{ in.}$$

$$A_s = \frac{11{,}640}{\left[60\left(27.5 - \dfrac{1.14}{2}\right)\right]} = 7.20 \text{ in}^2$$

$a = (7.20)(0.151) = 1.09$ in. Close enough.
Requirement for shrinkage reinforcement:
$0.0018(b_2)(h) = 0.0018 \times 156 \times 32 = 8.99 \text{ in}^2$
A_s required within the middle band (9 ft centered on centerline of column)

$$\frac{(A_s)(2)}{(\beta + 1)} = 8.99 \times \frac{2}{\left(\dfrac{13}{9} + 1\right)} = 7.35 \text{ in}^2$$

Use 10 #8 @ 12 in. in middle band, $A_s = 7.9 \text{ in}^2$ and one #8 in each edge band

4. Transfer of load at column base:

Capacity of bearing on footing concrete = $0.7(0.85 f'_c)\sqrt{A_2/A_1}(A_1)$
$$= 0.7 \times 0.85 \times 3 \times 2 \times (16)(18) = 1028 \text{ k}$$

Capacity of bearing on column concrete
$$= 0.7(0.85 f'_c)(16)(18) = 0.7 \times 0.85 \times 5 \times 16 \times 18 = 857 \text{ k} \leftarrow \text{controls.}$$

Load to be transferred by dowels = $1070 - 857 = 213$ k

Area of dowels required $= \dfrac{213}{(0.7)(60)} = 5.07 \text{ in}^2 \leftarrow \text{controls.}$

Minimum area of dowels = $0.005 \times 16 \times 18 = 1.44 \text{ in}^2$

Use four #10 dowels.

6.10 INTRODUCTION TO PRESTRESSED CONCRETE

We will limit our discussion to beams in this section. Prestressed concrete differs from the (conventionally) reinforced concrete in two main aspects: the type of reinforcing steel used and the way the reinforcing steel is stressed. In reinforced concrete, the reinforcement made of mild steel is just placed in the concrete passively to resist forces (mainly tension) resulting from externally applied loads ($DL + LL + I$, etc.). On the other hand, prestressed concrete introduces stresses in the concrete by prestressing the steel reinforcement made of high-strength steel to counteract the stresses from the externally applied loads. The stress-strain relationships of mild steel (deformed reinforcing bars) and high-strength steel (7-wire strand) are shown below:

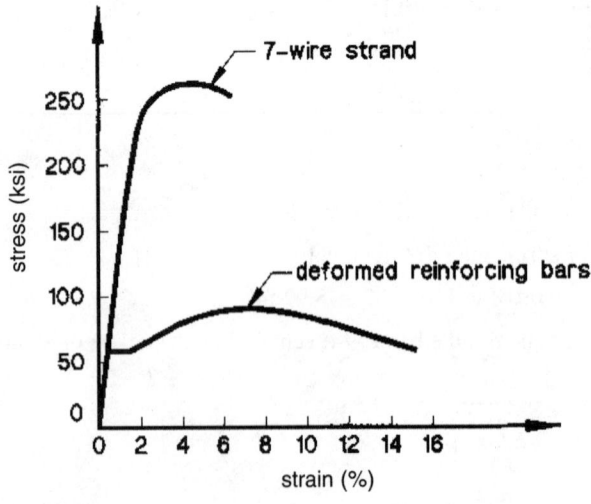

FIGURE 6.22

These two aspects are interrelated—that is, if we simply bury the high-strength steel in concrete as mild steel in reinforced concrete, then the advantage of the high strength can not be fully realized because the majority of the concrete will exist merely to hold the steel in place away from a small portion of highly stressed uncracked concrete. By prestressing the high-strength steel, we put the otherwise cracked concrete to work so that the entire concrete section is now fully utilized.

There are essentially two ways of introducing the prestressing force in the prestressed concrete: pre-tensioning and post-tensioning. For pre-tensioning, the tendons are tensioned (with anchors outside the concrete member) before concrete is cast in forms, and the prestressing force is primarily transferred to the concrete through bond. For post-tensioning, the tendons are tensioned (with anchors on the concrete member) after the concrete has attained required initial strength, and the prestressing force is primarily transferred through end anchorages.

The introduction of the prestressing practically transforms concrete members into an elastic material. The stresses in the concrete members then amount to superimposing stresses caused by the prestressing and by the externally applied loads:

$$f_t = \left(-\frac{P}{A} + \frac{Pey}{I}\right) - \frac{My_t}{I}$$

at top of beam

$$f_b = \left(-\frac{P}{A} - \frac{Pey}{I}\right) + \frac{My_b}{I}$$

at bottom of beam.

The stresses in the parentheses are associated with prestress, and the last term in the above equations is from external loads (i.e., $DL + LL + I$, etc.).

In addition to the final ultimate loading stage, the introduction of the prestressing operation also brings in some additional modes of failure which should be considered in the design. They include:

At jacking:

- The rupture of the prestressing tendon
- The failure of the tendon anchor

After prestress transfer and before all the time-dependent losses:

- The cracks of concrete due to high tension resulted from the high eccentric prestress
- Crushing of concrete near end of the prestressing tendon if prestress is too high

At service load after losses have occurred:

- The overstress of concrete in compression
- The tension (and induced cracking) in the precompressed tensile zone

To prevent these failure modes, ACI specifies allowable stresses at the different stages for prestressing tendon and for the concrete in Sections 18.4 and 18.5 respectively.

In a format somewhat similar to the conventionally reinforced concrete, ACI also provides nominal design strength calculations for flexure (ACI 18.7) and for shear (ACI 11.4).

A rectangular beam with the same outside dimensions as Example 6.1 is used here as an example to illustrate the general principles of prestressed concrete design and to provide a comparison between conventionally reinforced concrete and prestressed concrete.

EXAMPLE 6.10

For a section shown in Figure 6.23, assume that the member is prestressed by 1.53 in² of steel strand having a maximum acceptable initial tensile stress of 189 ksi. The prestressing strands are centered at 5 in. from the bottom. The concrete has f'_c = 4000 psi, and assume f'_{ci} = 3500 psi at the time of transfer. If the member is used on a 50-ft span, determine:

1. Stresses due to prestress immediately after transfer
2. Final stress after losses
3. Service live-load moment capacity according to the ACI Code, allowing for a 20 percent loss of prestress due to creep, shrinkage, and other sources (ACI 18.6 and commentary provide some additional information on the loss of prestress)
4. Nominal ultimate moment capacity of the cross section.

Solution

1. Temporary stress (prestress + DL) immediately after transfer

$$T_0 = f_{si} A_s = (189)(1.53) = 289 \text{ k}$$

$$f_{top} = -\frac{P}{A} + \frac{Pey}{I} - \frac{My_t}{I}$$

$$= -\frac{(289)(1000)}{(35)(12)} + \frac{(289)(1000)(12.5)(17.5)}{(35)^3} - \frac{(36)(50 \times 12)^2(17.5)}{(8)(35)^3}$$

$$= -688 + 1470 - 660$$

$$= 122 \text{ psi}$$

$$f_{bottom} = -688 - 1470 + 660 = -1498 \text{ psi}$$

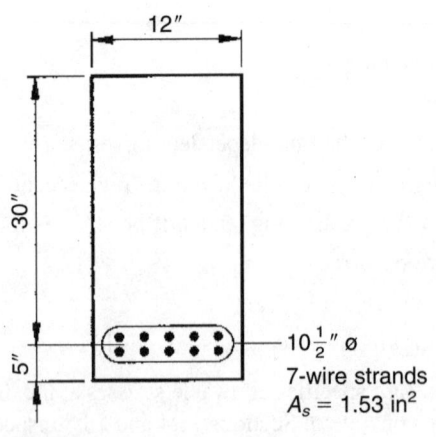

FIGURE 6.23

(Cross section: 12" wide × 30" tall; strands at 5" from bottom; 10½" ⌀ 7-wire strands, A_s = 1.53 in²)

CONCRETE DESIGN **6.39**

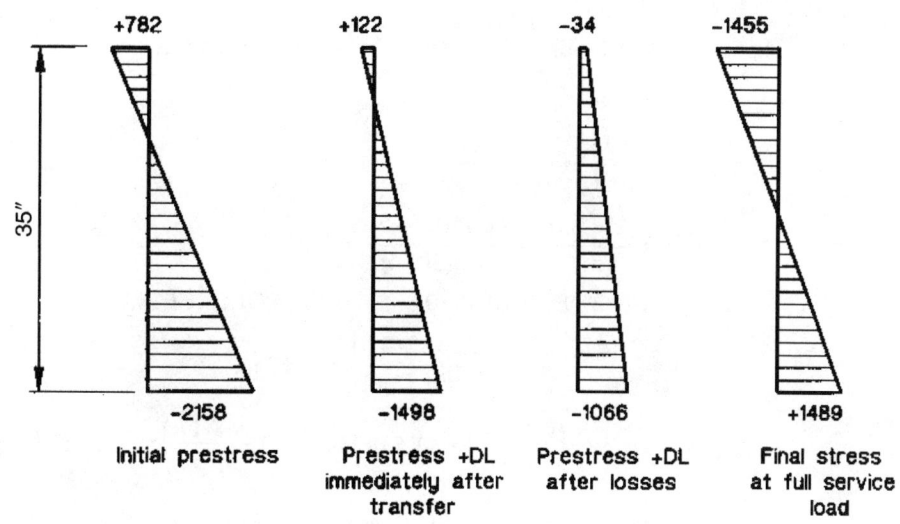

FIGURE 6.24

+782	+122	−34	−1455
−2158	−1498	−1066	+1489
Initial prestress	Prestress +DL immediately after transfer	Prestress +DL after losses	Final stress at full service load

2. Final stress (prestress + DL) after allowance for 20 percent prestress losses
   $f_{top} = 122 - 0.2(-688 + 1470) = -34$ psi $< 3\sqrt{f'_c}$ OK.
   $f_{bottom} = -1498 - 0.2(-688 - 1470) = -1066$ psi $< 0.60 f'_c$ OK.

3. Service live-load moment capacity
   $f_{top(max)} = 0.45 f'_c = (0.45)(4000) = 1800$ psi
   $f_{bottom(max)} = 6\sqrt{f'_c} = 379.5$ psi
   top: $-34 + f_{LL} = -1800$
   $f_{LL} = -1766$ psi
   bottom: $-1066 + f_{LL} = 379.5$
   $f_{LL} = 1445.5$ psi ← controls
   $$1445.5 = \frac{(M_{LL})(17.5)}{(35)^3}$$
   $\therefore M_{LL} = 3{,}541{,}500$ lb-in. $= 295$ k-ft

4. Nominal moment capacity
   For a fully prestressed member (with no non-prestressed tension or compression reinforcement), ACI Equation 18-3 reduces to
   $$f_{ps} = f_{pu}\left[1 - \left(\frac{\gamma_p}{\beta_1}\right)\rho_p\left(\frac{f_{pu}}{f'_c}\right)\right]$$
   $$= (270)\left[1 - \left(\frac{0.28}{0.85}\right)(0.00425)\left(\frac{270}{4}\right)\right]$$
   $= 244.5$ ksi

where $\gamma_p = 0.28$ for $\dfrac{f_{py}}{f_{pu}} > 0.90$ for low-relaxation strains

$\beta_1 = 0.85$ for $f'_c = 4000$ psi

$$\rho_p = \dfrac{A_{sp}}{(bd_p)} = \dfrac{1.53}{(12 \times 30)} = 0.00425$$

$$\omega_p = \dfrac{(A_{ps}f_{ps})}{(bd_p f'_c)} = \dfrac{(1.53)(244.5)}{(12)(30)(4)} = 0.26$$

$$\leq (0.36)\beta_1 = (0.36)(0.85) = 0.306 \quad \text{OK. ACI 18.8.1}$$

$$a = \dfrac{(A_{ps}f_{ps})}{(0.85bf'_c)} = \dfrac{(1.53)(244.5)}{(0.85)(12)(4)} = 9.17 \text{ in.}$$

$$M_n = A_{ps}f_{ps}\left(d_p - \dfrac{a}{2}\right) = (1.53)(244.5)\left(30 - \dfrac{9.17}{2}\right) = 9508 \text{ k-in.}$$

Compare the similarities and differences between this example and Example 6.1. Notice that the yield stress of the prestressing strands is about four times that of the deformed reinforcing bars, and the area is about $\dfrac{3}{4}$. What if we did not prestress the strands? What would be the crack width under service load condition for the concrete member with strands not prestressed? Notice little difference in the nominal moment strength between Examples 6.1 and 6.10.

6.11 SUMMARY

Item	Equations	ACI
Required Strength (Applied Factored Load)	$U = 1.4D + 1.7L$	9.2 (Eq. 9-1)
	$U = 0.75(1.4D + 1.7L + 1.7W)$	(Eq. 9-2)
	$U = 0.9D + 1.3L$	(Eq. 9-3)
	$U = 1.4D + 1.7L + 1.7H$	(Eq. 9-4)
	$U = 0.75(1.4D + 1.4T + 1.7L)$	(Eq. 9-5)
	$U = 1.4(D + T)$	(Eq. 9-6)
	Note: Compare these factors with those for Steel Building Design (i.e., AISC Code) and those for Bridge Design (i.e., AASHTO code)	
Design Strength (Reduced Nominal Capacity)	ϕ (Nominal Strength)	9.3
Strength Design Criteria	ϕ (Nominal Strength) $\geq U$	9.1.2
	$\phi M_n \geq M_u$	
	$\phi P_n \geq P_u$	
	$\phi V_n \geq V_u$	
	$\phi T_n \geq T_u$	

CONCRETE DESIGN 6.41

Item	Equations	ACI
Nominal Strength		
Flexure		
Rectangular Single Reinf. (Under-Reinforced)	$M_n = A_s f_y \left(d - \dfrac{a}{2}\right)$	
	$a = \dfrac{(A_s f_y)}{(0.85 f'_c b)}$	
	$\rho_b = \dfrac{0.85 f'_c \beta_1 \dfrac{87000}{f_y}}{(87{,}000 + f_y)}$	(Eq. 8-1)
Rectangular Double Reinf.	$\rho_{max} = 0.75 \rho_b$	
	$M_n = A_s f_y \left(d - \dfrac{a}{2}\right) + A'_s f'_s \left(\dfrac{a}{2} - d'\right)$	
Axial Load	$P_{n0} = A_g(0.85 f'_c) + A_{st}(f_y - 0.85 f'_c)$	(Eqs. 10-1,
	$A_{st} = A_s + A'_s$	10-2)
Combined Axial and Flexural Load (Eccentrically Loaded Column)	P-M Interaction Diagram	
Shear Associated with Bending	$V_n = V_c + V_s$	(Eq. 11-2)
	$V_c = 2\sqrt{f'_c}\, b_w d$ or	(Eq. 11-3)
	$V_c = \left(1.9\sqrt{f'_c} + 2500\rho_w \left\lvert \dfrac{V_u d}{M_u} \right\rvert \right) b_w d$	(Eq. 11-5)
	$\leq 3.5 \sqrt{f'_c}\, b_w d$	
	$V_s = A_v f_y d \dfrac{(\sin\alpha + \cos\alpha)}{s} < 8\sqrt{f'_c}\, b_w d$	(Eq. 11-16)
	when $\alpha = 90°$	
	$V_s = A_v f_y \dfrac{d}{s}$	(Eq. 11-15)
	$f_y \leq 60{,}000$ psi	11.5.2
Shear Friction	$V_n = A_v f_y \mu \leq 0.2 f'_c A_c$ or $800 A_c$	11.7.5
Strength Reduction Factors, ϕ		
Flexure, Without Axial Load	0.9	9.3.2.1
Axial Compression with Flexure,		9.3.2.2
For Spiral Reinf.	0.75	
For Tied Reinf.	0.70	
Shear and Torsion	0.85	9.3.2.3
Bearing on Concrete	0.70	9.3.2.4
Development Length	1.0	9.3.3
Cantilever Retaining Wall Design		
Preliminary Proportioning	Some empirical dimensions	
Check against Overturning, Bearing, and Sliding	F.S. (overturning) = 2.0	
	F.S. (sliding) = 1.5	

Item	Equations	ACI
Design of Heel Cantilever	Check both moment and shear at the back face of stem wall	
Design of Toe Cantilever	Check moment at the front face of the wall; check shear one d from the face	
Design of Wall Cantilever	Check moment at the base of the wall; check shear one d from the base	
Footing Design		
Flexure	Reinf. distribution in central band of short direction:	
	$\text{Total reinf. in short direction} \times \dfrac{2}{(\beta + 1)}$	(Eq. 15-1)
Shear	Beam shear	(Eq. 11-3)
	Punching shear	(Eq. 11-35)
Bearing/Dowels	Bearing	10.17
	Dowels	15.8
Prestressed Concrete Design		
Service Load Level		
Allowable Stresses in Prestressing Tendons		
	At jacking	18.5.1
	For stress relieved strand: $0.80 f_{pu}$	
	For low relaxation strand: $0.80 f_{pu}$	
	After transfer	
	For stress relieved strand: $0.74 f_{pu}$*	
	For low relaxation strand: $0.70 f_{pu}$	
	* Post-tensioned tendons limited to $0.70 f_{pu}$	
Loss of Prestress	Anchorage seating losses	18.6.1
	Elastic shortening of concrete	
	Creep of concrete	
	Shrinkage of concrete	
	Relaxation of tendon stress	
	Friction loss (curvature and wobble)	
Flexural Stresses in Concrete Resulting from Prestress and Applied Loads	$f = \left(-\dfrac{P}{A} \pm \dfrac{Pey}{I}\right) \mp \dfrac{My}{I}$	
Allowable Flexural Stresses in Concrete		
	At transfer before losses due to creep and shrinkage	18.4.2
	Compressive Stress	
	$0.60 f'_{ci}$	
	Tensile Stress	
	$3\sqrt{f'_c}$	
	$6\sqrt{f'_c}$ at simply supported ends	

CONCRETE DESIGN **6.43**

Item	Equations	ACI
	At service load after losses have occurred Compressive Stress $0.45 f_c'$ Tensile Stress $6\sqrt{f_{ci}'}$ in precompressed tensile zone	
Factored Load Level Flexural Nominal Strength	$M_n = A_s f_{ps}\left(d - \dfrac{a}{2}\right)$ $a = \dfrac{(A_s f_{ps})}{(0.85 f_c' b)}$ where $f_{ps} = f_{pu}\left\{1 - \left(\dfrac{\gamma_p}{\beta_1}\right)\left[\rho_p\left(\dfrac{f_{pu}}{f_c'}\right) + \left(\dfrac{d}{d_p}\right)(\omega - \omega')\right]\right\}$	(Eq. 18-3)
Shear Nominal Strength	$V_n = V_c + V_s$ *Simplified method* $V_c = \left[0.6\sqrt{f_{ci}'} + 700\left(\dfrac{V_u d}{M_u}\right)\right] b_w d$ *Detailed method* V_c = lesser of V_{ci} and V_{cw} $V_{ci} = 0.6\sqrt{f_c'}\, b_w d + V_d + \dfrac{(V_i M_{cr})}{M_{max}}$ $V_{cw} = \left(3.5\sqrt{f_c'} + 0.3 f_{pc}\right) b_w d + V_p$	(Eq. 11-2) (Eq. 11-9) (Eq. 11-10) (Eq. 11-12)

CHAPTER 7
STEEL DESIGN

7.1 INTRODUCTION

Chapter 22 of the 1997 Uniform Building Code (UBC) deals with the quality, testing, and design of steel used structurally in building and structures. The Load and Resistance Factor Design (LRFD) method will be used in steel design in this chapter.

Although UBC allows the use of the Allowable Stress Design (ASD) method as an alternative to LRFD, the structural design profession has been using LRFD in steel design for many years and the trend is accelerating. Steel-design courses in colleges and universities are devoted almost exclusively to LRFD. LRFD has the potential to save money compared with ASD, especially when the live loads are small compared with the dead loads. But immediate cost saving is not the reason AISC introduced LRFD. The primary objective is to provide a uniform reliability for steel structures under various loading conditions. The load and resistance factors in LRFD have been calibrated statistically to provide a uniform level of reliability.

Different load factors are applied to different types of loads. For example, a smaller load factor is used for dead loads, because we can estimate these loads with better accuracy, while the load factor for live loads is larger, because of the uncertainties involved in determining the future changes in live loads. In LRFD, the connections are designed stronger than the members to reflect the lessons learned from structural performance and failures. This uniformity cannot be obtained with ASD using the same factor of safety for all types of loads, members, and connections.

LRFD is written in a form that prompts the designers to consider serviceability and strength limit states systematically so a structure will have high performance throughout its 50 or more years of service life. High performance means few problems during construction, low maintenance and operations cost, and overall low life-cycle cost.

UBC adopted the Load and Resistance Factor Design Specifications for Structural Steel Buildings, Dec, 1, 1993, and it was published by the American Institute of Steel Construction, with modifications as set forth in Section 2207 of Division II of UBC. Therefore, we will be referring to the AISC Manual of Steel Construction, Load and Resistance Factor Design, Volumes I and II, Second Edition (LRFD manual) in this chapter on steel design. Volume I contains the LRFD specification and commentary, tables, and other design information for structural members in Part I, and the esssentials of LRFD in Part II. Volume II covers connections. This LRFD manual applies to buildings and not to bridges.

7.2 ATTRIBUTES OF STRUCTURAL STEELS

With the introduction of the Bessemer and open-hearth steel-making processes in the mid-1850s, steel became available at reasonable cost and in large quantity for building and bridge construction. In 1878, the all-steel Glasgow Bridge over the Missouri River in South Dakota was completed. In 1890, the Rand-McNally Building, the first all-steel-framed building, was completed in Chicago. In the same year, the all-steel Firth of Forth Railroad Bridge in Scotland was completed with a record span of 1700 feet.

Since the construction of the Rand-McNally Building and the Firth of Forth Bridge, steel production and fabrication have undergone significant and progressive changes. Today there is a wide selection of structural steels and shapes available to designers in the design and construction of modern buildings and bridges.

Structural steels have excellent properties for building and bridge construction. These properties include a combination of strength, ductility, uniformity, fracture toughness, fabricability, repairability, and cyclability. These properties, and other attributes of structural steels, are imparted to the steels during steel making when the steel is in a liquid state, and during processing when the steel is in a solid state. Recent advances in steel making and processing enable close control of melting, alloy additions, temperatures, and cooling rates to produce steels with outstanding properties for structural applications.

Table 7.1 summarizes the mechanical properties of structural steels. Complete information on availability and strengths is in Tables 1.1 and 1.2 of the LRFD manual.

7.3 TENSION MEMBERS

This section applies to prismatic members subject to axial tension caused by static forces acting through the centroidal axis. A prismatic member is a structural member having a straight longitudinal axis and constant cross section throughout its length. LRFD specification, Chapter D, contains provisions for the design of tension members.

7.3.1 Design Tensile Strength

LRFD specification (D1) stipulates that the design strength of tension members $\phi_t P_n$ shall be the lower value obtained according to the limit states of yielding in the gross section and fracture in the net section.

a. For yielding in the gross section, LRFD Equation D1-1 gives:

$$\phi_t = 0.9 \qquad P_n = F_y A_g \qquad (7.1)$$

$$P_u = \phi_t F_y A_g \qquad \text{(LRFD D1-1)}$$

TABLE 7.1 Minimum Mechanical Properties of Structural Steels

Type	Carbon	High-strength Low-alloy		$Q + T$ Low-Alloy	High-yield strength $Q + T$ low-alloy
ASTM	A36	A572	A588	A852	A514
Thickness of plate	Up to 8 in. inclusive	To $1\frac{1}{4}$ & 6 in. based on F_y	To 4 & 8 in. based on F_y	Up to 4 in. inclusive	$2\frac{1}{2}$ to 6 in. inclusive	Up to $2\frac{1}{2}$ in. inclusive
Shapes	All groups	All groups	All groups	Not applicable	Not applicable	Not applicable
Min. yield stress, F_y	36,000 psi	42,000 to 65,000 psi	42,000 to 50,000 psi	70,000 psi	90,000 psi	100,000 psi
Tensile stress, F_u	58,000 to 80,000 psi	60,000 to 80,000 psi	63,000 to 70,000 psi	90,000 to 110,000 psi	100,000 to 130,000 psi	110,000 to 130,000 psi

b. For fracture in the net section, LRFD Equation D1-2 gives:

$$\phi_t = 0.75$$
$$P_n = F_u A_e \quad (7.2)$$
$$P_u = \phi_t F_u A_e \quad \text{(LRFD D1-2)}$$

where A_e = Effective net area, in^2
A_g = Gross area of member, in^2
F_y = Specified minimum yield stress, ksi
F_u = Specified minimum yield stress, ksi
P_u = Nominal axial strength, kips

7.3.2 Gross Area, A_g

The gross area A_g of a member at any point is the sum of the products of the thickness and the gross width of each element measured normal to the axis of the member. For angles, the gross width is the sum of the widths of the legs less thickness. The gross areas of structural shapes are given in the LRFD manual.

7.3.3 Net Area, A_n

The term net area A_n refers to the gross area A_g minus the bolt holes. In computing the net area for tension and shear, the width of a hole is taken as $\frac{1}{16}$ in. larger than the nominal diameter of the hole. For punched holes, it is common practice to assume that $\frac{1}{16}$ in. more of the surrounding material is damaged. The width of a punched hole is then taken as $\frac{1}{8}$ in. larger than the nominal diameter of the hole.

EXAMPLE 7.1

Determine the net area of the $\frac{3}{8} \times 6$-in. plate shown in Figure 7.1. The plate is connected at its end with two lines of $\frac{3}{4}$ in. bolts. The bolt holes are drilled.

Solution

$$A_n = \left(\frac{3}{8}\right)(6) - (2)\left(\frac{3}{4} + \frac{1}{16}\right)\left(\frac{3}{8}\right) = 1.64 \text{ in}^2 \qquad \text{Answer}$$

When there is more than one row of bolt holes in a connection, it is desirable to stagger them to have a larger net area. When the holes are staggered, the net area is increased by the quantity $s^2/4g$ for each gage space in the chain of holes, where s is the spacing between any two consecutive holes and g is the gage distance between two lines of bolts as shown in Figure 7.2.

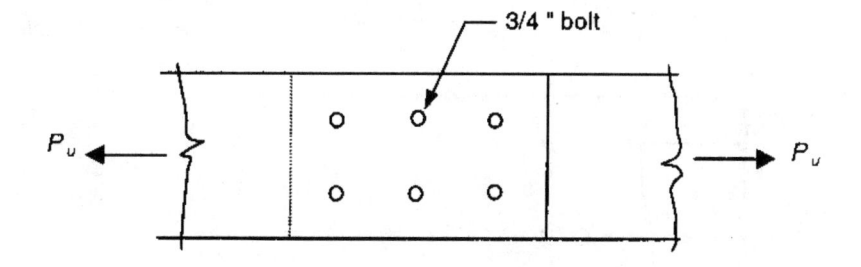

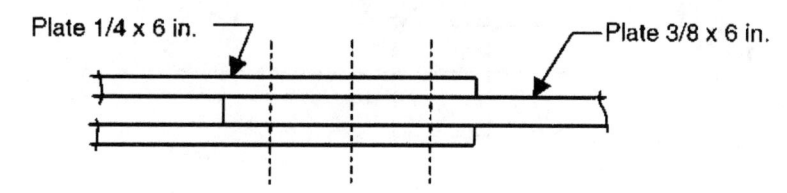

FIGURE 7.1

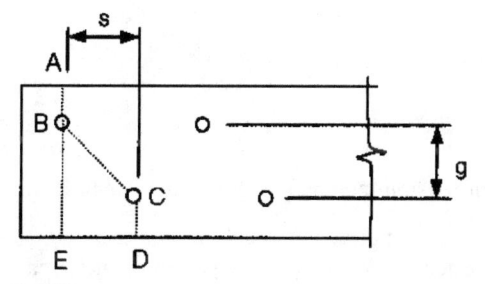

FIGURE 7.2

EXAMPLE 7.2

Determine the critical net area of the $\frac{1}{2}$ in. plate shown in Figure 7.3. The holes are punched for $\frac{3}{4}$ in. bolts.

Solution
The critical possible failure section could be ABCD, ABCEF or ABEF. Hole diameters to be subtracted are $\frac{3}{4} + \frac{1}{8} = \frac{7}{8}$ in. The net widths for each possible case are as follows:

$$ABCD = 9 - (2)\left(\frac{7}{8}\right) = 7.25 \text{ in.}$$

$$ABEF = 9 - (2)\left(\frac{7}{8}\right) + \frac{(3)^2}{(4)(6)} = 7.625 \text{ in.}$$

$$ABCEF = 9 - (3)\left(\frac{7}{8}\right) + \frac{(3)^2}{(4)(3)} = 7.125 \text{ in.} \quad (Controls)$$

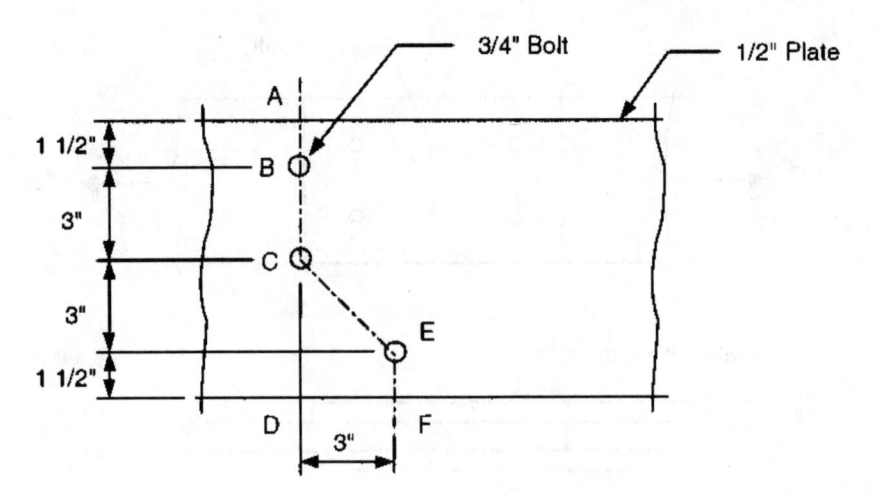

FIGURE 7.3

Hence the net area is given by $A_n = (7.125)\left(\dfrac{1}{2}\right) = 3.56$ in^2

7.3.4 Effective Net Area, A_e

The effective net area A_e for tension members is determined as follows:

1. When a tension load is transmitted directly to each of the cross-sectional elements by fasteners or welds, the effective net area A_e is equal to the net area A_n.
2. When a tension load is transmitted by bolts or rivets through some but not all of the cross-sectional elements of the member, the effective net area A_e is computed as

$$A_e = AU \tag{7.3}$$
(LRFD B3-1)

where A = Area as defined below
U = Reduction coefficient
   $= 1 - (\bar{x}/L) \leq 0.9$ or, as defined below
$\bar{x}$ = Conncection eccentricity, in.
L = Length of connection in the direction of loading

a. When the tension load is transmitted only by bolts or rivets:
$$A = A_n$$
= Net area of member, in^2

b. When the tension load is transmitted only by longitudinal welds to other than a plate member or by longitudinal welds in combination with transverse welds:
$$A = A_g$$
= Gross area of member, in^2

c. When the tension load is transmitted only by transverse welds:

A = Area of directly connected elements, in^2
$U = 1.0$

d. When the tension load is transmitted to a plate by longitudinal welds along both edges at the end of the plate for $l \geq w$: A = Area of plate, in.^{2}

For $l \geq 2w$ $U = 1.0$
For $2w > l \geq 1.5w$ $U = 0.87$
For $1.5w > l \geq w$ $U = 0.75$

where l = Length of weld, in.
       w = Plate width (distance between welds), in.

EXAMPLE 7.3

Determine the tensile strength of a W10 × 45 with two lines of $\frac{3}{4}$-in. diameter bolts in each flange using A588 steel with $F_y = 50$ ksi and $F_u = 70$ ksi. There are three bolts in each line 4 in. on center, and the holes are drilled but not staggered.

Solution

From Page 1-40 of AISC manual, the dimensions and properties of W10 × 45 may be found:

$$A_g = 13.3 \text{ in}^2, d = 10.10 \text{ in.}, b_f = 8.020 \text{ in.}, t_f = 0.620.$$

a. For the limit state of yielding in the gross section, Equation 7.1 gives

$$P_u = \phi_t F_y A_g = (0.90)(50)(13.3) = 598.5 \text{ k}$$

b. For the limit state of fracture in the net section, Equation 7.2 gives

$$P_u = \phi_t F_u A_e$$

$$A_n = 13.3 - (4)\left(\frac{13}{16}\right)(0.620) = 11.28 \text{ in}^2 = A$$

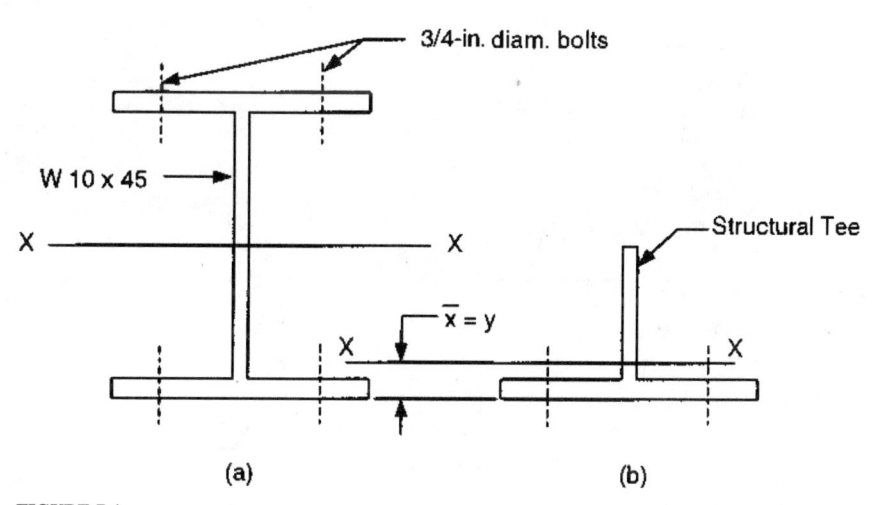

FIGURE 7.4

Next we will find the effective net area of the section. From the tables of the LRFD manual for half a W10 × 45 (or a WT5 × 22.5 on Page 1-82), $y = \bar{x} = 0.907$ in. Then

$$U = 1 - (\bar{x}/L) = 1 - 0.907/8 = 0.89$$

Hence

$$A_e = UA = (0.89)(11.13) = 9.91 \text{ in}^2$$

Therefore,

$$P = \phi_t F_u A_e = (0.75)(70)(9.91) = 520.3 \text{ k} \quad (Controls)$$

The limit state of fracture controls the design strength of the section.

EXAMPLE 7.4
Determine the design strength P_u of the member shown in Figure 7.5. The plates are of A572 Grade 50 steel with $F_y = 50$ ksi and $F_u = 65$ ksi.

Solution
The design strength is controlled by the smaller plate.

a. For the limit state of yielding in the gross section, Equation 7.1 gives

$$P_u = \phi_t F_y A_g = (0.90)(50)(0.75 \times 6) = 202.5 \text{ k}$$

b. For the limit state of fracture in the net section, Equation 7.2 gives

$$P = \phi_t F_u A_e$$

$$A = A_g = (0.75)(6.0) = 4.5 \text{ in}^2$$

$$1.5w = 9 \text{ in.} > l = 8 \text{ in.} > w = 6 \text{ in.}$$

Therefore,

$$U = 0.75$$

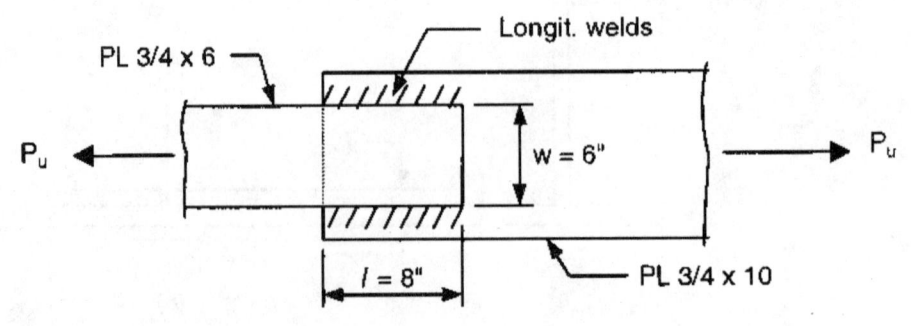

FIGURE 7.5

And,

$$A_e = UA = (0.75)(4.5) = 3.375 \text{ in}^2$$

Hence,

$$P_u = (0.75)(65)(3.375) = 164.5 \text{ k} \quad (Controls)$$

EXAMPLE 7.5
Determine the design strength P_u for the angle shown in Figure 7.6. The angle is welded on the end and sides of the 8-in. leg only. The angle plates are made of A36 structural steel with $F_y = 36$ ksi and $F_u = 58$ ksi.

Solution
The dimensions and section properties of the angle are given on Page 1-56 of the LRFD manual. $A_g = 9.94 \text{ in}^2$ and $\bar{x} = 1.56$ in. It is more efficient to weld the long leg of the angle than the short leg.

a. For the limit state of yielding in the gross section, Equation 7.1 gives
$$P_u = \phi_t F_y A_g = (0.90)(36)(9.94) = 322.1 \text{ k}$$

b. For the limit state of fracture in the net section, Equation 7.2 gives
$$P_u = \phi_t F_u A_e$$
$$A = A_g = 9.94 \text{ in.}$$
$$U = 1 - \frac{\bar{x}}{L} = 1 - \frac{1.56}{6.00} = 0.74$$

Hence,
$$A_e = UA = (0.74)(9.94) = 7.36 \text{ in}^2$$

Therefore,
$$P_u = (0.75)(58)(7.36) = 320.2 \text{ k} \quad (Controls)$$

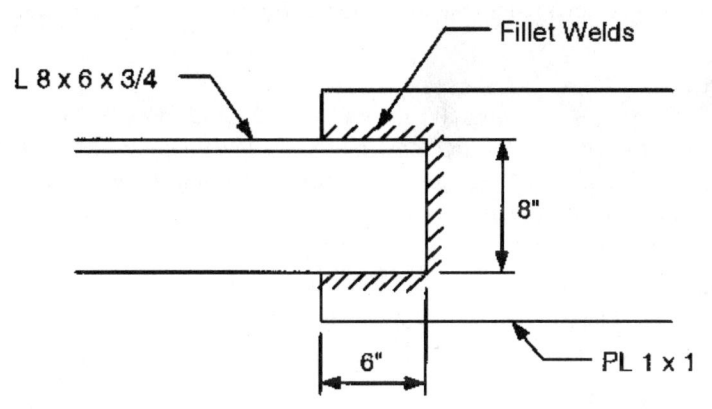

FIGURE 7.6

7.3.5 Design of Tension Members

In the previous section, we analyzed the members to determine the design strengths with known dimensions and properties. In the design of tension members, we have to select the member dimensions and properties to withstand design loads and load combinations. A design procedure involves determining minimum requirements and then trying a section that best meet the limit states, such as strength, deflection, and slenderness ratio. The LRFD specifications recommend a slenderness ratio L/r less than 300 to assure that tension members have sufficient stiffness.

a. To satisfy the limit state of yielding, we have

$$\text{Minimum } A_g = \frac{P_u}{\phi_t F_y} \quad (7.4)$$

b. To satisfy the limit state of fracture, we have

$$\text{Minimum } A_e = \frac{P_u}{F_u A_e}$$

Since $A_e = UA_n$ for a bolted member, the minimum value of A_n is

$$\text{Minimum } A_n = \text{Minimum } \frac{A_e}{U} = \frac{P_u}{F_u A_e U}$$

Hence,

$$\text{Minimum } A_g = \frac{P_u}{F_u A_e U} + \text{Estimated hole areas} \quad (7.5)$$

From the slenderness ratio L/r requirement, we have

$$\text{Minimum } r = \frac{L}{300} \quad (7.6)$$

Equations 7.5 and 7.6 can be used to estimate the size of member that will meet the gross area and net area requirements. Equation 7.6 can be used to check the L/r ratio.

EXAMPLE 7.6

Select a 32-ft-long wide-flange (W) section of A572 Grade 50 steel to support a tensile service dead load $P_D = 120$ k and a tensile service live load $P_L = 115$ k. The member has two lines of bolts in each flange for $\frac{7}{8}$-in. bolts as shown in Figure 7.7, and at least three in a line 4-in. on center.

Solution

Load factors and load combinations are given on Page 6-30 of the LRFD manual. The two load conditions for this design are

$P_u = 1.4P_D = (1.4)(120) = 168$ k
$P_u = 1.2P_D + 1.6P_L = (1.2)(120) + (1.6)(115) = 328$ k *(Controls)*

```
W-Section                    7/8-in. bolts at 4" ctrs.
                             At least 3 in a line
```

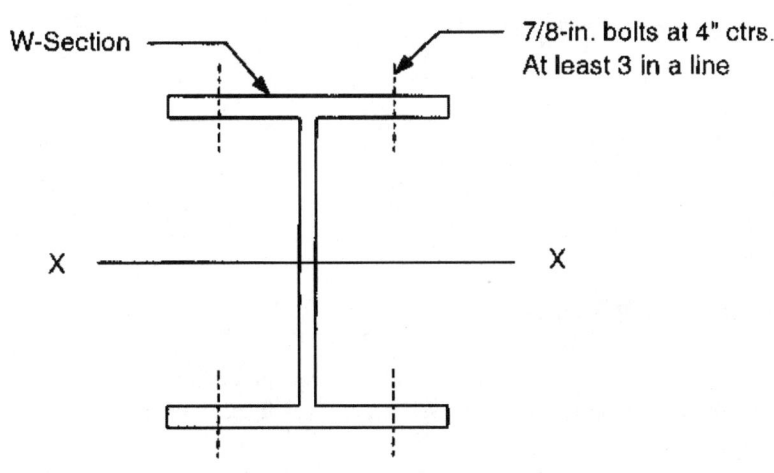

X ——————[I]—————— X

FIGURE 7.7

Determining the minimum A_g required

a. Minimum $A_g = \dfrac{P_u}{\phi_t F_y} = \dfrac{328}{(0.90)(50)} = 7.29$ in^2

b. Minimum $A_g = \dfrac{P_u}{F_u A_e U} +$ Estimated hole areas

Assume $U = 0.90$ and flange thickness of about 0.380 in. (after looking at the W Shapes tables in the LRFD manual to find a W shape that has about 7.29 in^2).

$$\text{Minimum } A_g = \dfrac{328}{(0.75)(65)(0.90)} + (4)\left(\dfrac{7}{8} + \dfrac{1}{8}\right)(0.380) = 9.0 \text{ in}^2$$

c. Minimum $r = \dfrac{L}{300} = \dfrac{(12)(32)}{300} = 1.28$

From the W Shapes tables in the LRFD manual, select a section that can meet the above minimum requirements with the lightest weight.

Try W10 × 33 giving $A_g = 9.71$ in^2, $d = 9.73$ in., $b_f = 7.960$ in., $t_f = 0.435$ in., and $r_y = 2.01$ in.
Checking

a. $P_u = \phi_t F_y A_g = (0.90)(50)(9.71) = 437.0$ k > 328 k   OK.
b. $P_u = \phi_t F_u A_e$

$\bar{x}$ for half of W10 × 33 or a WT5 × 16.5 = 0.869 in.
$L = (2)(4) = 8$ in.

Hence,

$$U = 1 - \dfrac{\bar{x}}{L} = 1 - \dfrac{0.869}{8} = 0.89$$

$$A_n = 9.71 - (4)(1)(0.435) = 7.97 \text{ in}^2$$

Therefore,

$$P_u = \phi_t F_u A_e = (0.75)(65)(0.89)(7.97) = 345.8 \text{ k} > 328 \text{ k} \quad \text{OK.}$$

c. $\dfrac{L}{r} = \dfrac{(12)(32)}{2.01} = 191 < 300 \quad \text{OK.}$

Use W10 × 33.

7.4 COMPRESSION MEMBERS

This section applies to compact and non-compact prismatic members subject to axial compression through the centroidal axis. Section 7.6 will deal with members subject to combined axial compression and flexure.

There are several types of compression members. The column is the most important compression member. Other types are the top chords of trusses and bracing members.

In this section we will be concerned mainly with two general modes of column failures: flexural buckling and local buckling. Flexural buckling, also known as the Euler buckling, occurs when axially loaded members are subject to flexure or bending as they become unstable. Column imperfection or out-of-straightness has significant effect on flexural buckling. Local buckling occurs when some part or parts of the cross section of an axially loaded member buckle locally in compression before flexural buckling can occur. The LRFD column formulas are developed for designing compression members to avoid the different modes of column failures.

The design strength of column is controlled by the slenderness ratio L/r of the column. The tendency to buckle is also affected by the types of end connections, or restraints. To account for this tendency, an effective length factor K is used to modify the column length L to give an effective length KL. LRFD manual Table C-C2.1, Page 6-184 provides the theoretical K values and the recommended design values for columns with different end conditions. An alignment chart also is given on Page 6-186 for more accurate determination of the K values for columns in frames braced against side sway and for frames not braced against side sway.

7.4.1 Classification of Steel Sections

Steel sections are classified as compact, noncompact, or slender-element sections. For a section to qualify as compact, its flanges must be continuously connected to the web or webs, and the width-thickness ratios of its compression elements must not exceed the limiting width-thickness ratios λ_p from Table B5.1, Pages 6-38 and 6-39 of the LRFD manual. If the width-thickness ratio of one or more compression elements exceeds λ_p, but does not exceed λ_r, the section is noncompact. If the width-thickness ratio of any element exceeds λ_r from Table B5.1, the section is referred to as a slender-element compression section.

For unstiffened elements that are supported along only one edge parallel to the direction of the compression force, the width shall be taken as follows:

a. For flanges of I-shaped members and tees, the width b is half the full-flange width, b_f.
b. For legs of angles and flanges of channels and zees, the width b is the full nominal dimension.

c. For plates, the width *b* is the distance from the free edge to the first row of fasteners, or line of welds.
d. For stems of tees, *d* is taken as the full nominal depth.

For stiffened elements that are supported along two edges parallel to the direction of the compression force, the width shall be taken as follows:

a. For webs of rolled or formed sections, *h* is the clear distance between flanges less the fillet or corner radius at each flange; h_c is twice the distance from the centroid to the inside face of the compression flange less the fillet or corner radius.
b. For webs of built-up sections, *h* is the distance between adjacent lines of fasteners or the clear distance between flanges when welds are used; h_c is twice the distance from the centroid to the nearest line of fasteners at the compression flange or the inside face of the compression flange when welds are used.
c. For flange or diaphragm plates in built-up sections, the width *b* is the distance between adjacent lines of fasteners or lines of welds.
d. For flanges of rectangle hollow structural sections, the width *b* is the clear distance between webs less the inside corner radius on each side. If the corner radius is not known, the width may be taken as the total section width minus three times the thickness.

Most of the dimensions of structural shapes can be found in Part 1 of the LRFD manual. The W, M, and S shapes listed in the AISC manual are compact for 36 ksi and 50 ksi yield stress steels unless otherwise indicated. Figure 7.8 shows various cases of stiffened and unstiffened elements.

(a) Unstiffened Elements

(b) Stiffened Elements

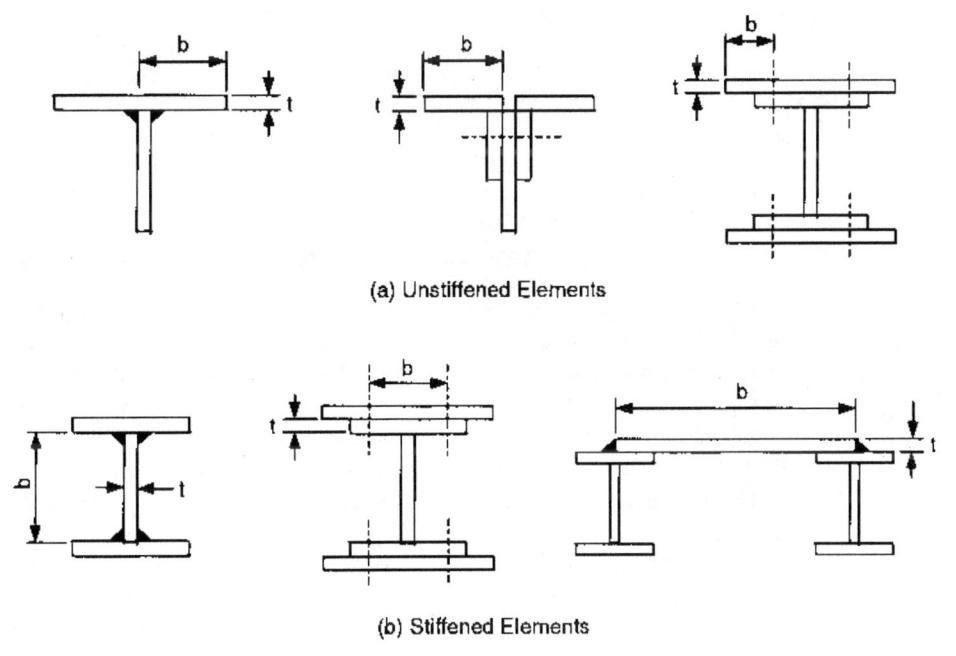

FIGURE 7.8

7.4.2 Column Formulas

The LRFD specification provides an empirical parabolic equation for short and intermediate columns and an equation (the Euler Equation) for long columns with inelastic buckling.

The design strength for flexural buckling of compression members whose elements have width-thickness ratios less than λ_r is $P_u = \phi_c P_n$:

$$\phi_c = 0.85$$
$$P_n = A_g F_{cr} \tag{7.7}$$

Therefore,

$$P_u = \phi_c A_g F_{cr} \tag{LRFD E2-1}$$

a. For $\lambda_c \leq 1.5$

$$F_{cr} = \left(0.658^{\lambda_c^2}\right) \tag{7.8}$$
$$\text{(LRFD E2-2)}$$

b. For $\lambda_c > 1.5$

$$F_{cr} = \left[\frac{0.877}{\lambda_c^2}\right] F_y \tag{7.9}$$
$$\text{(LRFD E2-3)}$$

where

$$\lambda_c = \frac{Kl}{r\pi}\sqrt{\frac{F_y}{E}} \tag{7.10}$$
$$\text{(LRFD E2-4)}$$

A_g = Gross area of member, in^2
F_y = Specified yield stress, ksi
E = Modulus of elasticity, ksi
K = Effective length factor
l = Laterally unbraced length of member, in
r = Governing radius of gyration about the axis of buckling, in.

LRFD specification states that the slenderness ratio Kl/r should not exceed 200 for members in which the design is based on compression.

LRFD manual Tables 3-36 and 3-50 of Part 6 are valuable aids for determining design strengths of columns with F_y = 36 ksi and 50 ksi, respectively. With known effective slenderness ratio Kl/r, the tables give values of design stress $\phi_c F_{cr}$, which can be multiplied by the cross-sectional area A of the member to obtain the design strength of the columns.

LRFD manual, Part 3 has column design tables giving design axial strength of structural shapes normally used as columns for F_y values of 36 ksi and 50 ksi, and for most of the commonly used effective lengths or KL values given in feet. These are very useful design aids for the design or selection of structural shapes for columns.

Once the readers are familiar with the use of these tables, the design and analysis of columns can be very fast and accurate.

STEEL DESIGN 7.15

EXAMPLE 7.7
  a. Determine the design strength of the $F_y = 50$ ksi axially loaded column shown in Figure 7.9.
  b. Repeat the problem using the column design stress values shown in Table 3-50, Part 6 of the LRFD manual.

Solution
  a. From Page 1-38 of the LRFD manual, we have

  W12 × 72, $A = 21.1$ in², $r_x = 5.31$ in. and $r_y = 3.04$ in. (controls buckling)

  From Table C-C2.1 on Page 6-184 of the LRFD manual, we obtain $K = 0.80$ for the end conditions shown for this problem.

  Hence,

  $$\frac{Kl}{r} = \frac{(0.8)(12 \times 15)}{(3.04)} = 47.37$$

  Therefore

  $$\lambda_c = \frac{Kl}{r\pi}\sqrt{\frac{F_y}{E}} = \frac{47.37}{\pi}\sqrt{\frac{50}{29{,}000}} = 0.626 \leq 1.5$$

  $$\lambda_c^2 = (0.626)^2 = 0.392$$

  Hence,

  $$F_{cr} = (0.658)^{0.392}(50) = 42.43 \text{ ksi}$$

  Therefore,

  $$P_u = \phi_c A_g F_{cr} = (0.85)(21.1)(42.43) = 761.0 \text{ k} \qquad \text{Answer}$$

  b. $KL = (0.80)(15) = 12$ ft

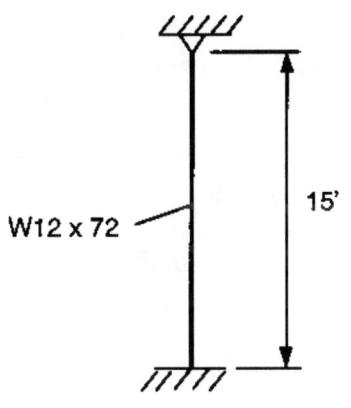

FIGURE 7.9

Entering values into the table on Page 3-24 of the LRFD manual for W12 × 72 and $KL = 12$, and $F_y = 50$ ksi, we have $P_u = \phi_c A_g F_{cr} = 761$ k.

EXAMPLE 7.8

a. Select the lightest W14 available for the service column loads of P_D 5 120 k and P_L 5 170 k, using F_y 5 36 ksi steel and a column length of 16 feet with fixed ends as shown in Figure 7.10.

b. Repeat the design problem using design tables in Part 3 of the LRFD manual to select the lightest W section available.

Solution

a. Design factored loads

$$P_u = 1.2P_D + 1.6P_L = (1.2)(120) + (1.6)(170) = 416 \text{ k}$$

Assume

$$\frac{KL}{r} = 50 \quad \text{(Any reasonable number will do for the first trial)}$$

From Table 3-36 of Part 6 of the LRFD manual (Page 6-147), we have

$$\phi_c F_{cr} = 26.83 \text{ ksi} \quad \text{(for } \frac{KL}{r} = 50\text{)}$$

Hence,

$$A \text{ required} = \frac{416}{26.83} = 15.50 \text{ in}^2$$

Try W14 × 53 having $A = 15.6$ in^2, $r_x = 5.89$ in., and $r_y = 1.92$ in (Controls)

$$\frac{KL}{r} = \frac{(0.65)(12 \times 16)}{(1.92)} = 65$$

From Table 3-36, we have

$$\phi_c F_{cr} = 24.50 \text{ ksi}$$
$$\phi_c P_n = (24.50)(15.6) = 382.2 \text{ k} < 416 \text{ k} \quad \text{No good.}$$

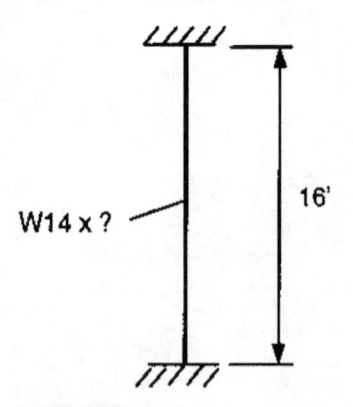

W14 x ?

16'

FIGURE 7.10

Try the next larger section W14 × 61, giving $A = 7.9$ in², min $r_y = 2.45$ in.

$$\frac{KL}{r} = \frac{(0.65)(12 \times 16)}{(2.45)} = 51$$

From Table 3-36, we have

$$\phi_c F_{cr} = 26.68 \text{ ksi}$$
$$\phi_c P_n = (26.68)(17.9) = 477 \text{ k} > 416 \text{ k} \quad \text{OK.}$$

Use W14 × 61.

b. $K_y L_y = (0.65)(16) = 10.4$ ft, and $P_u = 416$ k.
Enter tables in Part 3 of the LRFD manual (Pages 3-21, 3-25, 3-27, and 3-28). The lightest suitable section in each W series is shown below:

W14 × 61 $\phi_c P_n = 477$ k
W12 × 53 $\phi_c P_n = 418$ k   (Lightest section)
W10 × 54 $\phi_c P_n = 427$ k
W8 × 58 $\phi_c P_n = 435$ k

Use W12 × 53.

7.5 BEAMS

This section applies to compact and noncompact prismatic members subject to flexure and shear in accordance with Chapter F of the LRFD specification. Members subject to combined flexure and axial force will be covered in Section 7.6.

This section applies to homogeneous and hybrid shapes with at least one axis of symmetry and which are subject to simple bending about one principal axis. For simple bending, the beam is loaded in a plane parallel to a principal axis that passes through the shear center, or the beam is restrained against twisting at load points and supports. Only the limit states of yielding and lateral-torsional buckling are considered in this section. The lateral-torsional buckling provisions are limited to doubly symmetric shapes, channels, double angles, and tees.

7.5.1 Design for Flexure

The nominal flexural strength M_n is the lowest value obtained according to the limit state of (a) yielding; (b) lateral-torsional buckling; (c) flange local buckling; and (d) web local buckling. For laterally braced compact beams with $L_b \leq L_p$, only the limit state of yielding is applicable. L_b is the distance between points braced against lateral displacement of the compression flange, or between points braced to prevent twist of the cross section; L_p is the limiting unbraced length for full plastic bending capacity. For $L_b > L_p \leq L_r$, the limit states of lateral-torsional buckling, or inelastic buckling, are applicable. Lr is the limiting laterally unbraced length when elastic buckling will occur at a fair low load. When $L_b > L_r$, the section will buckle elastically before the yield stress is reached anywhere along the beam. Hence, three cases of flexural problems will be addressed, depending on L_b.

Case 1

For laterally braced compact beams with $L_b \leq L_p$, the flexural design strength of beams, determined by the limit state of yielding is

$$M_u = \phi_b M_n$$
$$\phi_b = 0.90$$
$$M_n = M_p = F_y Z \leq 1.5 M_y \qquad (7.11)$$
$$\text{(LRFD F1-1)}$$

where M_p = Plastic moment, kip-in
 M_y = Moment corresponding to onset of yielding at the extreme fiber from an elastic stress distribution ($= F_y S$ for homogeneous section and $F_{yf} S$ for hybrid sections), kip-in.

M_p is limited to no greater than $1.5 M_y$. For compact sections with a large shape factor, large inelastic deformations might occur when M_p is reached. To avoid large deformations in the design, the LRFD specification imposes $M_p \leq 1.5 M_y$.

For I-shaped members, including hybrid sections and channels, the limiting unbraced length for full plastic bending capacity, L_p, is given by

$$L_p = \frac{300 r_y}{\sqrt{F_{yf}}} \qquad (7.12)$$
$$\text{(LRFD F1-4)}$$

where r_y = Radius of gyration about y axis or weak axis
 F_{yf} = Yield stress of the flange

EXAMPLE 7.9

Determine the bending strength of a W10 × 22 member of $F_y = 50$ ksi and fully supported laterally.

Solution

From the Table of W Shapes Properties, Part 1 of the LRFD manual (Pages 1-40 and 1-41), we have for W10 × 22,

$$S_x = 23.2 \text{ in}^3, \quad Z_x = 26.0 \text{ in}^3 \quad \text{(Properties about major axis)}$$

Hence,

$$M_p = F_y Z_x = (50)(26.0) = 1300 \text{ kip-in.}$$

Checking,

$$1.5 M_y = 1.5 F_y S_x = (1.5)(50)(23.2) = 1740 \text{ kip-in.} > M_p \quad \text{OK.}$$

Therefore,

$$M_u = \phi_b M_n = \phi_b M_p = (0.90)(1300) = 1170 \text{ kip-in.} \qquad \textit{Answer}$$

EXAMPLE 7.10

Select a beam section to support dead moment M_D of 57.6 kip-ft and M_L of 165.4 kip-ft. Use steel with $F_y = 50$ ksi and assume full lateral support of compression flange.

Solution

$$M_u = 1.2M_D + 1.6M_L = (1.2)(57.6) + (1.6)(165.4) = 333.8 \text{ kip-ft}$$

Hence,

$$Z \text{ required} = \frac{M_u}{\phi_b F_y} = \frac{(12)(333.8)}{(0.90)(50)} = 89.0 \text{ in}^3$$

From the Tables of W Shapes Properties, Part 1 of the LRFD manual (Pages 1-30 to 1-35), we have

W24 × 55 gives $Z_x = 134.0$ in^3
W21 × 44 gives $Z_x = 95.4$ in^3 (Lightest section)
W18 × 46 gives $Z_x = 90.7$ in^3
W16 × 50 gives $Z_x = 92.0$ in^3

Select W21 × 44

Case 2:

For beams of doubly symmetric shapes and channels with $L_b > L_p \leq L_r$, the flexural design strength, determined by the limit state of lateral-torsional buckling, is

$$M_u = \phi_b M_n$$
$$\phi_b = 0.90$$

$$M_n = C_b \left[M_p - (M_p - M_r)\left(\frac{L_b - L_p}{L_r - L_p} \right) \right] \leq M_p \quad (7.13)$$
$$\text{(LRFD F1-2)}$$

where L_b = Distance between points braced against lateral displacement of the compression flange, or between points braced to prevent twist of the cross section, in.

Equation 7.13 (LRFD F1-2) can be rewritten as

$$\phi_b M_n = C_b [\phi_b M_p - BF(L_b - L_p)] \leq \phi_b M_p \quad (7.14)$$

where

$$BF = \frac{\phi_b (M_p - M_r)}{L_r - L_p}$$

BF is a factor given in the Load Factor Design Selection Table in Part 4 of the LRFD manual for each W shape. This helps us do the proportioning with a simple formula.

In the above equation, C_b is a modification factor for non-uniform moment diagrams where, when both ends of the beam segment are braced

$$C_b = \frac{12.5 M_{max}}{2.5 M_{max} + 3 M_A + 4 M_B + 3 M_C} \quad (7.15)$$
(LRFD F1-3)

where M_{max} = Absolute value of maximum moment in the unbraced segment, kip-in.
M_A = Absolute value of moment at quarter-point of the unbraced segment
M_B = Absolute value of moment at centerline of the unbraced beam segment
M_C = Absolute value of moment at three-quarter point of the unbraced beam segment

C_b is permitted to be conservatively taken as 1.0 for all cases. For cantilevers or overhangs where the free end is unbraced, $C_b = 1.0$.

The limiting laterally unbraced length L_r and the corresponding buckling moment M_r shall be determined as follows:

$$L_r = \frac{r_y X_1}{F_L} \sqrt{1 + \sqrt{1 + X_2 F_L^2}} \quad (7.16)$$
(LRFD F1-6)

$$M_r = F_L S_x \quad (7.17)$$
(LRFD F1-7)

where

$$X_1 = \frac{\pi}{S_x} \sqrt{\frac{EGJA}{2}} \quad (7.18)$$
(LRFD F1-8)

$$X_2 = 4 \frac{C_w}{I_y} \left(\frac{S_x}{GJ} \right)^2 \quad (7.19)$$
(LRFD F1-9)

S_x = Section modulus about major axis, in^3
E = Modulus of elasticity of steel (29,000 ksi)
G = Shear modulus of elasticity of steel (11,200 ksi)
F_L = Smaller of $(F_y - F_r)$ or F_{yw}
F_r = Compressive residual stress in flange; 10 ksi for rolled shape, 16.5 ksi for welded shapes
F_{yf} = Yield stress of flange
F_{yw} = Yield stress of web
I_y = Moment of inertia about y axis, in^4
C_w = Warping constant, in^6

The above equations are conservatively based on $C_b = 1.0$.

EXAMPLE 7.11
Determine the bending moment capacity of a W24 × 68 beam with (1) F_y = 36 ksi and (2) F_y = 50 ksi. Use L_b = 8 ft and C_b = 1.0.

Solution

1. $F_y = 36$ ksi. From the Load Factor Design Selection Table, Part 4 of the LRFD manual (Page 4-18), we have for W24 × 68:

$$BF = 12.1 \text{ kips}$$
$$L_r = 22.4 \text{ ft}$$
$$L_p = 7.8 \text{ ft}$$
$$\phi_b M_r = 300 \text{ kip-ft}$$
$$\phi_b M_p = 478 \text{ kip-ft}$$

We have $L_b = 8 > L_p = 7.8 < L_r = 22.4$, therefore Case 2 applies and $\phi_b M_n$ can be determined from Equation 7.14:

$$\phi_b M_n = C_b[\phi_b M_p - BF(L_b - L_p)]$$
$$= 1.0[478 - 12.1(8 - 7.8)]$$
$$= 475.6 \text{ kip-ft}$$

2. $F_y = 50$ ksi. From the Load Factor Design Selection Table, Part 4 of the LRFD manual (Page 4-18), we have

   For W24 × 68:

$$BF = 18.7 \text{ kips}$$
$$L_r = 17.4 \text{ ft}$$
$$L_p = 6.6 \text{ ft}$$
$$\phi_b M_r = 462 \text{ kip-ft}$$
$$\phi_b M_p = 664 \text{ kip-ft}$$

We have $L_b = 8 > L_p = 6.6 < L_r = 17.4$, therefore Case 2 applies and $\phi_b M_n$ can be determined from Equation 7.14:

$$\phi_b M_n = C_b[\phi_b M_p - BF(L_b - L_p)] = 1.0[664 - 18.7(8 - 6.6)]$$
$$= 637.8 \text{ kip-ft} \qquad \text{Answer}$$

Case 3

For beams of doubly symmetric shapes and channels with $L_b > L_r$, the flexural design strength is $M_u = \phi_b M_n$:

$$\phi_b = 0.90$$
$$M_n = M_{cr} \leq M_p$$

where M_{cr} = Critical elastic moment, determined as follows:

$$M_{cr} = C_b \frac{\pi}{L_b} \sqrt{EI_y GJ + \left(\frac{\pi E}{L_b}\right)^2 I_y C_w}$$

$$= \frac{C_b S_x X_1 \sqrt{2}}{L_b/r_y} \sqrt{1 + \frac{X_1^2 X_2}{2(L_b/r_y)^2}}$$

(7.20)
(LRFD F1-13)

The values for X_1 and X_2 are shown for W shapes in the tables for W Shapes Properties in Part 1 of the LRFD manual.

EXAMPLE 7.12
Determine the flexural strength of a W18 × 97 section of $F_y = 36$ ksi and the unbraced length L_b is 45 ft. Assume $C_b = 1.0$.

Solution
For $F_y = 36$ ksi: From the Load Factor Design Selection Table, Part 4 of the LRFD manual (Page 4-18), we have for W18 × 97:

$$L_r = 38.1 \text{ ft}$$
$$L_p = 11.0 \text{ ft}$$

Hence,

$$L_b = 45 > L_p = 11.0 > L_r = 38.1$$

Case 3 applies and the flexural design strength $\phi_b M_n$ is given by (7.20):

$$\phi_b M_n = \phi M_{cr}$$

where

$$M_{cr} = C_b \frac{\pi}{L_b} \sqrt{EI_y GJ + \left(\frac{\pi E}{L_b}\right)^2 I_y C_w}$$

From tables of W Shapes Properties and Torsion Properties W Shapes in Part 1 of the LRFD manual (Pages 1-33 and 1-148), we have

$I_y = 201$ in^4
$J = 5.86$ in^4
$C_w = 15{,}800$ in^6

Therefore,

$$M_{cr} = (1.0)\left(\frac{\pi}{12 \times 45}\right)\sqrt{(29{,}000)(201)(11{,}200)(5.86) + \left(\frac{\pi \times 29{,}000}{12 \times 45}\right)^2 (201)(15{,}800)}$$
$$= 4001 \text{ kip-in.} = 333.4 \text{ kip-ft.}$$

Hence flexural design strength

$$M_u = \phi M_{cr} = (0.90)(333.4) = 300 \text{ kip-ft.} \qquad \textit{Answer}$$

7.5.2 Beam Design Charts

Part 4 of the LRFD manual contains beam-design charts to reduce the tedious work of designing and analyzing beams with laterally unbraced lengths exceeding L_p and L_r. These charts cleverly and simply cover three cases of flexural problems discussed in the previous section. The charts show the design moment $\phi_b M_n$ for structural W and M shapes of

$$F_y = 36 \text{ ksi} \quad \text{and} \quad F_y = 50 \text{ ksi steels}, \qquad \text{using } \phi = 0.9 \quad \text{and} \quad C_b = 1.0.$$

The design moment $\phi_b M_n$ in kip-ft. is plotted with respect to the unbraced length in feet, with no consideration of the moment due to weight of the beam. Design moments are shown for unbraced lengths in feet, starting at spans less than L_p, spans between L_p and L_r, and for spans beyond L_r. For each of the shapes, L_p is indicated with a solid symbol, ●, while L_r is indicated by a hollow symbol, ○. The solid portion of each curve indicates the most economical section by weight. The dashed portion of each curve indicates ranges in which a lighter weight beam will satisfy the loading conditions.

To select a member, it is only necessary to enter the chart with the unbraced length L_b and the factored design moment M_u. For example, we want to select a beam with $F_y = 36$ ksi, $L_b = 20$ ft. and design moment $M_u = \phi_b M_n = 600$ kip-ft. We enter the beam-design moments in Part 4 of the LRFD manual for $F_y = 36$ ksi, and design moment = 600 kip-ft (Page 4-124 for this example). With unbraced length equal to 20 feet on the bottom scale, we proceed upward to intersect the horizontal line corresponding to the design moment = 600 kip-ft on the left-hand scale. Any beam listed above and to the right of the intersection point satisfies the design-moment requirement. In this example, the nearest section to the right is W21 × 101, but it is on the dashed portion of the curve, meaning that it is not the most efficient section. The lightest section is a W30 × 99, which is the nearest solid line upward and to the right.

EXAMPLE 7.13
Select the lightest W section of A572 Grade 50 steel for the beam shown in Figure 7.11. The beam has lateral bracing provided for the compression flange only at the ends. Assume $C_b = 1.0$.

Solution
Assume beam weight = 70 lb/ft

$$W_u = (1.2)(1.070) + (1.6)(2) = 4.484 \text{ lb/ft}$$

$$M_u = \frac{(4.484)(25)^2}{8} = 350.3 \text{ kip-ft}$$

Enter the beam-design moments charts (Page 4-157) with

$$L_b = 25 \text{ ft}, \quad M_u = \phi_b M_n = 350.3 \text{ kip-ft} \quad \text{and} \quad F_y = 50 \text{ ksi}$$

Select W14 × 74

D = 1 klf (not incl. beam weight)
L = 2 klf

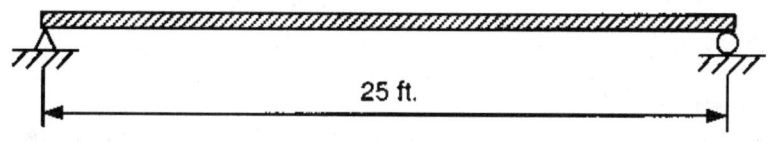

25 ft.

FIGURE 7.11

7.5.3 Design Shear Strength

This section applies to unstiffened webs of singly or doubly symmetric beams, including hybrid beams, and channels subject to shear in the plane of the web.

The design shear strength of unstiffened webs, with $h/t_w \leq 260$, is $\phi_v V_n$,

where $\phi_v = 0.90$
V_n = Nominal shear strength defined as follows

Case 1
Shear strength is controlled by web yielding. Almost all rolled-beam sections in the LRFD manual fall under this classification.

For
$$\frac{h}{t_w} \leq \frac{418}{\sqrt{F_{yw}}}$$

$$V_n = 0.6 F_{yw} A_w \qquad (7.21)$$
$$\text{(LRFD F2-1)}$$

The web area A_w is taken as the overall depth d times the web thickness t_w.

Case 2
Shear strength is controlled by the inelastic buckling of web.

For
$$\frac{418}{\sqrt{F_{yw}}} < \frac{h}{t_w} \leq \frac{523}{\sqrt{F_{yw}}}$$

$$V_n = \frac{0.6 F_{yw} A_w \left(\dfrac{418}{\sqrt{F_{yw}}}\right)}{\left(\dfrac{h}{t_w}\right)} \qquad (7.22)$$
$$\text{(LRFD F2-2)}$$

Case 3
Shear strength is controlled by the elastic buckling of web.

For
$$\frac{523}{\sqrt{F_{yw}}} < \frac{h}{t_w} \leq 260$$

$$V_n = \frac{132{,}000 A_w}{\left(\dfrac{h}{t_w}\right)^2} \qquad (7.23)$$
$$\text{(LRFD F2-3)}$$

EXAMPLE 7.14
Determine the design shear strength of a W14 × 74 beam of $F_y = 50$ ksi.

Solution
From the W Shape Dimensions Table on Page 1-36 of Part 1, LRFD manual, we have $d = 14.17$ in., $t_w = \frac{7}{16}$ in., $k = 1\frac{9}{16}$ in. (distance to web toe fillet)
Hence,

$$A_w = (14.17)\left(\frac{7}{16}\right) = 6.20 \text{ in}^2$$

$$h = d - 2k = (14.17) - \left(2 \times 1\frac{9}{16}\right) = 12.61 \text{ in.}$$

$$\frac{h}{t_w} = \frac{12.61}{\frac{7}{16}} = 28.8 < \frac{418}{\sqrt{50}} = 59.1$$

Therefore,

$$\phi_v V_n = (0.90)(0.6)(F_{yw})(A_w)$$
$$= (0.90)(0.6)(50)(6.20) = 167.4 \text{ k}$$

7.5.4 Deflections of Beams

The LRFD specification does not cover the maximum deflections allowed for structural steel members. However, UBC requires that structural member shall not exceed the values set forth in Table 16-D, based on the factors set forth in Table 16-E. The 1989 AASHTO specifications limit deflections to $\frac{1}{800}$ of span length for live load plus impact and $\frac{1}{1000}$ of span length if bridge is also used for pedestrians.

The code provisions for deflections are intended to ensure there is adequate stiffness in the members to avoid:

1. Damage of materials attached to or supported by the beam. For example, cracking of plaster or board attached to ceiling or floor.
2. Discomfort experienced by persons using the buildings or bridges.
3. Ponding on flat roofs because of little or no runoff.

The designers can use any methods suitable for computing the deflections of beams. Part 4 of the LRFD manual (Page 4-29) has simple formula for determining deflections for structural steel shapes under different loading conditions. The simple formula takes the form

$$\Delta = \frac{ML^2}{C_1 I_x}$$

where Δ = Maximum vertical deflection at center of span, in.
M = Maximum service load moment, kip-ft
L = Span length, ft
C_1 = Loading constant (See Page 4-29 of LRFD manual)
I_x = Moment of inertia, in^4

EXAMPLE 7.15
Find the maximum deflection due to the factored dead and live loads.

Solution
In this problem,

$$M = 350.3 \text{ kip-ft}$$
$$L = 25 \text{ ft}$$
$$C = 161 \text{ (uniform loads)}$$
$$I_x = 796 \text{ in}^4$$

Therefore,

$$\Delta = \frac{ML^2}{C_1 I_x}$$

$$= \frac{(350.3)(25)^2}{(161)(796)} = 1.71 \text{ in.} \qquad \text{Answer}$$

7.6 BENDING AND AXIAL FORCE

This section applies to prismatic members subject to axial force and flexure about one or both axes of symmetry.

Case 1
Doubly and singly symmetric members in flexure and tension.

The interaction of flexure and tension in symmetric shapes is limited by the following LRFD equations:

a. For $\dfrac{P_u}{\phi P_n} \geq 0.2$

$$\frac{P_u}{\phi P_n} + \frac{8}{9}\left(\frac{M_{ux}}{\phi_b M_{nx}} + \frac{M_{uy}}{\phi_b M_{ny}} \right) \leq 1.0 \qquad (7.24)$$
$$\text{(LRFD H1-1a)}$$

b. For $\dfrac{P_u}{\phi P_n} < 0.2$

$$\frac{P_u}{2\phi P_n} + \left(\frac{M_{ux}}{\phi_b M_{nx}} + \frac{M_{uy}}{\phi_b M_{ny}} \right) \leq 1.0 \qquad (7.25)$$
$$\text{(LRFD H1-1b)}$$

where P_u = Required tensile strength, kips
P_n = Nominal tensile strength determined in accordance with Section 7.2 of this chapter, kips
M_u = Required flexural strength determined in accordance with Section 7.4 of this chapter, kip-in.
M_n = Nominal flexural strength determined in accordance with Section 7.4 of this chapter, kip-in.

x = Subscript relating symbol to strong axis bending
y = Subscript relating symbol to weak axis bending
$\phi = \phi_t$ = Resistance factor for tension (see Section 7.2)
ϕ_b = Resistance factor for flexure = 0.9

EXAMPLE 7.16
Check the adequacy of a W12 × 45 tension member with no holes, consisting of 50 ksi steel and subjected to a factored tensile force P_u of 100 k and a factored bending moment M_{ux} of 90 kip-ft. The member has an L_b = 8 ft. Assume C_b = 1.0.

Solution
For the W12 × 45 section:

$$A = 13.2 \text{ in}^2, \quad Z_x = 64.7 \text{ in}^4, \quad L_p = 6.9 \text{ ft}, \quad L_r = 20.3 \text{ ft}$$
$$\phi_t P_n = \phi_t F_y A_g = (0.90)(50)(13.2) = 594 \text{ k}$$

$$\therefore \frac{P_u}{\phi_t P_n} = \frac{100}{594} = 0.17 < 0.2$$

Hence, Equation 7.24 (LRFD Equation H1-1b) applies.
Next, $L_b = 8$ ft $> L_p = 6.9$ ft $< L_r = 20.3$ ft from Load Factor Design Selection Table (Page 4-19, Part 4 of the LRFD manual), we have

$$\phi_b M_p = 243 \text{ kip-ft}, \quad BF = 5.07$$

$$\therefore \phi_b M_n = C_b[\phi_b M_p - BF(L_b - L_p)]$$
$$= 1.0[243 - 5.07(8.0 - 6.9)]$$
$$= 237.4 \text{ kip-ft}$$

From Equation 7.24 (LRFD H1-1b), we have

$$\frac{P_u}{2\phi P_n} + \left(\frac{M_{ux}}{\phi_b M_{nx}} + \frac{M_{uy}}{\phi_b M_{ny}}\right) = \frac{100}{(2)(594)} + \left(\frac{90}{237.4} + 0\right) = 0.46 < 1.0$$

The section is adequate. However, it is overdesigned. A smaller and more economical section can be selected.

Case 2
Doubly and singly symmetric members in flexure and compression.
The interaction of flexural and compression in symmetric shapes is limited by the same interaction equations for members in flexure and tension. These equations are repeated here for ready reference:

a. For $\dfrac{P_u}{\phi P_n} \geq 0.2$

$$\frac{P_u}{\phi P_n} + \frac{8}{9}\left(\frac{M_{ux}}{\phi_b M_{nx}} + \frac{M_{uy}}{\phi_b M_{ny}}\right) \leq 1.0 \quad\quad\quad (7.24)$$
$$\text{(LRFD H1-1a)}$$

b. For $\dfrac{P_u}{\phi P_n} < 0.2$

$$\dfrac{P_u}{2\phi P_n} + \left(\dfrac{M_{ux}}{\phi_b M_{nx}} + \dfrac{M_{uy}}{\phi_b M_{ny}}\right) \leq 1.0 \qquad (7.25)$$
(LRFD H1-1b)

where P_u = Required tensile strength, kips
P_n = Nominal tensile strength determined in accordance with Section 7.2 of this chapter, kips
M_u = Required flexural strength determined in accordance with Section 7.4 of this chapter, kip-in.
M_n = Nominal flexural strength determined in accordance with Section 7.4 of this chapter, kip-in.
x = Subscript relating symbol to strong axis bending
y = Subscript relating symbol to weak axis bending
$\phi = \phi_c$ = resistance factor for compression = 0.85 (see Section 7.3)
ϕ_b = Resistance factor for flexure = 0.9

Before the above interaction equations can be applied, we need to find the second order ($P\Delta$) effects in the design frame. When a beam column is subject to flexural and compression between its unbraced length, the beam will deflect laterally in the plane of bending, a secondary moment is caused by the compression load times the lateral deflection ($P\Delta$) as shown in Figure 7.12. This secondary moment causes additional deflection and a larger column moment, which, in turn, causes more deflection and moment until equilibrium is reached.

In structures designed on the basis of elastic analysis, M_u for beam-columns, connections, and connected members shall be determined from a second-order elastic analysis or from the following approximate second-order analysis procedure provided by the LRFD specification:

$$M_u = B_1 M_{nt} + B_2 M_{lt} \qquad (7.26)$$
(LRFD C1-1)

where M_{nt} = Required flexural strength in member assuming there is no lateral translation of the frame, kip-in.
M_{lt} = Required flexural strength in member as a result of lateral translation of the frame only, kip-in.

$$B_1 = \dfrac{C_m}{1 - \dfrac{P_u}{P_{e1}}} \geq 1 \qquad (7.27)$$
(LRFD C1-2)

$P_{e1} = \dfrac{A_g F_y}{\lambda_c^2}$ where λ_c is the slenderness parameter, in which the effective length factor K in the plane of bending is taken as unity.

$\lambda_c = \dfrac{Kl}{r\pi}\sqrt{\dfrac{F_y}{E}}$

$P_{e1} = \dfrac{\pi^2 EI}{(Kl)^2}$

(a) Braced Against Side-sway

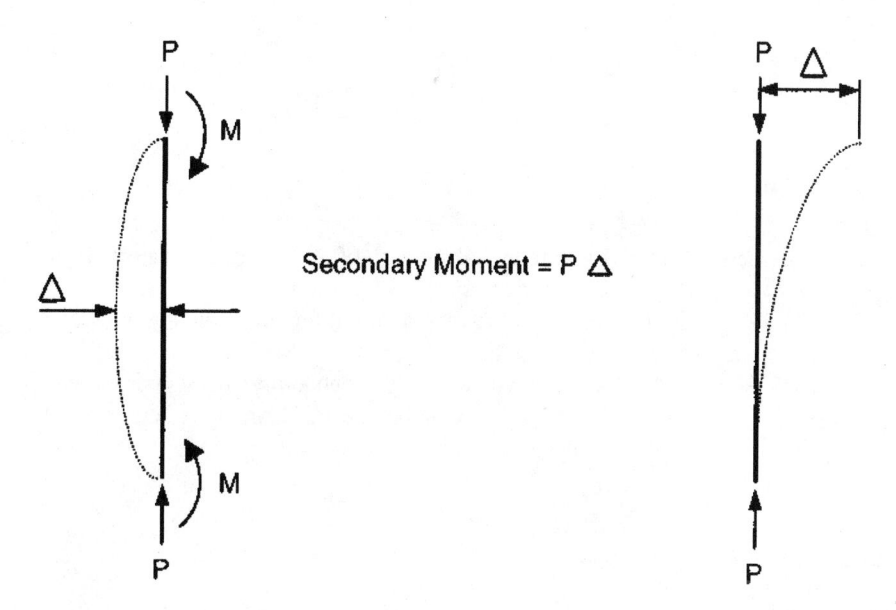

(b) Unbraced Against Side-sway

FIGURE 7.12

P_u = Required axial compressive strength for the member under consideration, kips
C_m = Coefficient based on elastic first-order analysis assuming no lateral translation of the frame. The value of C_m is taken as follows:

a. For compression members not subject to transverse loading between their supports in the plane of bending

$$C_m = 0.6 - 0.4\left(\frac{M_1}{M_2}\right) \quad (7.28)$$
$$\text{(LRFD C1-3)}$$

where M_1/M_2 is the ratio of the smaller to larger moments at the ends of that portion of the member unbraced in the plane of bending under consideration. M_1/M_2 is positive when the member is bent in reverse curvature, negative when bent in single curvature.

b. For compression members subjected to transverse loading between their supports, the value of C_m is determined either by rational analysis or by the use of the following values:

for members whose ends are restrained, $C_m = 0.85$
for members whose ends are unrestrained, $C_m = 1.00$

$$B_2 = \frac{1}{1 - \sum P_u \left(\dfrac{\Delta_{oh}}{\sum HL}\right)} \quad (7.29)$$
$$\text{(LRFD C1-4)}$$

or

$$B_2 = \frac{1}{1 - \frac{\sum P_u}{\sum P_{e2}}} \qquad (7.30)$$
(LRFD C1-5)

where $\sum P_u$ = Required axial strength of all columns in a story, kips
Δ_{oh} = Lateral inter-story deflection, in.
$\sum H$ = Sum of all story horizontal forces producing, kips
L = Story height, in.
$P_{e2} = A_g F_y / \lambda_c^2$ where λ_c is the slenderness parameter, in which the effective length factor K in the plane of bending is determined by structural analysis and is given by

$$\lambda_c = \frac{Kl}{r\pi}\sqrt{\frac{F_y}{E}}$$

$$P_{e2} = \frac{\pi^2 EI}{(Kl)^2}$$

EXAMPLE 7.17

A 13-ft-long W12 × 96 of A588 steel is used as a beam column in a braced frame. It is bent in single curvature, with equal but opposite end moments, and is not subjected to intermediate transverse loads. Determine whether this member is adequate for $P_u = 550$ k and first-order moment $M_{ntx} = 150$ kip-ft. See Figure 7.13.

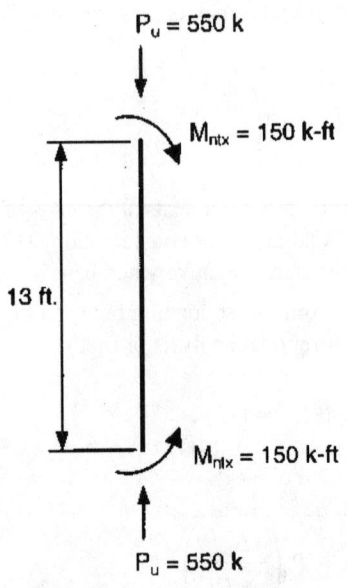

FIGURE 7.13

Solution

From Table of W Shapes Dimensions of Part 1, LRFD manual (Pages 1-38 and 1-39), we have

$$A = 28.2 \text{ in}^2, \quad I_x = 833 \text{ in}^4, \quad r_x = 5.44 \text{ in.}$$

From the Load Factor Design Selection Table of Part 4, the LRFD manual (Page 4-18), we have for $F_y = 50$ ksi

$$L_p' = 10.9 \text{ ft}, \quad L_r = 41.3 \text{ ft}, \quad BF = 5.20 \text{ kips}, \quad \phi_b M_p = 551 \text{ kip-ft}$$

Since

$$L_b = 13 \text{ ft} > L_p = 10.9 \text{ ft}, \quad \phi_b M_{nx} = Cb[\phi_b M_p - BF(L_b - L_p)]$$
$$= 1.0[551 - 5.2(13.0 - 10.9)] = 540.1 \text{ kip-ft}$$

Next, for a braced frame, $K = 1.0$

$$K_x L_x = K_y L_y = (1.0)(13.0) = 13 \text{ ft}$$

From Column Tables in Part 3 of the LRFD manual (Page 3-24), we have $\phi_c P_n = 995$ k (for $KL = 13$ ft and $F_y = 50$ ksi.)
Hence,

$$\frac{P_u}{\phi_c P_n} = \frac{550}{995} = 0.553 > 0.2$$

∴ Must use interaction Equation 7.24 (LRFD Equation H1-1a).

However, before applying the interaction equation, we must find the magnification factor B_1. $B_2 = 0$, since there is no side-sway.

$$\therefore M_{ux} = B_1 M_{ntx}$$

$$C_m = 0.6 - 0.4\left(\frac{M_1}{M_2}\right) = 0.6 - 0.4\left(-\frac{150}{150}\right) = 1.0$$

$$P_{e1} = \frac{\pi^2 E I_x}{(Kl)^2} = \frac{(\pi^2)(29{,}000)(833)}{(12 \times 13)^2} = 9805 \text{ k}$$

P_{e1} can be estimated from the bottom portion of the Column Tables in Part 3, LRFD manual (Page 3-24). The value is given as $P_{ex}(KL)^2/10^4 = 23{,}900$, which gives $P_{ex} \cong 9820$ k for $KL = 12 \times 13$.

The value of P_{e1} also can be obtained from Table 8 in Part 6 of the LRFD manual (Page 6-154). The values in Table 8 are based on $P_e/A_g = \pi^2 E /(Kl/r)^2$. We have to find $K_x L_x/r_x$ and then pick P_e/A_g from the table. For this example, $K_x L_x/r_x = 12 \times 13/5.44 = 28.7$. By interpolation, $P_e/A_g = 347.75$, giving $P_e = 347.75 \times A_g = 347.75 \times 28.2 = 9807$ k

$$B_1 = \frac{C_m}{1 - \dfrac{P_u}{P_{e1}}} = \frac{1.0}{1 - \dfrac{550}{9805}} = 1.059$$

$$\therefore M_{ux} = B_1 M_{ntx} = (1.059)(150) = 158.9 \text{ kip-ft}$$

Now we can apply Equation 7.27 (LRFD Equation H1-1a) to check the adequacy of the section:

$$\frac{P_u}{\phi P_n} + \frac{8}{9}\left(\frac{M_{ux}}{\phi_b M_{nx}} + \frac{M_{uy}}{\phi_b M_{ny}}\right) \leq 1.0$$

$$= \frac{550}{995} + \frac{8}{9}\left(\frac{158.9}{540.1} + \frac{0}{\phi_b M_{ny}}\right) = 0.815 < 1.$$

$\therefore$ The member is adequate.

EXAMPLE 7.18

A 13-ft-long W12 × 96 of A588 steel is used as a beam column in an unbraced frame. It is bent in reverse curvature with equal but opposite end moments, and is not subjected to intermediate transverse loads. Determine whether this member is adequate for $P_u = 550$ k and first order moment $M_{ntx} = 150$ kip-ft, and $M_{ltx} = 100$ kip-ft. Assume that the total factored gravity load ΣP_u for this case is 5500 k, and $\Sigma P_{ex} = 20{,}000$ k $K_x = 1.2$.

Solution

From Table of W Shapes Dimensions of Part 1 in the LRFD manual (Pages 1-38 and 1-39), we have $A = 28.2$ in², $I_x = 833$ in⁴, $r_x = 5.44$ in.

From Load Factor Design Selection Table of Part 4 in the LRFD manual (p 4-18), we have, for $F_y = 50$ ksi,

$$Z_x = 147 \text{ in}^3, \quad L_p = 10.9 \text{ ft}, \quad L_r = 41.3 \text{ ft}, \quad BF = 5.20 \text{ kips}, \quad \phi_b M_p = 551 \text{ kip-ft}$$

$$K_x L_x = (1.2)(13) = 15.6 \text{ ft}$$

From Column Tables in Part 3 in the LRFD manual (Page 3-24), we have $\phi_c P_n = 916.4$ k (for $KL = 15.6$ ft and $F_y = 50$ ksi).

Hence,

$$\frac{P_u}{\phi_c P_n} = \frac{550}{916.4} = 0.600 > 0.2$$

$\therefore$ Must use interaction Equation 7.24 (LRFD Equation H1-la).

However, before applying the interaction equation, we must find magnification factors B_1 and B_2.

$$\therefore M_{ux} = B_1 M_{ntx} + B_2 M_{ltx}$$

$$C_m = 0.6 - 0.4\left(\frac{M_1}{M_2}\right) = 0.6 - 0.4(+1.0) = 0.2$$

$$P_{elx} = \frac{\pi^2 EI_x}{(Kl)^2} = \frac{(\pi^2)(29{,}000)(833)}{(12 \times 13)^2} = 9805 \text{ k}$$

$$B_{1x} = \frac{C_m}{(1 - P_u/P_{elx})} = \frac{0.2}{1 - 550/9805} = 0.212 < 1.0 \quad \text{Use } 1.0$$

$$B_{2x} = \frac{1}{1 - \dfrac{\sum P_u}{\sum P_{e2x}}} = \frac{1}{1 - \dfrac{5500}{20{,}000}} = 1.379$$

Hence,

$$\therefore M_{ux} = B_1 M_{ntx} + B_2 M_{ltx}$$
$$= (1.0)(150) + (1.379)(100) = 287.9 \text{ k}$$

$$L_b = 13 \text{ ft} > L_p = 10.9 \text{ ft, but} < L_r = 41.3 \text{ ft}$$

$$\phi_b M_{nx} = Cb[\phi_b M_p - BF(L_b - L_p)]$$
$$= (0.2\,[551 - 5.2(13.0 - 10.9)] - 108.1 \text{ kip-ft}$$

Applying the interaction Equation 7.24 (LRFD Equation H1-1a), we have

$$\frac{P_u}{\phi P_n} + \frac{8}{9}\left(\frac{M_{ux}}{\phi_b M_{nx}} + \frac{M_{uy}}{\phi_b M_{ny}}\right) \le 1.0$$

$$\frac{550}{916.4} + \frac{8}{9}\left(\frac{287.9}{108.1} + \frac{0}{\phi_b M_{ny}}\right) = 2.967 \gg 1.0$$

$\therefore$ The member is not adequate. A larger section will have to be selected.

7.7 BOLTED CONNECTIONS

This section covers the use of high-strength bolts in structural connections. Bolted connections require only a two-person crew for installation; there is no need for skilled labor. Bolted connections can be made in the field at a much faster pace than riveting and welding. High-strength bolts must conform to the provisions of the Load and Resistance Factor Design Specifications for Structural Joints Using ASTM A325 or A490 Bolts, as approved by the Research Council on Structural Connections. The specification is given in Part 6 of the LRFD manual, starting on Page 6-371. High-strength bolts are available in $\frac{1}{2}$-in. to $1\frac{1}{2}$-in. diameters. However, the nominal sizes of $\frac{3}{4}$-in., $\frac{7}{8}$-in. and 1-in. diameters usually are preferred for availability and practical applications.

The LRFD manual, Volume 1, Part 6, Chapter J, outlines the provisions for connections, joints, and fasteners. The readers should refer to Chapter J for the specifications covering the materials, design, and installation of high-strength bolts. The LRFD manual, Volume 2, Part 8, contains general information, design considerations, examples, and design aids for the design of high-strength bolts. The examples and the design aids are especially helpful to understand the specifications and designing bolted connections simply and quickly. They will help solve problems expeditiously and score points in the PE examination.

In the LRFD manual, various types of high-strength bolts are abbreviated as follows:

A325-SC = Slip-critical or fully tensioned A325 bolts

A325-N = Snug-tight or bearing A325 bolts with threads included in the shear planes

A325-X = Snug-tight or bearing A325 bolts with threads excluded from the shear planes.

Similar abbreviations are used for A490 bolts. We will discuss the applications of high-strength in the following sections.

7.7.1 General Provisions

Bolted connections consist of the elements of the connected members, such as beam webs, gussets, angles, brackets, and the bolts. These components must be proportioned so their design strength is equal to or greater than the required strength determined by design and analysis. The connections can be simple or moment-resistant. Simple connections of beams must be designed as flexible beam connections that can accommodate end rotations of unrestrained beams. To accomplish end rotation, some inelastic deformation in the connection is permitted. The connections can be proportioned for the reaction shears only. Moment-resistant or moment connections are designed for the combined effect of forces resulting from moment and shear induced by the rigidity of the connections.

7.7.2 Snug-Tight and Full-Tensioned Bolts

High-strength bolts can be tightened to *snug-tight* or *full-tensioned*. Snug-tight is a term referring to the tightening of the bolts by the full effort of a person using a spud wrench, or the tightness obtained after a few impacts of an impact wrench. This is not a numerical measure of the force required to tighten the bolts, but a general indication of the tightness of the bolts. Snug-tight bolts must be marked clearly on the design plans and the erection drawings. Full-tensioned bolts are high-strength bolts tightened further, after reaching snug-tight condition, by one of the following methods.

1. Turn-of-the-nut method: The bolts are brought to a snug-tight condition and then, with an impact wrench or hand wrench, given from one-third to one full turn, depending on the length and the slope of the surfaces under the heads and nuts. The actual amount of turn is given in Table 5 Nut Rotation from Snug-Tight Condition in Part 6 of the LRFD manual (Page 6-385). The amount of turn can be controlled by marking the snug-tight condition with paint or crayon.

2. Calibrated wrench method: The bolts are tightened to the specified tension by a calibrated wrench that halts tightening after reaching the specified tension. The wrench must be calibrated daily, and a hardened washer must be used under the element turned. The calibrated wrench is set to provide a tension not less than 5 percent in excess of the minimum tension specified in Table 4 in Part 6 of the LRFD manual (Page 6-380).
3. Direct-tension indicator tightening: The direct-tension indicator consists of a hardened washer which has protrusions on one face. As the bolt is tightened, the protrusions flatten. The gap between the washer and the connecting element gives a measure of the bolt tension. The manufacturer of the indicator furnishes a feeler gage for measuring the gap to achieve fully tensioned bolts.

7.7.3 Types of Connections

Bolted joints may be divided into bearing-type connections and slip-critical connections. Bearing-type connections are divided further into snug-tightened bearing-type connections and fully tensioned bearing-type connections. However, the friction between the connected parts is not a consideration in either the snug-tight or fully tensioned bearing-type connections. As the name implies, bearing-type connections rely on bearings for the strength of the connections. There are practical cases where slip of the connection is undesirable for service-load conditions or for the serviceability of the structure, such as joints subject to fatigue stresses, heavy impact loads, and severe vibrations. Slip-critical connections rely on the clamping force and the friction between the faying surfaces for slip resistance. High-strength bolts used in slip-critical connections must be fully tensioned, and the coefficient of friction μ of the faying surfaces must be as specified in the LRFD specification or determined by tests. The LRFD specification defines three classes of surfaces with their corresponding slip coefficients. Clean-mill scale with no coating is defined as a Class A surface with $\mu = 0.33$. Blast-cleaned surfaces with no coatings are defined as a Class B surface with $\mu = 0.50$. Hot-dip galvanized and roughened surfaces are defined as Class C surface with $\mu = 0.40$.

In bolted connections subject to shear, the load is transferred between the connected parts by friction up to a certain level of force that is dependent upon the total clamping force on the faying surfaces and the coefficient of friction of the faying surfaces. At this point, the bolts are not subject to shear, nor is the connected material subject to bearing stress. As loading is increased to a level in excess of the frictional resistance between the faying surfaces, slip occurs, but rupture does not occur. As a higher load is applied, the load is resisted by shear in the fastener and bearing on the connected material plus some uncertain amount of friction between the faying surfaces. The final failure will be by shear failure of the bolts, tear-out of the connected material, or unacceptable ovalization of the holes. Final failure load is independent of the clamping force provided by the bolts.

The size of the bolt hole affects the performance and strength of a joint. A standard bolt hole is $\frac{1}{16}$ in. larger in diameter than the bolt. There are three types of enlarged holes: oversized, short-slotted, and long-slotted. The nominal hole dimensions are shown in Table J3.3 in Part 6 of the LRFD manual (Page 6-82). Oversized holes are allowed in any or all plies of slip-critical connections, but not allowed in bearing-type connections. Short-slotted holes are allowed in any or all plies of slip-critical or bearing-type connections. Long-slotted holes are allowed in only one of the connected parts of either a slip-critical or bearing-type connection at an individual faying surface. Long-slotted holes are permitted to be used without regard to direction of loading in slip-critical connections, but must be normal to the direction of load in bearing-type connections.

The design of high-strength bolted connections considers the strength required to prevent premature failure by shear of the bolts and bearing failure of the connection material. In the case of slip-critical connections, resistance to slip load is checked.

Table J3.2 in Part 6 of the LRFD manual (Page 6-81) shows the design strength of fasteners, and Table J3.6 in Part 6 of the LRFD manual (Page 6-84) shows the slip-critical nominal resistance to shear of high-strength bolts.

7.7.4 Minimum Spacing and Edge Distance

The distance between centers of standard, oversized, or slotted holes, must not be less than $2\frac{2}{3}$ times the nominal diameter of the fastener; a distance of 3d is preferred. The distance from the center of a standard hole to an edge of a connected part must not be less than the applicable value from Table J3.4 in Part 6 of the LRFD manual (Page 6-82). The distance from the center of an oversized, or slotted, hole to an edge of a connected part must not be less than that required for a standard hole plus the applicable increment C_2 from Table J3.8 in Part 6 of the LRFD manual (Page 6-86).

7.7.5 Maximum Spacing and Edge Distances

The LRFD specification imposes maximum spacing and edge distances to ensure the joint is watertight. Moisture causes failure of the coating in the faying surfaces. Corrosion will develop and cause separation of the connected parts, resulting in further moisture penetration and corrosion. The presence of moisture in the joint of weathering steel is even more critical. The constant moisture in the joint will disrupt the weathering process and the weathering steel will continue to corrode.

The maximum distance from the center of any bolt to the nearest edge of parts in contact is limited to 12 times the thickness of the connected part under consideration, but must not exceed six inches. The longitudinal spacing of connectors between elements in continuous contact consisting of a plate and a shape or two plates is limited as follows:

a. For painted members, or unpainted members not subject to corrosion, the spacing must not exceed 24 times the thickness of the thinner plate or 12 inches.

b. For unpainted members of weathering steel subject to atmospheric corrosion, the spacing shall not exceed 14 times the thickness of the thinner plate or seven inches.

7.7.6 Minimum Strength of Connections

Bolted connections must be designed to provide sufficient strength to support a factored load of not less than 10 kips.

7.7.7 Design Tension or Shear Strength

The design tension or shear strength of a high-strength bolt or threaded part is $\phi F_n A_b$

where ϕ = Resistance factor = 0.75
F_n = Nominal tensile strength F_t or shear strength, F_v, tabulated in Table J3.2, ksi.
A_b = Nominal unthreaded body area (gross area) of bolt, in^2

For example, for a 1-in. A325 bolt, $A_b = 0.785$ in^2, $F_n = F_t = 90$ ksi (from Table J3.2) and $F_n = F_v = 48$ ksi when threads are not excluded from shear planes and 60 ksi when threads are excluded from shear planes.

7.7.8 Combined Tension and Shear in Bearing-Type Connections

The design strength of a bolt subject to combined tension and shear is $\phi F_t A_b$, where $\phi = 0.75$. The nominal tension stress F_t must be computed from the equations in Table J3.5 in Part 6 of the LRFD manual as a function of f_v, the required shear stress produced by the factored loads. The design shear strength ϕF_v, tabulated in Table J3.2, must equal or exceed the shear stress, f_v.

EXAMPLE 7.19
The bearing-type connection in Figure 7.14 is subject to a load $P_u = 200$ kips, which passes through the center of gravity of the bolt group. Is the connection adequate to resist the load?

Solution
Resolving the load P_u into vertical component P_v and horizontal component P_h we have $A_b = 0.442$ in^2 (gross area of $\frac{3}{4}$-in. bolt, Table 8-7 of LRFD manual, Volume 2 (Page 8-17)

$$P_v = \frac{3}{5} \times 200 = 120 \text{ k}$$

$$P_h = \frac{4}{5} \times 200 = 160 \text{ k}$$

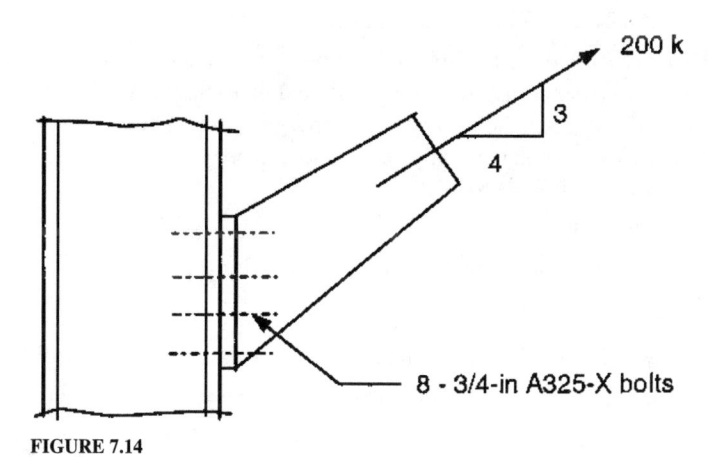

FIGURE 7.14

Hence,

$$\text{Shear stress}, f_v = \frac{120}{8} \times 0.442 = 33.94 \text{ ksi}$$

$$\text{Tensile stress}, f_t = \frac{160}{8} \times 0.442 = 45.25 \text{ ksi}$$

From Table J3.5, the limiting tensile stress

$$\phi F_t = 0.75[117 - (1.50)(33.94)]$$
$$= 49.57 \text{ ksi} > 45.25 \text{ ksi} \quad \text{OK}.$$

From Table J3.2,

$$\text{The design shear strength } \phi F_v = 60 \text{ ksi} > 33.94 \text{ ksi} \quad \text{OK}.$$

The connection is adequate.

7.7.9 Bearing Strength at Bolt Holes

The design bearing strength at bolt holes is ϕR_n,

where $\phi = 0.75$
R_n = Nominal bearing strength

Bearing strength must be checked for both bearing-type and slip-critical connections. The use of oversize holes and short- and long-slotted holes parallel to the line of force is restricted to slip-critical connections as outlined in Section 7.7.3.

The following terms are used in cases (a) and (b) described below:

L_e = Distance (in.) along the line of force from the edge of the connected part to the center of a standard hole or the center of a short- and long-slotted hole perpendicular to the line of force. For oversize holes and short- and long-slotted holes parallel to the line of force, L_e must be increased by increment C_2 of Table J3.8 in Part 6 of the LRFD manual (Page 6-86).

s = Distance (in.) along the line of force between centers of standard holes, or between centers of short- and long-slotted holes perpendicular to the line of force. For oversize holes, and short- and long-slotted holes parallel to the line of force, s must be increased by spacing increment C_1 of Table J3.7 in Part 6 of the LRFD manual (Page 6-86).

d = Diameter of bolt, in.

F_u = Specified minimum tensile strength of the critical part, ksi

T = Thickness of the critical connected part, in.

a. When $L_e \geq 1.5d$ and $s \geq 3d$ and there are two or more bolts in line of force:

When deformation around the bolt holes is a design consideration

$$R_n = 2.4dtF_u \tag{7.30}$$
$$\text{(LRFD J3-1a)}$$

When deformation around the bolt holes is not a design consideration, for the bolt nearest the edge

$$R_n = 2.4dtF_u \leq 3dtF_u \qquad (7.31)$$
$$(\text{LRFD J3-1b})$$

and for the remaining bolts

$$R_n = (s - d/2)tF_u \leq 3.0dtF_u \qquad (7.32)$$
$$(\text{LRFD J3-1c})$$

For long-slotted bolt holes perpendicular to the line of force:

$$R_n = 2.0dtF_u \qquad (7.33)$$
$$(\text{LRFD J3-1d})$$

b. When $L_e < 1.5d$ and $s < 3d$ or for a single bolt in the line of force:

For a single bolt hole or the bolt hole nearest the edge when there are two or more bolt holes in the line of force

$$R_n = L_e tF_u \leq 2.4dtF_u \qquad (7.34)$$
$$(\text{LRFD J3-2a})$$

For the remaining bolt holes

$$R_n = (s - d/2)tF_u \leq 2.4dtF_u \qquad (7.35)$$
$$(\text{LRFD J3-2b})$$

For long-slotted bolt holes perpendicular to the line of force:

For a single bolt hole or the bolt hole nearest the edge where there are two or more bolt holes in the line of force

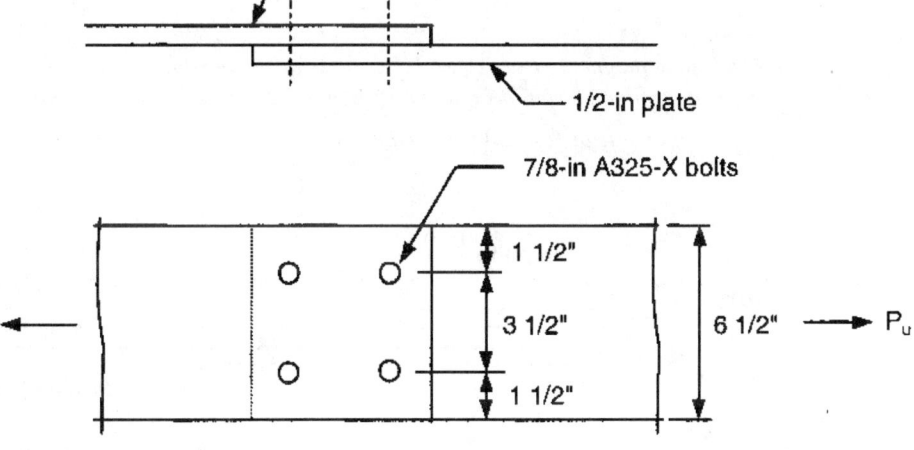

$$R_n = L_e tF_u \leq 2.4dtF_u \qquad (7.36)$$
$$(\text{LRFD J3-2c})$$

FIGURE 7.15

For the remaining bolt holes
$$R_n = (s - d/2)tF_u \leq 2.0dtF_u \qquad (7.37)$$
$$\text{(LRFD J3-2d)}$$

EXAMPLE 7.20
Compute the design strength P_u of the bearing-type connection shown in Figure 7.15. The steel is A588 with $F_y = 50$ ksi and $F_u = 70$ ksi, the bolts are $\frac{7}{8}$-in. diameter A325-X and standard size holes. The edge distances are $> 1\frac{1}{2}d$ and the distance center-to-center of the holes is $> 3d$.

Solution
Design strength of plates:

$$A_g = \left(\frac{1}{2}\right)(10) = 5.0 \text{ in}^2$$

$$A_n = 5.0 - (2)\left(\frac{7}{8} + \frac{1}{8}\right)\left(\frac{1}{2}\right) = 4.0 \text{ in}^2 = A_e$$

$$P_u = \phi_t F_y A_g = (0.9)(50)(5.0) = 225 \text{ k}$$

$$P_u = \phi_t F_u A_e = (0.75)(70)(4.0) = 210 \text{ k}$$

Bolts in single shear:

$$P_u = \phi F_n A_b = (0.75)(60)(0.601)(4 \text{ bolts}) = 108.2 \text{ k}$$

Bolts in bearing on $\frac{1}{2}$-in. plate:

$$P_u = \phi 2.4 dt F_u = (0.75)(2.4)\left(\frac{7}{8}\right)\left(\frac{1}{2}\right)(70)(4 \text{ bolts})$$
$$= 220.5 \text{ k}$$

Hence, design $P_u = 108.2$ k.   *Controls by shear*

EXAMPLE 7.21
Determine the number of 1-in. A325-X bolts in standard holes required for the bearing-type connection shown in Figure 7.16 Use A572 Grade 50 steel and edge distances $= 1\frac{1}{2}d$ and the distances center-to-center of bolts $= 3d$.

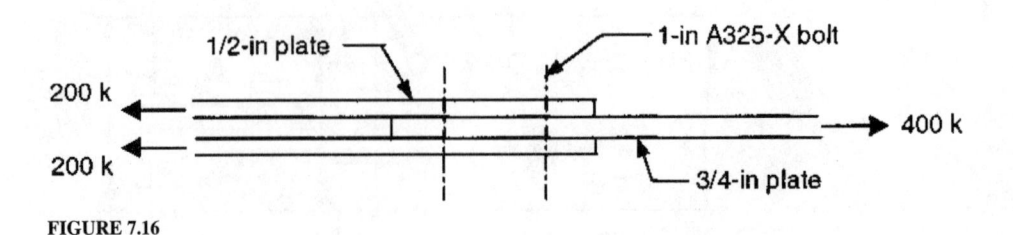

FIGURE 7.16

Solution

Gross area A_b of 1-in. bolt = 0.785 in^2

The bolts are in double shear.

∴ Design shear strength per bolt = (0.75)(2 × 0.785)(60) = 70.65 k

(The same value may be obtained from Table 8-11 of LRFD manual.)

The bolts are in bearing on $\frac{3}{4}$-in. plate.

∴ Design bearing strength per bolt = (0.75)(2.4)(1)(3/4)(65) = 87.75 k

The design shear strength controls the design of the connection.

(The same value may be obtained from Table 8-13 of LRFD manual.)

Hence,

$$\text{Number of bolts required} = \frac{400}{70.65} = 5.7$$

Use six bolts for the connection.

7.7.10 Slip-Critical Connections Designed at Service Loads

The design for shear of high-strength bolts in slip-critical connections is given below, and must be checked for bearing at factored loads.

The design resistance to shear of a bolt in a slip-critical connection is

$$\phi R_n = \phi F_v A_b \tag{7.38}$$

where $\phi = 1.0$ for standard, oversized, short-slotted, and long-slotted holes when the long slot is perpendicular to the line of force

$\phi = 0.85$ for long-slotted holes when the long slot is parallel to the line of force

F_v = Nominal slip-critical shear resistance tabulated in Table J3.6 of the LRFD manual (Page 6-84), ksi. The values for F_v in Table J3.6 are based on Class A (slip coefficient 0.33), clean-mill scale and blast-cleaned surfaces with Class A coatings. For other classes of finishes, the values must be adjusted accordingly.

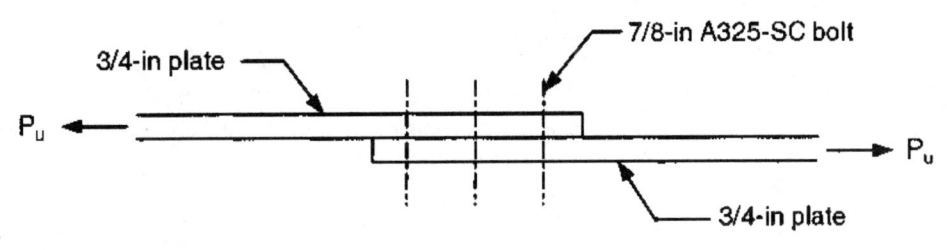

FIGURE 7.17

The design resistance to shear must equal or exceed the shear on the bolt due to service loads. When the loading combination includes wind loads, in addition to dead and live loads, the total shear on the bolt due to combined load effects, at service load, can be multiplied by 0.75.

EXAMPLE 7.22

Design a slip-critical connection for the $\frac{3}{4}$-in. A588 ($F_y = 50$ ksi and $F_u = 70$ ksi) steel plates shown in Figure 7.17 to resist the axial service loads $P_D = 40$ k and $P_L = 50$ k, using $\frac{7}{8}$-in. A325-SC bolts with threads excluded from the shear plane and with standard size holes. $L_e \geq 1\frac{1}{2}d$ and distance center-to-center of bolts $\geq 3d$.

Solution

Slip-Critical Design at Service Loads: bolts in shear, the design shear resistance per bolt is given by

$$\phi F_v A_b = (1.0)(17)(0.601) = 10.2 \text{ k per bolt}$$

$$\text{Number of bolts required} = \frac{(40 + 50)}{10.2} = 8.8; \text{ use 9 bolts.}$$

Check as bearing-type connection at factored load.

$$\text{Factored load } P_u = (1.2)(40) + (1.6)(50) = 128 \text{ k}$$

$$\text{Bolts in single shear} = (0.75)(0.601)(60) = 27.0 \text{ k}$$

$$\text{Bolts in bearing} = (0.75)(2.4)\left(\frac{7}{8}\right)\left(\frac{3}{4}\right)(70) = 82.9 \text{ k}$$

$$\text{Number of bolts required} = \frac{128}{27} = 4.5; \text{ (5 bolts)}$$

Use 9 bolts. OK.

7.7.11 Design Rupture Strength

a. Shear Rupture Strength. The design strength for the limit state of rupture along a shear failure path in the effected elements of connected members can be taken as ϕR_n

where $\phi = 0.75$
$R_n = 0.6 F_u A_{nv}$ (7.39)
(LRFD J4-1)

A_n = Net area subject to shear, in^2

b. Tension Rupture Strength. The design strength for the limit state of rupture along a tension path in the affected elements of connected members can be taken as ϕR_n

where $\phi = 0.75$
$R_n = F_u A_{nt}$ (7.40)
(LRFD J4-2)

A_{nt} = Net area subject to tension, in^2

STEEL DESIGN

c. Block Shear Rupture Strength. Block shear is a limit state in which the resistance is determined by the sum of the shear strength on a failure path and the tensile strength on a perpendicular segment. It must be checked at beam end connections where the top flange is coped and in similar situations, such as tension members and gusset plates. When ultimate rupture strength on the net section is used to determine the resistance on one segment, yielding on the gross section shall be used on the perpendicular segment.

The block shear rupture design strength ϕR_n can be determined as follows:

1. When $F_u A_{nt} \geq 0.6 F_u A_{nv}$

$$\phi R_n = \phi[0.6 F_y A_{gv} + F_u A_{nt}] \qquad (7.41)$$
$$(\text{LRFD J4-3a})$$

2. When $0.6 F_u A_{nv} > F_u A_{nt}$

$$\phi R_n = \phi[0.6 F_u A_{nv} + F_y A_{gt}] \qquad (7.42)$$
$$(\text{LRFD J4-3b})$$

where $\phi = 0.75$
A_{gv} = Gross area subject to shear, in^2
A_{gt} = Gross area subject to tension, in^2
A_{nv} = Net area subjected to shear, in^2
A_{nt} = Net area subjected to tension, in^2

EXAMPLE 7.23

A W14 × 43 beam is connected to A36 gusset plates as shown in Figure 7.18. Standard holes use $\frac{7}{8}$-in. A325-X bearing-type bolts. Compute (a) the tensile strength of the W-section and the gusset plates, (b) the strength of the bolts in single shear and bearing, and (c) the block shear strength of the W-section.

Solution

For A36 W14 × 43 beam, $A_g = 12.6$ in^2, flange width, $b_f = 7.995$ in, $t_f = 0.53$ in.

$$F_y = 36 \text{ ksi}, \qquad F_u = 58 \text{ ksi}$$

a. Tensile design strength of W-beam

$$P_u = \phi_t F_y A_g = (0.9)(36)(12.6) = 408 \text{ k} > 340 \text{ k} \quad \text{OK.}$$

$$A_n = 12.6 - (4)\left(\frac{7}{8} + \frac{1}{8}\right)(0.53) = 10.48 \text{ in.}$$

$$U = 1 - \frac{\bar{x}}{L} = 1 - \frac{1.31}{6} = 0.78$$

$$\therefore A_e = (0.78)(10.48) = 8.17 \text{ in}^2$$

$$P_u = \phi_t F_u A_e = (0.75)(58)(8.17) = 355 \text{ k} > 340 \text{ k} \quad \text{OK.}$$

Tensile design strength of gusset plates

$$P_u = \phi_t F_y A_g = (0.9)(36)\left(\frac{1}{2} \times 11\right)(2) = 356.4 \text{ k} > 340 \text{ k} \quad \text{OK.}$$

$$A_n = 2\left[\left(\frac{1}{2} \times 11\right) - (2 \times 1 \times 1/2)\right] = 9 \text{ in}^2$$

$$0.85 A_g = (0.85)\left(\frac{1}{2}\right)(11)(2) = 9.35 \text{ in}^2$$

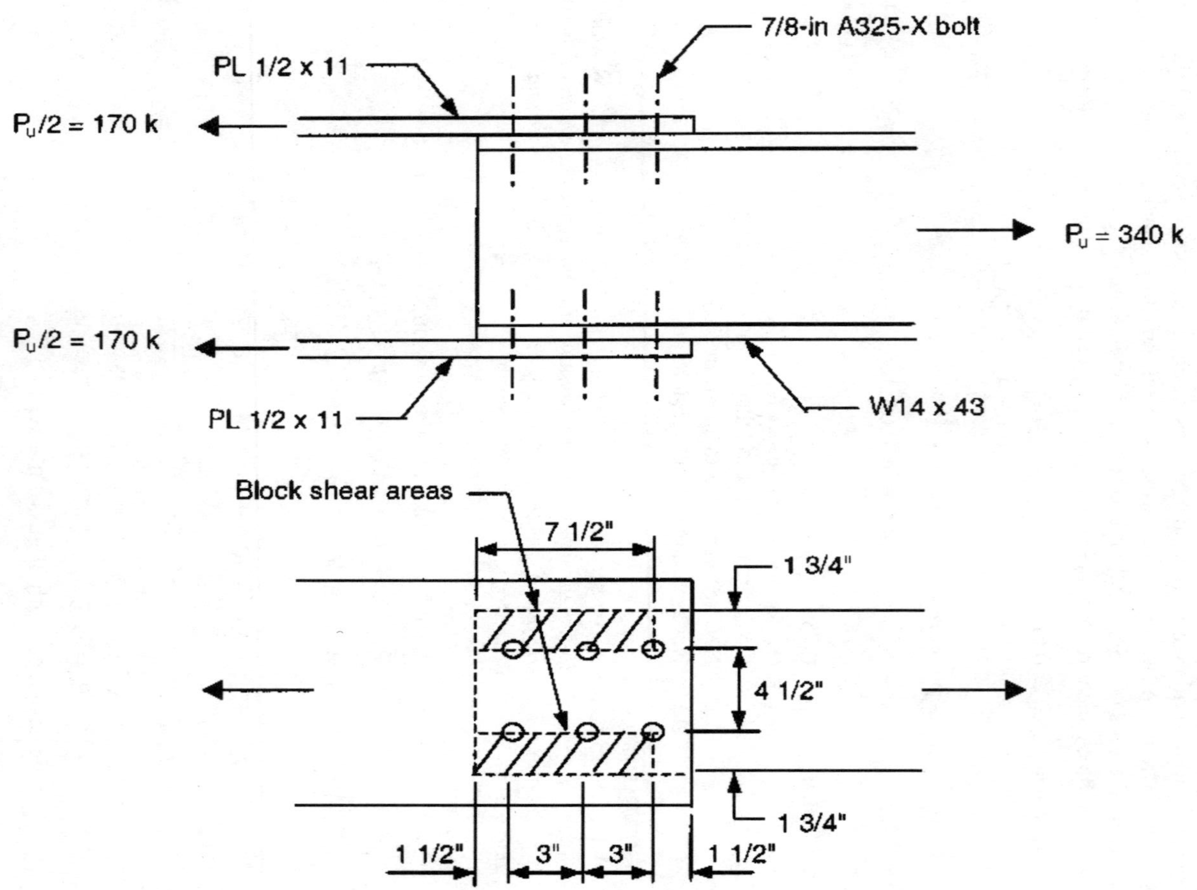

FIGURE 7.18

Hence,
$$P_u = \phi_t F_u A_n = (0.75)(58)(9) = 391.5 \text{ k} > 340 \text{ k} \quad \text{OK.}$$

b. Strength of bolts in single shear and bearing

Bolts in single shear (total 12 bolts):
$$P_u = \phi_t F_n A_b = (0.75)(60)(0.601)(12 \text{ bolts}) = 324.5 \text{ k} < 340 \text{ k} \quad \text{No good}$$

Bolts in bearing on $\frac{1}{2}$-in plate:
$$P_u = \phi 2.4 dt F_u = (0.75)(2.4)\left(\frac{7}{8}\right)\left(\frac{1}{2}\right)(58)(12 \text{ bolts})$$
$$= 548.1 \text{ k} > 340 \text{ k} \quad \text{OK.}$$

c. Blocking shear strength of W-section

The block shear areas are shown shaded in Figure 7.18.

The total areas for the top and bottom flanges are:

A_{gv} = gross area subject to shear, in^2
    = (4)(0.53)(7.5) = 15.9 in^2

A_{gt} = gross area subject to tension, in^2
    = (4)(0.53)(1.75) = 3.71 in^2

A_{nv} = net area subjected to shear, in^2
    = (4)(7.5 − 2.5 × 1.0)(0.53) = 10.6 in^2

A_{nt} = net area subjected to tension, in^2
    = (4)(1.75 − 0.5 × 1)(0.53) = 2.65 in^2

Hence, $F_u A_{nt} = (58)(2.65) = 153.7$ k

$0.6 F_u A_{nv} = 0.6(58)(10.6) = 368.9$ k

Therefore, $0.6 F_u A_{nv} > F_u A_{nt}$

Use Equation 7.42 (LRFD Equation J4-3b)
$$\phi R_n = \phi[0.6 F_u A_{nv} + F_y A_{gt}]$$
$$= 0.75[(0.6)(58)(10.6) + (36)(3.71)]$$
$$= 376.8 \text{ k} > 340 \text{ k} \quad \text{OK.}$$

7.8 WELDED CONNECTIONS

Welding offers many advantages over bolted connections. Some of them are:

- Welding makes simple and clean-looking connections without the need for gusset and splice plates, and there are no bolt heads and nuts in the joints. These result in significant savings in steel weight and cost.
- Welding can be used to join smaller components in the shop or in the field, such as pipes and tubes.

- Welded structures are more rigid, because parts are connected directly and little or no slip takes place. It is a good method to provide moment resisting connection, and seismic-resistant structures.
- Welding can be used efficiently in field connection and repair of members in steel structures.
- Welding produces less noise than bolting using pneumatic tools. This is an important feature where noise control is a requirement of the building permit or city ordinance.

Part 8 of the LRFD manual, Volume 2, has a good discussion on Welded Construction, covering weldability of steel, welding materials and processes, inspection, economical considerations, design strength of welds, prequalified welded joints, design examples, and tables.

7.8.1 Welding Code

The American Welding Society's Structural Welding Code is the most widely used standard for welding of steel buildings and bridges. AWS D1.1 is used for steel buildings and AWS D1.5 is used for welded steel bridges. Readers are encouraged to refer to the AWS code for design of welded connections, welding procedures, selection of electrodes, weld and welder qualifications, inspection, and acceptance of welds.

7.8.2 Types of Welding

Several types of welding are available. Arc welding is the predominant type of welding used in structural steel work. A good source of information on arc welding is the *Procedure Handbook of Arc Welding Design and Practice*, published by the Lincoln Electric Co., Cleveland, Ohio.

7.8.3 Types of Welds

Four types of welds—fillet, groove, plug, and slot welds—are used usually in structural steel work. The fillet and groove welds are the two main structural welds. We will discuss these two types of welds, and show some examples on the design and application of these welds.

7.8.4 Fillet Weld

The profile of a fillet weld is shown in Figure 7.19. A slight convex surface is preferred over a concave surface. As the weld shrinks on cooling, high tensile stress is set up in the concave surface, resulting in a tendency to cause cracking in the weld.

The strength of a fillet weld is controlled by the shear strength and is equal to the design shear stress times the theoretical throat area of the weld. For a 45° fillet weld, the throat dimension is 0.707 times the leg of the weld. The size of a fillet weld is indicated by the symbol shown in Figure 7.20. The size or leg of the weld is $\frac{5}{16}$ in. and the throat is

$$0.707 \times \frac{5}{6} \text{ in.} = 0.221 \text{ in.}$$

FIGURE 7.19

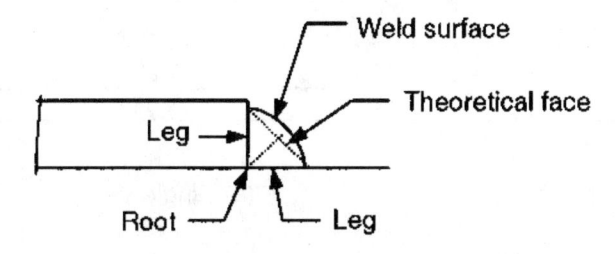

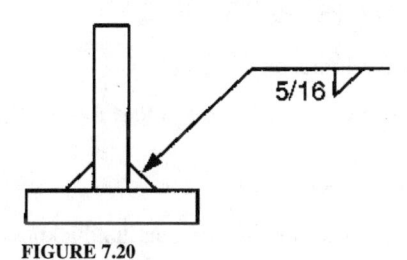

(a) Convex Weld Surface

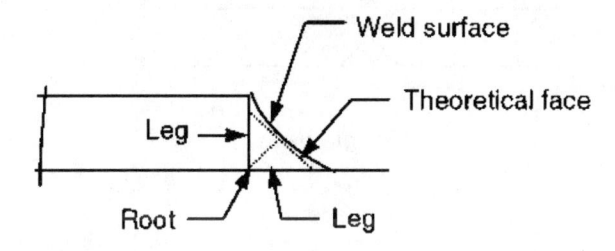

(b) Concave Weld Surface

FIGURE 7.20

7.8.5 Complete Penetration Groove Weld

Complete penetration groove welds are shown in Figure 7.21. Square groove welds are used to joint plates no thicker than $\frac{1}{4}$ in. Single- and double-vee groove welds are used to connect thicker plates. The strength of full-penetration groove welds is equal to or greater than the strength of the base metal.

7.8.6 Nominal Strength of Weld

The LRFD specification makes recommendations on the choice of electrodes to meet the required weld strength level. Table J2.5 in Part 6 of the LRFD manual (Page 6-78) gives the design strength of welds for the various types of welds. The design strength of a weld is taken as the lower value of $\phi F_w A_w$ and $\phi F_{BM} A_{BM}$

(a) Square

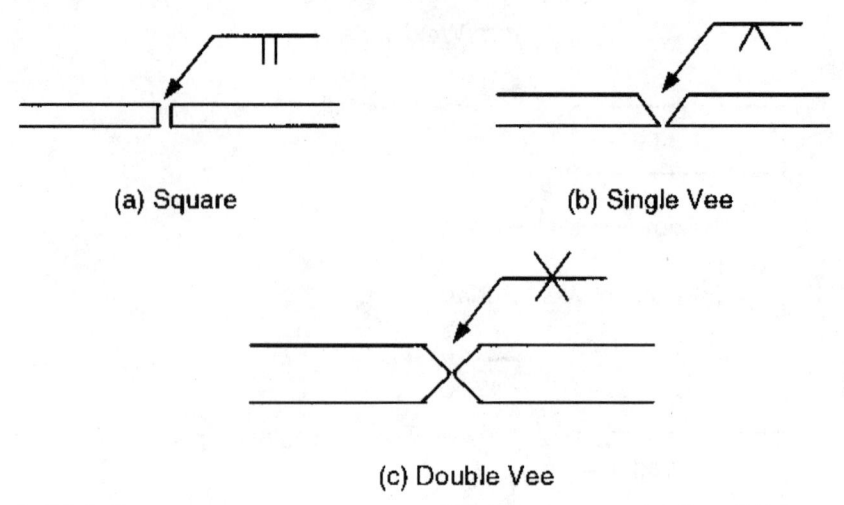

(b) Single Vee

(c) Double Vee

FIGURE 7.21

where ϕ = Resistance factor = 0.75
F_w = Nominal strength of the weld electrode, ksi
A_w = Effective cross-sectional area of the weld, in^2
F_{BM} = Nominal strength of the base material, ksi
A_{BM} = Cross-sectional area of the base material, in^2

The LRFD specification imposes some limits on fillet welds:

- The minimum size of fillet welds is shown in Table J2.4 in Part 6 of the LRFD manual (Page 6-75).
- The maximum size of fillet welds of connected parts is:

  a. Along edges of plate less than $\frac{1}{4}$-in. thick, not greater than the thickness of the plate.

  b. Along edges of plate $\frac{1}{4}$-in. or more in thickness, not greater than the thickness of the plate minus $\frac{1}{16}$-in.

  c. For flange-web welds and similar connections, the actual weld size need not be larger than that required to develop the web capacity, and the requirements of Table J2.4 need not apply.

- The minimum effective length of fillet welds design on the basis of strength must be not less than four times the nominal size, or else the size of the weld must be considered not to exceed $\frac{1}{4}$ of its effective length. If longitudinal fillet welds are used alone in end connections of flat-bar members, the length of each fillet weld must not be less than the perpendicular distance between them.

- The maximum effective length of fillet welds loaded by forces parallel to the weld, such as lap splices, must not exceed 70 times the filled weld leg. A uniform stress distribution may be assumed throughout the maximum effective length.

- In lap joints, the minimum amount of lap is five times the thickness of the thinner part joined, but not less than 1 in.
- Fillet welds must either be returned continuously around the ends or sides respectively for a distance of not less than two times the nominal weld size or terminate not less than the nominal weld size from the sides or ends. End returns must be indicated on the design and detail drawings.

EXAMPLE 7.24
Find the design strength of a unit length of $\frac{5}{16}$-in. fillet weld using E70 electrodes with a minimum tensile strength of 70 ksi and assume the load is applied parallel to the weld.

Solution

Effective throat thickness = $(0.707)\left(\frac{5}{16}\right) = 0.221$ in.

From Table J2.5 in Part 6 of LRFD manual (Page 6-78), nominal strength of weld

$F_w = 0.60 \times F_{EXX} = (0.60)(70) = 42$ ksi

Hence, design strength per inch = $\phi F_w A_w = (0.75)(42)(0.221 \times 1) = 6.96$ k/in.

EXAMPLE 7.25
Design a welded connection using fillet welds to resist a factored load of 95 kips on the $\frac{1}{2} \times 6$ in. A36 steel member shown in Figure 7.22.

Solution

$P_u = 95$ k

Minimum fillet weld size = $\frac{3}{16}$ in. (from Table J2.4, p. 6-75 of LRFD manual)

Maximum fillet weld size = $\frac{1}{2}$ in. $- \frac{1}{16}$ in. = $\frac{7}{16}$ in. (LRFD specification). Try $\frac{5}{16}$ in. weld.

Effective throat thickness = $(0.707)\left(\frac{5}{16}\right) = 0.221$ in.

Nominal strength of weld = $(0.60)(70) = 42$ ksi

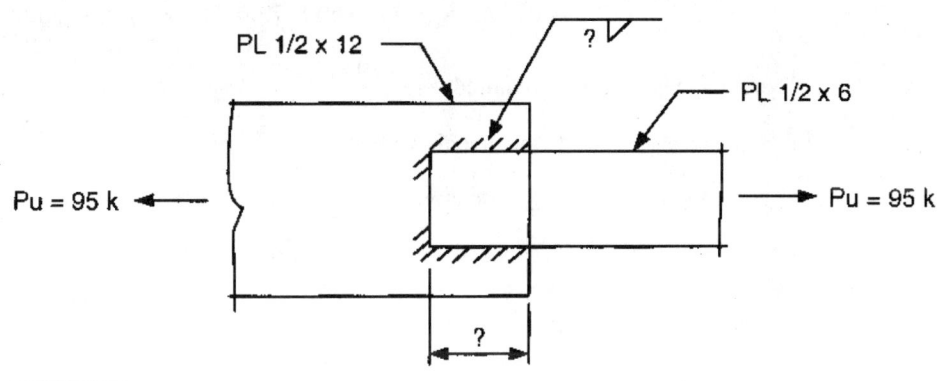

FIGURE 7.22

Hence, design strength per inch = $\phi F_w A_w$ = (0.75)(42)(0.221 × 1) = 6.96 k/in.
Length of weld required = $\frac{95}{6.96}$ = 13.6 in. (use 14 in.)

Use end returns not less than 2 × $\frac{5}{16}$ = $\frac{5}{8}$ in. (use 1 in.)
Leaving 14 − 2 = 12 in. Use 6 in. on each side.
LRFD specification requires minimum effective length of fillet welds ≥ perpendicular distance between the two longitudinal fillet welds = 6 in. < 7 in.   OK.

EXAMPLE 7.26

A ∠ 6 × 4 × $\frac{1}{2}$ of A36 is used as a member of a roof truss. Design a connection to the bottom chord of the truss using E70 electrodes for fillet welds for the full tension capacity of the angle as shown in Figure 7.23. Place the welds to avoid any eccentricity in the connection.

Solution

Tension capacity of angle $P_u = \phi_t F_y A_g$ = (0.9)(36)(4.75) = 153.9 k   (Controls)

or

$P_u = \phi_t F_u A_e$ assuming U = 0.9 ($A_e = A_g$ in welded connection)
(0.75)(58)(0.9)(4.75) = 186.0 k

Minimum weld size = $\frac{3}{16}$ in.

Maximum weld size = $\frac{1}{2} - \frac{1}{16} = \frac{7}{16}$ in.

Try $\frac{7}{16}$-in. fillet weld:

Effective throat thickness = $(0.707)\left(\frac{7}{16}\right)$ = 0.309

$\frac{\text{Strength of weld}}{\text{inch}}$ = $\phi F_w A_w$ = (0.75)(0.60 × 70)(0.309 × 1) = 9.73 k/in.

Total length of weld required = $\frac{153.9}{9.73}$ = 15.8   Use 16 in.

Locate c.g. of welds to coincide with c.g. of section:

Taking moment about L_2 to determine force P_1 (See Figure 7.22):

$$(153.9)(4.01) - (P_1 \times 6.0) = 0$$

Hence,

P_1 = 102.9 k
P_2 = 153.9 − 102.9 = 51.0 k

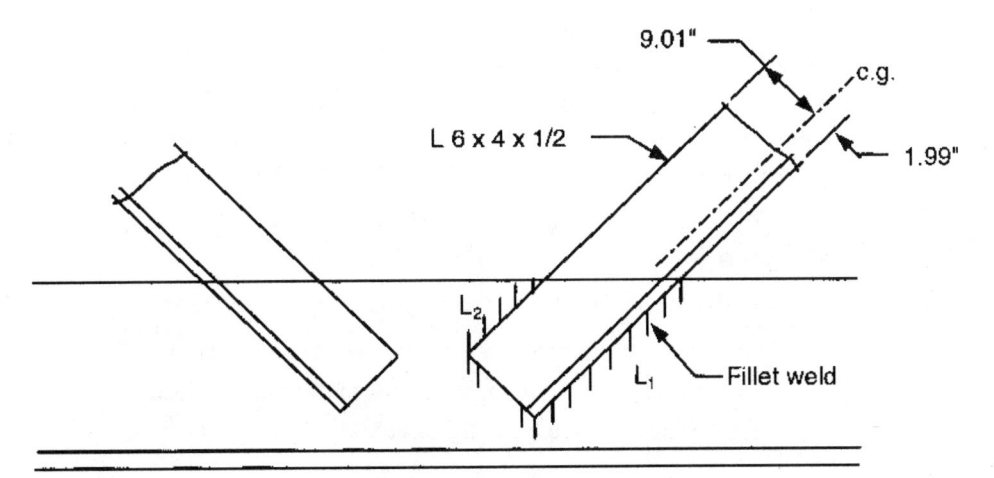

FIGURE 7.23

$$L_1 = \frac{102.9}{9.73} = 10.5 \text{ in.}$$

$$L_2 = 16.0 - 10.5 = 4.5 \text{ in.}$$

Use end returns of $2 \times \frac{7}{16} = \frac{7}{8}$ in. Use 1 in. end returns.

∴ Side weld lengths are 9.5 in. and 3.5 in.

7.9 COMPOSITE BEAMS

This section applies to composite steel beams supporting a reinforced concrete slab so interconnected that the beams and the slab act together to resist bending. Simple and continuous composite beams with shear connectors and concrete beams, constructed with or without temporary shoring, are included.

Elastic analysis will be assumed in the design of composite beams. In determining forces in members and connections of a structure, consideration is given to the effective sections at the time each increment of load is applied. Strains in steel and concrete are assumed to be directly proportional to the distance from the neutral axis. The stress is equal to strain times the modulus of elasticity of steel E_s, or modulus of elasticity for concrete E_c.

Tensile strength of concrete is neglected. Maximum stress in the steel is limited to F_y. Maximum compressive stress in the concrete is limited to $0.85f'_c$, where f'_c is the specified compressive strength of the concrete. In composite hybrid beams, the maximum stress in the steel flange is limited to F_{yf}, but the strain in the web can exceed the yield strain with a corresponding stress taken as F_{yw}.

Only fully composite beams are considered in this section. Shear connectors are provided in sufficient numbers to develop the maximum flexural strength of the composite beam. No slip is assumed.

There are several advantages of composite construction. It takes advantage of the high compressive strength of concrete by putting a large part of the concrete in compression, and a larger part of the steel section in tension. This results in less steel for the same loads and spans, or longer spans for the same section. Composite sections have much higher modulus of section than noncomposite sections, resulting in a much smaller deflection. Another advantage of composite construction is the possibility of reducing the thickness of the slab and/or the depth of beam. This is a very important cost-saving feature of composite construction.

7.9.1 Effective Width

The effective width of the concrete slab on each side of the beam centerline is limited to:

a. One-eighth of the beam span, center to center of supports
b. One-half the distance to the centerline of the adjacent beam
c. The distance to the edge of the slab.

7.9.2 Strength of Beams with Shear Connectors

The positive design flexural strength $\phi_b M_n$ is determined as follows:

a. For $\dfrac{h}{t_w} \leq \dfrac{640}{\sqrt{F_{vf}}}$

$\phi_b = 0.85$; M_n is determined from the plastic stress distribution on the composite section.

b. For $\dfrac{h}{t_w} \geq \dfrac{640}{\sqrt{F_{vf}}}$

$\phi_b = 0.90$; M_n shall be determined from the superposition of elastic stresses, considering the effects of shoring.

The negative design flexural strength $\phi_b M_n$ is determined for the steel section alone in accordance with procedures and requirements outlined in Section 7.5 of this chapter.

Alternatively, the negative design flexural strength $\phi_b M_n$ may be computed with: $\phi_b = 0.85$ and M_n determined from the plastic stress distribution on the composite section, provided that:

1. Steel beam is an adequately braced compact section.
2. Shear connectors connect the slab to the steel beam in the negative moment region.
3. Slab reinforcement parallel to the steel beam, is properly developed within the effective width of the slab.

The following may be assumed for determining the plastic stress distribution for negative moment area:

1. Tensile stress F_{yr} in all adequately developed longitudinal reinforcing bars within the effective width of the concrete slab.
2. Uniformly distributed steel stress F_y throughout the tension zone and throughout the compression zone in the structural section.

3. The net compressive force in the steel section is equal to the total tensile force in the reinforcing steel.

7.9.3 Strength During Construction

When temporary shores are not used during construction, the steel section alone must have adequate strength to support all loads applied prior to the concrete attaining 75 percent of its specified strength F_c'. The design flexural strength of the steel section is determined in accordance with the procedures and requirements of Section 7.5 of this chapter.

7.9.4 Design Shear Strength

The design shear strength of composite beams shall be determined by the shear strength of the steel web.

7.9.5 Shear Connectors

This section applies to the design of steel headed-studs. Headed-studs must not be less than four stud diameters in length after installation.

Strength of Headed-Stud Shear Connectors:
The nominal strength Q_n in kips of one headed-stud shear connector embedded in a solid concrete slab is

$$Q_n = 0.5 A_{sc} \sqrt{f_c' E_c} \leq A_{sc} F_u \qquad (7.43)$$
$$\text{(LRFD I5-1)}$$

where A_{sc} = Cross-sectional area of a headed-shear connector, in²
 f_c' = Specified compressive strength of concrete, ksi
 F_u = Minimum specified tensile strength of a headed-stud shear connector, ksi
 E_c = Modulus of elasticity of concrete, ksi.

A set of Q_n values based on the above equation is given in Table 5-1, Part 5 of the LRFD manual for $\frac{3}{4}$-in. headed studs made from A36 steel and embedded in concrete slabs with different concrete strengths and weights.

7.9.6 Required Number of Shear Connectors

The number of shear connectors required between the section of maximum bending moment, positive or negative, and the adjacent section of zero moment is equal to the horizontal shear force divided by the nominal strength Q_n of one shear connector.

The total horizontal shear force between the point of maximum positive moment and the point of zero moment is taken as the smallest of the following:

a. $0.85f'_c A_c$
b. $A_s F_y$
c. ΣQ_n

where f'_c = Specified compressive strength of concrete, ksi
A_c = Area of concrete slab within effective width, in²
A_s = Area of steel cross section, in²
F_y = Minimum specified yield stress, ksi
ΣQ_n = Sum of nominal strengths of shear connectors between the point of maximum positive moment and the point of zero moment, kips.

In continuous composite beams where longitudinal reinforcing steel in the negative moment regions is considered to act compositely with the steel beam, the total horizontal shear force between the point of maximum negative moment and the point of zero moment is taken as the smaller of

a. $A_r F_{yr}$
b. ΣQ_{ny}

where A_r = Area of adequately developed longitudinal reinforcing steel within the effective width of the concrete slab, in³
F_{yr} = Minimum specified yield stress of the reinforcing steel, ksi
ΣQ_n = Sum of nominal strengths of shear connectors between the point of maximum negative moment and the point of zero moment, kips.

7.9.7 Shear Connector Placement and Spacing

Shear connectors required on each side of the point of maximum bending moment, positive or negative, must be distributed uniformly between that point and the adjacent points of zero moment. However, the number of shear connectors placed between any concentrated load and the nearest point of zero moment must be sufficient to develop the maximum moment required at the concentrated load point.

Shear connectors must have at least one inch of lateral concrete cover. Unless located over the web, the diameter of studs must not be greater than 2.5 times the thickness of the flange to which they are welded. The minimum center-to-center spacing of stud connectors shall be six diameters along the longitudinal axis of the supporting composite beam and four diameters transverse to the longitudinal axis of the supporting composite beam. The maximum center-to-center spacing of shear connectors must not exceed eight times the total slab thickness.

7.9.8 Neutral Axis in Concrete Slab

Load tests have shown that the nominal moment capacity M_n of a composite section can be estimated very accurately with the plastic theory. It is assumed that the steel section at failure is fully yielded, concrete is stressed to $0.85 f_c'$, and any part of the concrete in the tension side of the neutral axis is cracked and incapable of carrying stress.

The plastic neutral axis (PNA) can fall in the slab, or in the flange or web of the steel section. We will discuss only the more common case where PNA falls in the concrete slab.

The concrete slab compression stresses are assumed to be uniform with a value of $0.85 f_c'$ over an area of depth a and an effective width b_e. The steel section is assumed to be fully yielded. With these assumptions, the nominal or plastic moment capacity of a composite section can be determined as shown in Figure 7.24.

Total compression C in the concrete is given by $C = 0.85 f_c' a b_e$

Total tensile force T in the steel section is given by $T = A_s F_y$

Equating the equal but opposite forces, we have $0.85 f_c' a b_e = A_s F_y$

$$a = \frac{A_s F_y}{0.85 f_c' b_e} \qquad (7.44)$$

When a is equal to or less than slab thickness t, PNA falls in the slab and the nominal moment capacity M_n or plastic moment capacity M_p is given by

$$M_n = M_p = A_s F_y \left(\frac{d}{2} + t - \frac{a}{2} \right) \qquad (7.45)$$

where A_s = Area of steel section
F_y = Specified yield stress of steel section
d = Depth of steel beam
t = Thickness of concrete slab
a = Depth of concrete stress block as determined in Equation 7.44

EXAMPLE 7.27

Compute the moment capacity M_u for the composite section shown in Figure 7.25, using $f_c' = 4$ ksi and $F_y = 36$ ksi.

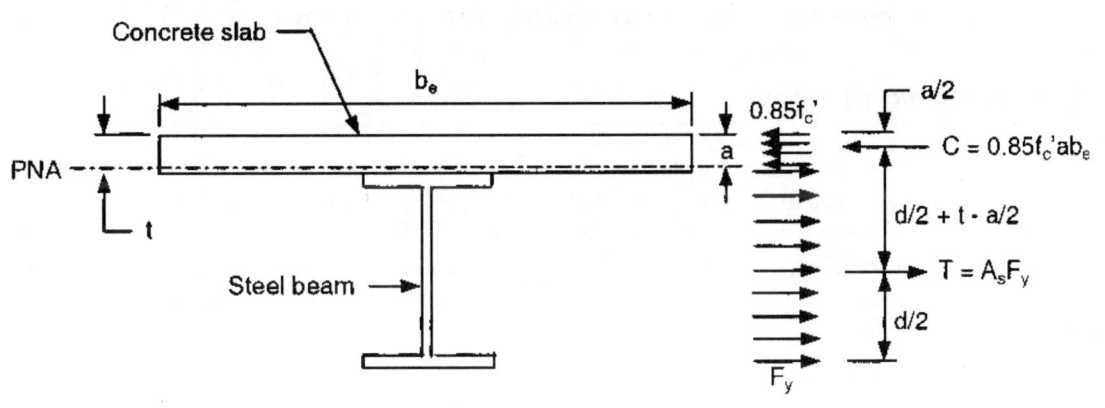

FIGURE 7.24

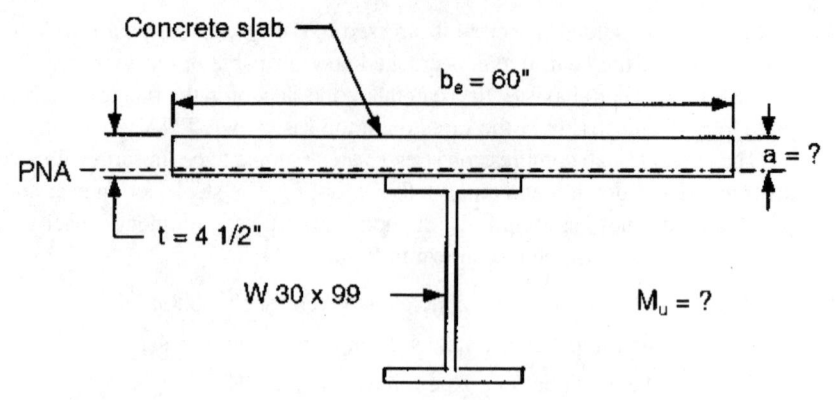

FIGURE 7.25

Solution
In this example, $A_s = 29.1$ in^2, $F_y = 36$ ksi, $f'_c = 4.0$ ksi, $d = 30$ in., $t = 4.5$ in., $b_e = 70$ in.

$$a = \frac{A_s F_y}{0.85 f'_c b_e} = \frac{29.1 \times 36}{0.85 \times 4.0 \times 70} = 4.40 \text{ in.} < 4.5 \text{ in.}$$

$$M_n = M_p = A_s F_y \left(\frac{d}{2} + t - \frac{a}{2} \right)$$

$$= 29.1 \times 36 \left(\frac{30}{2} + 4.5 - \frac{4.40}{2} \right) = 18{,}123 \text{ k-in.} = 1510 \text{ k-ft}$$

$$M_u = \phi M_n = (0.85)(1510) = 1284 \text{ k-ft}$$

7.9.9 Deflection of Composite Section

Deflection of composite beams is computed using the same methods as for other types of beams. However, the stages of loadings must be considered. For example, for unshored construction, the dead load of the concrete slab is applied to the steel section alone; other dead loads and live loads are applied to the composite section. For shored construction, all loads are applied to the composite section after the shoring is removed.

7.10 BEARING PLATES

Column base plates distribute the forces at the base of the column to an area of foundation large enough to prevent crushing the concrete. Base plate thicknesses should be specified in multiples of $\frac{1}{8}$-in. up to $1\frac{1}{4}$-in. and in multiples of $\frac{1}{4}$-in. thereafter. Typical column plates are shown in Figures 11-14 and 11-15 in Part 11 of the LRFD manual, Volume II (Pages 11-55).

In accordance with the LRFD specification, the design bearing loads on concrete can be taken as $\phi_c P_p$:

a. On the full area of a concrete support

$$P_p = 0.85 f'_c A_1 \tag{7.46}$$

b. On less than the full area of a concrete support

$$P_p = 0.85 f'_c A_1 \sqrt{\frac{A_2}{A_1}} \tag{7.47}$$

where $\phi_c = 0.60$
A_1 = Area of steel concentrically bearing on a concrete support, in
A_2 = Maximum area of the portion of the supporting surface that is geometrically similar to and concentric with the loaded area, in.

$$\sqrt{\frac{A_2}{A_1}} \leq 2$$

The LRFD specification does not require a particular method for designing column base plates. However, Part 11 of the LRFD manual suggests that maximum moments in a base plate occur at distances 0.80 b_f and 0.95 d apart. The bending moment can be calculated at each of these sections and the larger value used to determine the plated thickness needed. The variables are illustrated in Figure 11-20, Part 11 of the LRFD manual (Page 11-61). The LRFD manual contains several examples on the design of column base. The method consists of the following steps and can best be illustrated by an example.

EXAMPLE 7.28
A W12 × 96 column (F_y = 50 ksi) with a factored load of 800 kips bears on a concrete 48 in. × 48 in. footing. The concrete has a specified compressive strength f'_c = 3.0 ksi and the base plate has F_y = 36 ksi. Design the base plate to distribute the column load.

Solution
W12 × 96:

$$d = 12.71 \text{ in.}, \quad b_f = 12.16 \text{ in.}$$
$$t_w = 0.55 \text{ in.}, \quad t_f = 0.90 \text{ in.}$$

Step 1: Calculate required base-plate area

$$\sqrt{\frac{A_2}{A_1}} > 2, \text{ use 2 maximum}$$

$$A_1 = \frac{P_u}{\phi_c(0.85 f'_c)\sqrt{\frac{A_2}{A_1}}}$$

$$= \frac{800}{0.6(0.85 \times 3)(2)} = 261.4 \text{ in}^2$$

Step 2: Optimize base-plate dimensions

The plate dimensions B and N now can be optimized.

$$\Delta = \frac{0.95d - 0.8b_f}{2} = \frac{0.95(12.71) - 0.8(12.16)}{2} = 1.17 \text{ in.}$$

$$N = \sqrt{A_1} + \Delta = \sqrt{261.4} + 1.17 = 17.3 \text{ in.}$$

Try $N = 18$ in. and $B = 15$ in.

Step 3: Calculate required base-plate thickness

The required base-plate thickness may be calculated as

$$t_{req} = l\sqrt{\frac{2P_u}{0.9F_y BN}}$$

In the above equation, l is the larger of m, n, and $\lambda n'$ where

$$n = \frac{B - 0.8b_f}{2} = \frac{15 - 0.8(12.16)}{2} = 2.64 \text{ in.}$$

$$m = \frac{N - 0.95d}{2} = \frac{18 - 0.95(12.71)}{2} = 2.96 \text{ in.}$$

$$\phi_c P_p = 0.85 f'_c A_1 \sqrt{\frac{A_2}{A_1}}$$

$$= 0.6(0.85)(3)(18 \times 15)(2) = 826.2 \text{ k}$$

$$n' = \frac{\sqrt{d b_f}}{4} = \frac{\sqrt{(12.71)(12.16)}}{4} = 3.11$$

$$X = \left(\frac{4 d b_f}{(d + b_f)^2}\right)\left(\frac{P_u}{\phi_c P_p}\right) = \left(\frac{4(12.71)(12.16)}{(12.71 + 12.16)^2}\right)\left(\frac{800}{826.2}\right) = 0.97$$

$$\lambda = \frac{2\sqrt{X}}{1 + \sqrt{1 - X}} \leq 1$$

Hence, $\therefore \lambda = \dfrac{2\sqrt{0.97}}{1 + \sqrt{1 - 0.97}} = 1.68$ Use 1

$$\lambda n' = (1)(3.11) = 3.11$$

Now, l = Maximum of m, n, $\lambda n'$
= Maximum of 2.96, 2.64, 3.11
= 3.11

Finally,

$$t_{req} = l\sqrt{\frac{2P_u}{0.9F_y BN}} = 3.11\sqrt{\frac{2(800)}{0.9(36)(18)(15)}} = 1.33 \text{ in.}$$

Step 4: Select base-plate

Select PL $1\frac{1}{2} \times 15 \times$ 1 ft 6 in.

The same method and steps may be used for designing either heavily or lightly loaded base plates.

CHAPTER 8
MASONRY DESIGN

8.1 INTRODUCTION

Use of masonry materials can be traced back to ancient times. Around 2500 B.C., Egyptian used stone to build pyramids in Giza. Other masonry structures include the temples, monuments, lighthouses, and other magnificent structures of Greece, Egypt, and Rome. Despite these successful uses, architecture turned its creative focus away from masonry after the invention of portland cement in 1824, refinements in iron production in the early 19th century, and the development of the Bessemer furnace in 1854. As building height increased, masonry construction with no tensile strength was considered inadequate for resisting lateral loads from wind, blasts, and earthquakes. This condition remained unchanged until the advent of modern reinforced masonry.

This chapter will help the engineer prepare for masonry design questions on the Civil Professional Engineering Examination. With this in mind, the following subjects are included:

1. Basic materials, including masonry units, mortar, grout, and reinforcing accessories. Also, moduli of elasticity of concrete masonry unit and steel, design data, and section properties

2. General design requirements

3. Design of unreinforced masonry structures

4. Design of reinforced masonry structures for flexure, shear, axial forces, combined flexure and axial forces, and in-plane shear forces

5. Design-reinforced masonry lintels

6. Design-reinforced masonry columns and pilasters

8.2 MATERIALS

8.2.1 Masonry Units

Masonry units are available in many different materials, shapes, sizes, and finishes. The masonry units considered in this chapter are concrete masonry units, including hollow and solid units. These units are made from portland cement, water, and normal-weight or light-weight aggregate, with or without the inclusion of other materials. A *hollow masonry unit* is a unit whose net cross-sectional area (solid area) in any plane parallel to the surface containing the cores or cells is less than 75 percent of its gross cross-sectional area measured in the same plane.

Based on ASTM C90-95, UBC Standard 21-4 classifies concrete masonry units into the following two grades, according to their unit weight and use:

1. Grade N: Units having a weight classification of 85 pcf or greater, for general use in walls below and above grade that may or may not be exposed to moisture penetration or the weather
2. Grade S: Units having a weight classification of less than 85 pcf; uses limited to walls not exposed to the weather and above-grade installation in exterior walls with weather-protective coatings

Within each grade, there are two types of concrete masonry units, according to moisture content requirements:

1. Type I: Moisture-controlled units, which must conform to the moisture requirements given in UBC Standard Table 21-4-A
2. Type II: Non-moisture-controlled units, which need not conform to the moisture requirements

Concrete masonry units are available in a wide variety of sizes and shapes. Solid units are typically $7\frac{5}{8}$ in. high and are available in several lengths and widths. Hollow units are typically $15\frac{5}{8}$ in. long; either $7\frac{5}{8}$ in. or $3\frac{5}{8}$ in. high; and $7\frac{5}{8}$ in., $5\frac{5}{8}$ in., or $3\frac{5}{8}$ in. wide. When specifying a block size, it is common to give the nominal dimensions in the order of width, height, and length. For instance, the 8 × 4 × 16 standard block is 8 in. wide × 4 in. high × 16 in. long. For convenient wall layout, the actual width, height, and length dimensions are $\frac{3}{8}$ in. less than the nominal dimensions to allow for typical mortar joint thickness. Minimum face-shell and web thicknesses are listed in UBC Standard Table 21-4-C.

Material properties of concrete masonry units that affect the structural performance of installed masonry include absorption, linear shrinkage, moisture content, compressive strength, stiffness, and tensile strength. The strength and absorption requirements for concrete masonry units are listed in UBC Standard Table 21-4-B. The compressive strength of masonry units is established in accordance with UBC Standard 21-17.

8.2.2 Mortar

The purposes of mortar are to fill the irregularities between masonry units, to provide resistance to the penetration of light, wind, and water, and to provide bonding strength to bond masonry units into an assemblage that acts as an integral element having desired functional performance characteristics. Mortar is a mixture of cementitious material, aggregate, and water. A major distinction between concrete and mortar is illustrated by the manner in which they are handled during construction: Concrete usually is placed in nonabsorbent metal or treated timber forms so that most of the water will be retained. Mortar usually is placed between absorbent masonry units, to which the mortar loses water as soon as contact is made. Compressive strength is a prime consideration in concrete, but it is only one of several important factors in mortar. Bond strength is generally more important in mortar, as well as good workability and water retentivity. Both are required for maximum bond. Flexure strength is also important because it measures the ability of mortar to resist cracking.

There are four types of mortar: M, S, N, and O, in order of descending strength, given in ASTM C270. The compressive strength of mortar depends largely upon the cement content and water-cement ratio. The average compressive strength of mortar should typically be weaker than the masonry units, so that any cracks will occur in the mortar joints, where they can be more easily repaired. Thus, the selection of mortar type is based on the type of masonry units to be used, as well as the building code and engineering practice standard requirements, such as allowable design stresses and lateral support. Table 8.1 is a general guide for the selection of mortar type for various masonry wall constructions.

In seismic zone 2, however, UBC 2106.1.12.3 specifies that Type O mortar shall not be used as part of the vertical or lateral load-resisting systems. In seismic zones 3 and 4, design and construction of masonry structures must conform to the requirements for seismic zone 2, with additional restrictions specified in UBC 2106.1.12.4, such as that Type N mortar shall not be used as part of the vertical or lateral load-resisting system. Normally, the Type S mortar should be specified. The mortar compressive strength is typically measured by the aver-

TABLE 8.1 Guide for the Selection of Masonry Mortars* (ASTM C270 Table X1.1)

		Mortar Type	
Location	Building Segment	Recommended	Alternative
Exterior, above Grade	Load-bearing wall	N	S or M
	Non-load-bearing wall	O[†]	N or S
	Parapet wall	N	S
Exterior, at or below Grade	Foundation wall, retaining wall, manholes, sewers, pavements, walks, and patios	S[‡]	M or N[‡]
Interior	Load-bearing wall	N	S or M
	Non-bearing partitions	O	N

* This table does not provide for many specialized mortar uses, such as chimney, reinforced masonry, and acid-resistant mortars.

[†] Type O mortar is recommended for use where the masonry is unlikely to be frozen when saturated, or unlikely to be subjected to high wind or other significant lateral loads. Type N or S mortar should be used in other cases.

[‡] Masonry exposed to weather in a nominally horizontal surface is extremely vulnerable to weathering. Mortar for such masonry should be selected with due consideration.

TABLE 8.2 Average Compressive Strength for Mortar

Mortar	Type	Average compressive strength of 2-inch cubes at 28 days (min. psi)	Water retention (min., %)	Air content (max., %)
Cement-lime or mortar cement	M	2500	75	12
	S	1800	75	12
	N	750	75	14
	O	350	75	14
Masonry cement	M	2500	75	18
	S	1800	75	18
	N	750	75	18
	O	350	75	18

age compressive strength of three 2-inch cubes of mortar. Table 8.2 gives the minimum strength, water retention, and maximum air content for the mortar type specified.

8.2.3 Grout

Grout is a mixture of cementitious materials and small size aggregate, such as sand and pea gravel. Sufficient water must be added to permit grout to be fluid enough to fill all voids in the grout space without segregation of the materials, and to completely bond the steel reinforcement, ties, and anchors to form a unified composite structure. UBC 2103.4 specifies that the compressive strength of grout f_g' at 28 days old shall not be less than 2000 psi. Depending on the maximum size of aggregate used, grout used in masonry construction may be either fine grout or coarse grout. Grout proportions by volume are summarized in UBC Table 21-B.

8.2.4 Reinforcing Accessories

UBC Standard 21-10 Part I covers joint reinforcements fabricated from cold drawn steel wire. These joint reinforcements are suitable for placement in mortar joints between masonry courses.

Steel bars used to reinforce masonry are the same types of steel bars used in reinforced concrete. Reinforcement must conform to one of the following specifications:

1. ASTM A615, Specifications for Deformed and Plain Billet-steel Bars for Concrete Reinforcement
2. ASTM A616, Specifications for Rail-steel Deformed and Plain Bars for Concrete Reinforcement.

3. ATM A617, Specifications for Axle-steel Deformed and Plain Bars for Concrete Reinforcement
4. ASTM A706, Specifications for Low-alloy Steel Deformed Bars for Concrete Reinforcement.
5. ASTM A767, Specifications for Zinc-Coated (Galvanized) Steel Bars for Concrete Reinforcement
6. ASTM A775, Specifications for Epoxy-Coated Reinforcing Steel Bars

8.2.5 Modulus of Elasticity of Materials (UBC 2106.2.12)

1. Modulus of elasticity of concrete masonry unit

$$E_m = 750 f'_m \quad 3{,}000{,}000 \text{ psi maximum}$$

This value must be reduced by one-half when no special inspection is provided (UBC 2107.1.2). When the modulus of elasticity of masonry is established by test, however, the value shall not be reduced.

2. Modulus of elasticity of steel

$$E_s = 29{,}000{,}000 \text{ psi}$$

3. Shear modulus of masonry

$$G = 0.4 E_m$$

8.2.6 Design Data and Section Properties

TABLE 8.3 Assumed Dimensions of Hollow Concrete Units, in.

Nominal thickness of unit W	Design width	Face-shell thickness	Webs	Equivalent web thickness
6	$5\frac{5}{8}$	1	1	$2\frac{1}{4}$
8	$7\frac{5}{8}$	$1\frac{1}{4}$	1	$2\frac{1}{4}$
10	$9\frac{5}{8}$	$1\frac{3}{8}$	$1\frac{1}{8}$	$2\frac{1}{2}$
12	$11\frac{5}{8}$	$1\frac{1}{2}$	$1\frac{1}{8}$	$2\frac{1}{2}$

Example: 8-in. block, grouted at 16 in. oc. Calculate in-plane direct shear area.

For 8-in. block. Wall thickness = 7.625 in., Length of block = $15\frac{5}{8}$ in.
Face-shell thickness $t_s = 1\frac{1}{4}$ in., Web thickness, $t_w = 1$ in.
  Cell Size 6 in. × 5 in.
In-plane shear area = $(2 \times 1\frac{1}{4} \times 16) + 6(5) + (1 + 1\frac{1}{4})(5) = 81.25$ in^2/16 in.
Or 60.9 in^2/ft

TABLE 8.4 Average Weight for Grouted CMU, lb/ft²

Spacing of grout cells, in.	Medium weight 125-105 lb/ft³			Normal weight 125 lb/ft³ or more			
	Wall thickness, in.			Wall thickness, in.			
	6	8	12	6	8	10	12
Solid grouted	56	77	118	68	92	116	140
16 in. o.c.	46	60	90	58	75	92	111
24 in. o.c.	42	53	79	53	69	85	102
32 in. o.c.	40	50	73	51	65	78	93
40 in. o.c.	38	47	70	50	62	75	89
48 in. o.c.	37	46	68	49	60	72	85
No grout	31	35	50	43	50	59	69

Note: Average weights are based on medium weight and normal-weight units having average concrete weights of 120 lb/ft³ and 138 lb/ft³. A small quantity has been added to include the weight on bond beams and reinforcing.

TABLE 8.5 Equivalent Wall Thickness for Computing Compression, in.

Spacing of grout cells, in.	Equivalent solid thickness*, in.				In-plane shear area[†], in²/ft
	Wall thickness, in.				Wall thickness, in.
	6	8	10	12	8
Solid grouted	5.6	7.6	9.6	11.6	88.5
16 in. o.c.	4.5	5.8	7.2	8.5	60.9
24 in. o.c.	4.1	5.2	6.3	7.5	50.5
32 in. o.c.	3.9	4.9	5.9	7.0	45.3
40 in. o.c.	3.8	4.7	5.7	6.7	42.3
48 in. o.c.	3.7	4.6	5.5	6.5	40.5
No grout	3.4	4.0	4.7	5.5	30.0

* Equivalent solid thickness means the calculated thickness of the wall if there were no hollow cores, and is obtained by dividing the volume of the solid material in the wall by the face area of the wall. This Equivalent Solid Thickness (EST) is for the determination of area for structural design only, e.g., $f_a = P/(EST)b$.

† In-plan direct shear area = (total length of mortared face shells) × (face-shell thickness) + grouted core area + (total mortared web and end wall thickness adjacent to grout core) × (distance between face shells).

8.3 GENERAL DESIGN REQUIREMENTS

Three design methods are given in the Uniform Building Code, Section 2106, for the design of masonry structures. Method 1 is the working stress design method, method 2 is the strength-design method, and method 3 is the empirical design method.

TABLE 8.6 Radius of Gyration for Concrete Masonry Units (UBC Table 21-H-1)

Spacing of grout cells, in.	Nominal width of wall, in.				
	4	6	8	10	12
Solid grouted	1.04	1.62	2.19	2.77	3.34
16 in. o.c.	1.16	1.79	2.43	3.04	3.67
24 in. o.c.	1.21	1.87	2.53	3.17	3.82
32 in. o.c.	1.24	1.91	2.59	3.25	3.91
40 in. o.c.	1.26	1.94	2.63	3.30	3.97
48 in. o.c.	1.27	1.96	2.66	3.33	4.02
No grout	1.35	2.08	2.84	3.55	4.29

8.3.1 Working Stress Design Method

In working stress design, the stresses in concrete masonry under service loads must not exceed the pre-designated allowable values. The basic assumptions for the working stress design of masonry structures are as follows (UBC 2107.1.4):

1. Plane sections before bending remain plane after bending. This means that the unit strains in the cross section both above and below the neutral axis are directly proportional to the distance from the neutral axis.

2. Stress is proportional to strain; i.e. linear elastic material following Hooke's law is assumed. The modulus of elasticity is constant throughout the member.

3. Masonry elements combine to form a homogeneous member.

In addition to the assumptions outlined above, the working stress design of reinforced masonry is based on the following assumptions (UBC 2107.2.3):

1. Masonry carries no tensile stress. The reinforcing steel is assumed to carry all tensile stresses.

2. Reinforcement is completely surrounded by and bonded to masonry material, so that they work together as a homogeneous material within the range of allowable working stresses. The strain in masonry ϵ_m at a given load is equal to the strain in the reinforcing steel, ϵ_s, at the same location.

$$\epsilon_m = \frac{f_m}{E_m} = \epsilon_s = \frac{f_s}{E_s}$$

where f_m = Stress in the masonry, psi
 f_s = Stress in steel, psi
 E_m = Modulus of elasticity of masonry, psi
 E_s = Modulus of elasticity of steel, psi.

$$f_s = E_s \epsilon_s = \frac{E_s}{E_m} f_m = n f_m$$

where $n = \dfrac{E_s}{E_m}$ is called the modulus ratio.

8.3.1.1 Allowable Masonry Stresses (UBC 2107.1.2)

The compressive strength of concrete masonry units along with the mortar strength provides the basis for the specified design compressive strength of the masonry assemblage. Table 8.7 gives the specified compressive strength of masonry f'_m, based on the compressive strength of masonry units and the type of mortar used.

In seismic zones 3 and 4, the value of f'_m for concrete masonry shall be limited to 1500 psi when quality assurance provisions do not include requirements for special inspection as prescribed in UBC Section 1701, unless the value of f'_m is verified by tests in accordance with UBC 21.5.3.4.

8.3.1.2 Allowable Masonry Stresses for Unreinforced Masonry

One or more special inspectors are required to inspect the construction of structural masonry, as outlined in UBC section 1701, to ensure the quality of masonry units, mortar, grout, and reinforcing steel, and the conformance of construction to the approved plans, details, specifications, and applicable workmanship. Otherwise, the allowable stresses for masonry in the following sections shall be reduced by 50 percent (UBC 2107.1.2).

1. Allowable axial compressive stresses (UBC 2107.3.2)
   The allowable compressive stress F_a is

$$F_a = 0.25 f'_m \left[1 - \left(\dfrac{h'}{140r}\right)^2\right] \quad \text{for } \dfrac{h'}{r} \leq 99 \quad (8.1)$$

$$F_a = 0.25 f'_m \left(\dfrac{70r}{h'}\right)^2 \quad \text{for } \dfrac{h'}{r} > 99 \quad (8.2)$$

where f'_m = Specified compressive strength of masonry at 28 days, psi
       h' = Effective height of wall or column, in.
       r = Radius of gyration in inch, based on specified unit dimension or UBC Tables 21-H-1, 21-H-2, and 21-H-3 for partially grouted hollow unit masonry.

TABLE 8.7 Specified Compressive Strength of Masonry f'_m, psi, Based on Specifying the Compressive Strength of Masonry Units (UBC Table 21-D)

Compressive strength of concrete masonry units, psi	Specified compressive strength of masonry f'_m	
	Type M or S mortar	Type N mortar
4800 or more	3000	2800
3750	2500	2350
2800	2000	1850
1900	1500	1350
1250	1000	950

2. Allowable flexural compressive stresses (UBC 2107.3.3)

$$F_b = 0.33 f'_m \quad 2000 \text{ psi maximum} \quad (8.3)$$

3. Combined compressive stresses, unit formula (UBC 2107.3.4)

   Elements subjected to both axial compression and bending stresses shall be designed in accordance with accepted principles or shall satisfy the following formula:

$$\frac{f_a}{F_a} + \frac{f_b}{F_b} \le 1 \quad (8.4)$$

   where f_a = Computed axial compressive stress due to design axial load, psi
   F_a = Allowable axial compressive stress in columns for centroidally applied axial load only (Equations 8.1 or 8.2), psi
   f_b = Computed flexure stress in extreme fiber due to design bending loads only, psi
   F_b = Allowable flexure compressive stress in members subjected to bending load only (Equation 8.3), psi.

4. Allowable tensile stress (UBC 2107.3.5)

   Elements subject to both axial tension and bending stresses shall not exceed the allowable flexure tensile stress F_t given in Table 8.8.

   Values in Table 8.8 are for tension normal to the head joints in running bond masonry. No tension is allowed across head joints in stack bond masonry. These values shall not be used for horizontal flexure members. Running bond and stack bonds, as shown in Figure 8.1, are two of the most commonly used concrete block wall patterns. In a running bond wall pattern, the head joints of the masonry units are staggered. In the stack bond pattern, the units are all laid up with the head joints aligned. Running bond is the strongest bond pattern and should be used unless a stack bond pattern is essential to the architectural treatment of the building.

TABLE 8.8 Allowable Flexure Tension, psi (UBC Table 21-I)

Unit type	Cement-lime and mortar cement M or S	Cement-lime and mortar cement N	Masonry cement M or S	Masonry cement N
Normal to bed joints*				
Solid	40	30	24	15
Hollow	25	19	15	9
Normal to head joints				
Solid	80	60	48	30
Hollow	50	38	30	18

* Bed joint is the mortar joint that is horizontal at the time the masonry units are placed. Head joint is the mortar joint having a vertical transverse plane.

(a) Common or running bond

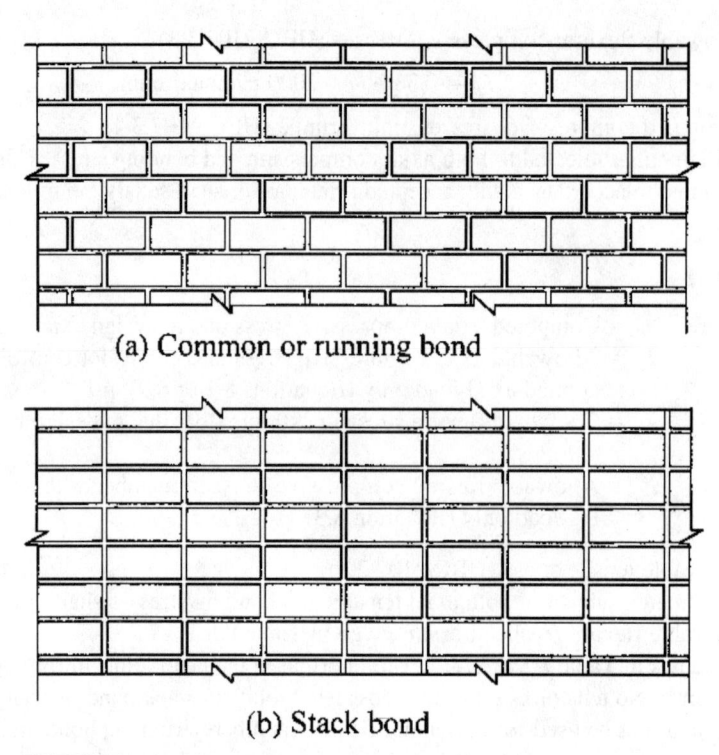

(b) Stack bond

FIGURE 8.1

5. Allowable shear stress in flexure members (UBC 2107.3.6)
   The allowable shear stress F_v in flexure members is

   $$F_v = 1.0\sqrt{f'_m} \qquad 50 \text{ psi maximum} \tag{8.5}$$

   Except at a distance of $\frac{1}{16}$th the clear span beyond the point of inflection, the maximum stress shall be 20 psi.

6. Allowable shear stress in shear walls (UBC 2107.3.7)
   The allowable shear stress F_v in shear wall is

   Concrete units with Type M or S mortar, $F_v = 34$ psi maximum
   Concrete units with Type N mortar, $F_v = 23$ psi maximum

   The allowable shear stress may be increased by $0.2f_{md}$, where f_{md} is computed as compressive stress due to dead load only.

7. Allowable bearing stress (UBC 2107.3.8)
   When a member bears on the full area of a masonry element, the allowable bearing stress F_{br} is

   $$F_{br} = 0.26f'_m \tag{8.6}$$

   When a member bears on one-third or less of a masonry element and the least dimension between the edges of the loaded and unloaded areas is a minimum of one quarter of the parallel side dimension, the allowable bearing stress F_{br} is

$$F_{br} = 0.38 f'_m \qquad (8.7)$$

Values between the full loaded area and one-third or less loaded area shall be interpolated.

8.3.1.3 Flexural Design for Unreinforced Masonry (UBC 2107.3.11)
Design flexure stress f_b shall be computed by

$$f_b = \frac{Mc}{I} \qquad (8.8)$$

where M = Design moment, in.-lb
   c = Distance from neutral axis to extreme fiber, in.
   I = Moment of inertia about neutral axis of cross-sectional area, in^4

8.3.1.4 Shear Design for Unreinforced Masonry (UBC 2107.3.12)
Design shear f_v shall be computed based on the formula

$$f_v = \frac{V}{A_e} \qquad (8.9)$$

where V = Design shear force, lb
   A_e = Effective area of masonry, in^2

The effective area of hollow masonry unit used in design is the minimum bedded area that is dependent upon the thickness of the face shells and the cross-webs, or the gross area of solid units plus any grouted area.

8.3.1.5 Minimum Reinforcement for Stack Bond Unreinforced Masonry (UBC 2107.3.14)
Masonry units laid in stack bond shall have longitudinal reinforcement of at least 0.00027 times the vertical cross-sectional area of the wall placed horizontally in the bed joints or in bond beams spaced vertically not more than 48 inches apart.

8.3.1.6 Allowable Masonry Stresses for Reinforced Masonry
The service load stresses of reinforced concrete masonry shall not exceed the values given in this section. When there are no special inspections as outlined in UBC section 1701, the allowable stresses for masonry in the following sections shall be reduced by 50 percent (UBC 2107.1.2).

1. Allowable stresses in reinforcement (UBC 2107.2.11)
   a. Tensile stress
      Deformed bars: $F_s = 0.5 f_y$, 24,000 psi maximum
      Wire reinforcement: $F_s = 0.5 f_y$, 30,000 psi maximum
      Ties, anchors, and smooth bars: $F_s = 0.4 f_y$, 20,000 psi maximum
   b. Compressive stress
      Deformed bars in columns: $F_{sc} = 0.4 f_y$, 24,000 psi maximum
   c. Deformed bars in flexural members: $F_s = 0.5 f_y$, 24,000 psi maximum

d. Deformed bars in shear walls that are confined by lateral ties throughout the distance where compression reinforcement is required, and where such lateral ties are not less than $\frac{1}{4}$-inch in diameter and spaced not farther apart than 16 bar diameters or 48 tie diameters: $F_{sc} = 0.4f_y$, 24,000 psi maximum

2. Allowable axial compressive force (UBC 2107.2.5)

The allowable compressive stress F_a for members other than reinforced masonry column is

$$F_a = 0.25 f'_m R \qquad (8.10)$$

For reinforced masonry columns, the allowable compressive force P_a is a combination of masonry capacity and steel capacity. The masonry capacity P_m is $0.25 f'_m A_e$, and the steel capacity P_s is $0.65 A_s F_{sc}$. Therefore, the allowable compressive force P_a is

$$P_a = [0.25 f'_m A_e + 0.65 A_s F_{sc}] R \qquad (8.11)$$

where f'_m = Specified compressive strength of masonry at 28 days, psi
A_e = Effective area of masonry, in^2
A_s = Effective cross-sectional area of reinforcement, in^2
F_{sc} = Allowable compressive stress in column reinforcement, psi
R = Load reduction factor to account for the added moment introduced by the lateral displacement of the deflected column shape ($P\Delta$ effect).

$$R = 1 - \left(\frac{h'}{140r}\right)^2 \quad \text{for } \frac{h'}{r} \leq 99 \qquad (8.12)$$

$$R = \left(\frac{70r}{h'}\right)^2 \quad \text{for } \frac{h'}{r} > 99 \qquad (8.13)$$

where h' = Effective height of wall or column, in., is the distance between pinned ends (Euler column formula, as mentioned in Chapter 3). Therefore, the effective column height in column with other conditions of end restraint becomes the distance between points of inflection (equivalent to pinned ends). For members not supported at top, normal to the axis considered, the effective height is twice the height of the member above the support.
r = Radius of gyration, in., based on specified unit dimension, or UBC Tables 21-H-1, 21-H-2, and 21-H-3 for partially grouted hollow unit masonry.

3. Allowable flexural compressive stresses (UBC 2107.2.6)

$$F_b = 0.33 f'_m \qquad 2000 \text{ psi maximum} \qquad (8.14)$$

4. Combined compressive stresses, unit formula (UBC 2107.2.7)

Elements subjected to both axial compression and bending stresses shall be designed in accordance with accepted principles, or shall satisfy the following formula

$$\frac{f_a}{F_a} + \frac{f_b}{F_b} \leq 1 \tag{8.15}$$

5. Allowable tensile stress (UBC 2107.2.3)
   The reinforcing steel is assumed to carry all tensile stresses. Masonry carries no tensile stress.
6. Allowable shear stress in flexure members (UBC 2107.2.8)
   The allowable shear stress F_v in flexure members without shear reinforcement is

$$F_v = 1.0\sqrt{f'_m} \quad \text{50 psi maximum} \tag{8.16}$$

Except at a distance of $\frac{1}{16}$th the clear span beyond the point of inflection, the maximum stress shall be 20 psi.

The allowable shear stress F_v in flexure members with shear reinforcement is

$$F_v = 3.0\sqrt{f'_m} \quad \text{150 psi maximum} \tag{8.17}$$

7. Allowable shear stress in shear walls (UBC 2107.2.9)
   The allowable shear stress F_v in shear walls, where in-plane flexure reinforcement is provided and masonry is used to resist all shear, is as follows

For $\dfrac{M}{Vd} < 1$

$$F_v = \frac{1}{3}\left(4 - \frac{M}{Vd}\right)\sqrt{f'_m} \tag{8.18}$$

With a maximum value of $F_{v(\max)} = \left(80 - 45\dfrac{M}{Vd}\right)$

For $\dfrac{M}{Vd} \geq 1$

$$F_v = 1.0\sqrt{f'_m} \quad \text{35 psi maximum} \tag{8.19}$$

where M = Design moment, in.-lb
       V = Design shear force, lb
       d = Distance from extreme compression fiber of flexure member to centroid of longitudinal tensile reinforcement, in.

The allowable shear stress F_v in shear wall, where shear reinforcement is provided to take all shears, is as follows:

For $\dfrac{M}{Vd} < 1$

$$F_v = \frac{1}{2}\left(4 - \frac{M}{Vd}\right)\sqrt{f'_m} \tag{8.20}$$

With a maximum value of $F_{v(max)} = \left(120 - 45\dfrac{M}{Vd}\right)$

For $\dfrac{M}{Vd} \geq 1$

$$F_v = 1.5\sqrt{f'_m} \qquad 75 \text{ psi maximum} \tag{8.21}$$

8. Allowable bearing stress (UBC 2107.3.8)
    When a member bears on the full area of a masonry element, the allowable bearing stress F_{br} is

$$F_{br} = 0.26 f'_m \tag{8.22}$$

When a member bears on one-third or less of a masonry element, and the least dimension between the edges of the loaded and unloaded areas are a minimum of one quarter of the parallel side dimension of the loaded area, the allowable bearing stress F_{br} is

$$F_{br} = 0.38 f'_m \tag{8.23}$$

Values between the full loaded area and one-third or less loaded area shall be interpolated.

8.3.1.7 Flexural Design for Rectangular Flexure Elements (UBC 2107.2.15)
In the design of reinforced rectangular section, as shown in Figure 8.2, the first step is to locate the neutral axis. This is found by equating the moment of the transformed steel area

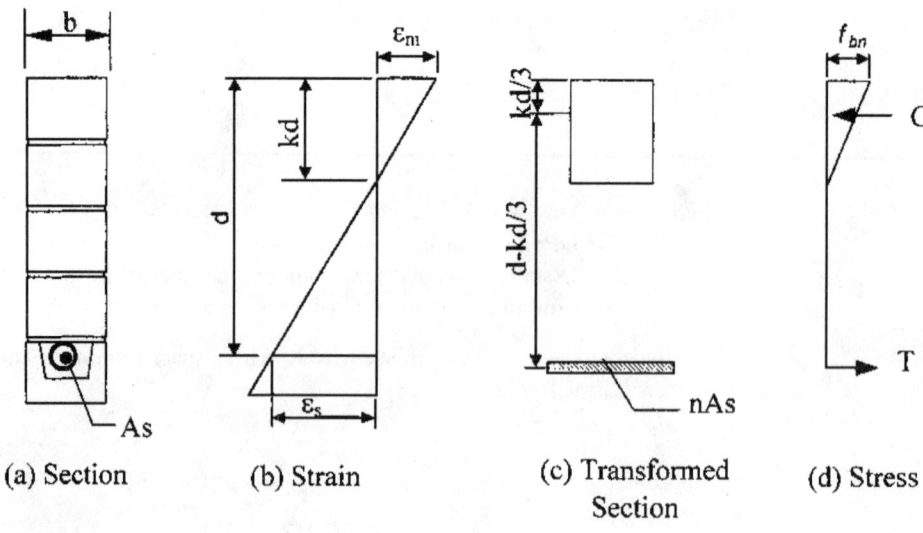

(a) Section   (b) Strain   (c) Transformed Section   (d) Stress

FIGURE 8.2

about the centroidal axis of the cross section to the moment of the compression area about the centroidal axis as follows:

$$b(kd)\left[\frac{kd}{2}\right] = nA_s(d - kd)$$

The coefficient k is the ratio of depth of compressive stress block to the total depth from the compression face to the reinforcing steel d, and b is the width of the member, in.

Rearranging

$$\frac{b(kd)^2}{2} - nA_s(d - kd) = 0$$

Substituting ρbd for A_s, where ρ is the steel ratio, we obtain

$$\frac{b(kd)^2}{2} - n\rho bd(d - kd) = 0$$

Dividing through by bd^2 and multiplying by 2;

$$k^2 - 2\rho n(1 - k) = 0$$

Solving for k

$$k = \sqrt{2\rho n + (\rho n)^2} - \rho n \tag{8.24}$$

Therefore, the location of the neutral axis for the section is determined by the modulus ratio $n = E_s/E_m$ and the steel ratio $\rho = A_s/bd$. This equation, however, applies only for rectangular beams without compressive reinforcement. It can be used to analyze the moment capacity of a beam with a given cross section for which the modular ratio n and the steel ratio ρ are known, or for calculating unit stresses in the steel and masonry when a known moment is applied.

The compression stresses in the beam vary linearly from zero at the neutral axis to a maximum stress of f_m at extreme compression fiber. The total compressive force is computed as

$$C = \frac{1}{2}f_m bkd \tag{8.25}$$

This force acts at the centroid of the triangular stress block, or $kd/3$ from the top. The level arm jd is

$$jd = d - \frac{kd}{3} = d\left(1 - \frac{k}{3}\right) \quad \text{or} \quad j = 1 - \frac{k}{3} \tag{8.26}$$

The coefficient j is the ratio of the distance between the resultant compressive force and the centroid of the tensile force to the distance d.

If the moment at service loads is M, taking moment about T, we can write

$$M = Cjd = \frac{1}{2}f_m bkd(jd) = \frac{1}{2}f_m kjbd^2 \tag{8.27}$$

The compressive stress in the masonry becomes

$$f_m = \frac{M}{bd^2}\left(\frac{2}{jk}\right) \tag{8.28}$$

Similarly, taking moment about C gives the following:

$$M = Tjd = A_s f_s (jd) \tag{8.29}$$

The tensile stress in the longitudinal reinforcement is

$$f_s = \frac{M}{jd} \tag{8.30}$$

Since ρ is usually unknown, the expression for k in Equation 8.24 cannot be used in design. For design purposes, the location of neutral axis is expressed in terms of allowable stresses in steel and masonry from the geometry of the strain diagram, as follows:

$$\frac{\epsilon_m}{\epsilon_s} = \frac{kd}{d - kd}$$

Substituting $\epsilon_m = \dfrac{f_m}{E_m}$ and $\epsilon_s = \dfrac{f_s}{E_s}$, we obtain

$$\left(\frac{f_m}{E_m}\right)\left(\frac{E_s}{f_s}\right) = \frac{k}{1 - k}$$

Since the modular ratio $n = \dfrac{E_s}{E_m}$, solve

$$k = \frac{1}{1 + \dfrac{f_s}{nf_m}} \tag{8.31}$$

This expression applies in design cases only.

The balanced steel ratio ρ_b is given by the equilibrium of tensile and compressive forces at the section, as follows:

$$A_s f_s = \frac{f_m}{2} b(kd)$$

Dividing both sides by bd and solving for ρ_b,

$$\rho_b = \frac{k}{\left(\dfrac{2f_s}{f_m}\right)}$$

Substituting the equation for k in the above equation gives

$$\rho_b = \frac{n}{2\left(\dfrac{f_s}{f_m}\right)\left[n + \left(\dfrac{f_s}{f_m}\right)\right]} \tag{8.32}$$

If the actual steel ratio ρ is less than ρ_b, the steel stress will reach its allowable value before the masonry reaches its allowable limit; thus the design moment will be governed by the allowable steel stress. If the actual steel ratio ρ is greater than ρ_b, the masonry will reach its allowable stress first, and it will dictate the permissible moment on the beam.

8.3.1.8 Flexural Design for T beams (partially grouted hollow unit masonry)

In order to minimize the amount of grout used and to reduce the weight of wall, grouts may be placed only at the locations of the reinforcing steel. Wall grouted only at the reinforcing steel may develop a rectangular or a T stress block depending on the location of the neutral axis. If the compression area were within the face shells, the wall would be analyzed or designed as a rectangular element. If the neutral axis were below the face shell, the section would have a T section stress block. Therefore, as in the design of rectangular sections, the first step in the design of reinforced T sections is to locate the neutral axis. See the T section of a partially grouted wall section as shown in Figure 8.3. Both the face shell flange and part of the web resist the compression force C.

The compressive force on the flange is

$$C_f = \frac{1}{2}\left(f_m + \frac{kd - t_f}{kd}f_m\right)bt_f = f_m\frac{2kd - t_f}{2kd}bt_f \tag{8.33}$$

The compressive force on the web is

$$C_w = \frac{f_m}{2}\left(\frac{kd - t_f}{kd}\right)[b_w(kd - t_f)] \tag{8.34}$$

However, the contribution of the portion of the web in compression is generally small and can be neglected. Therefore,

$$T \approx C \quad \text{or} \quad \rho b d f_s = f_m\frac{2kd - t_f}{2kd}bt_f \tag{8.35}$$

From the strain compatibility relationship, it can be determined that

$$k = \frac{n}{n + \left(\dfrac{f_s}{f_m}\right)}$$

FIGURE 8.3

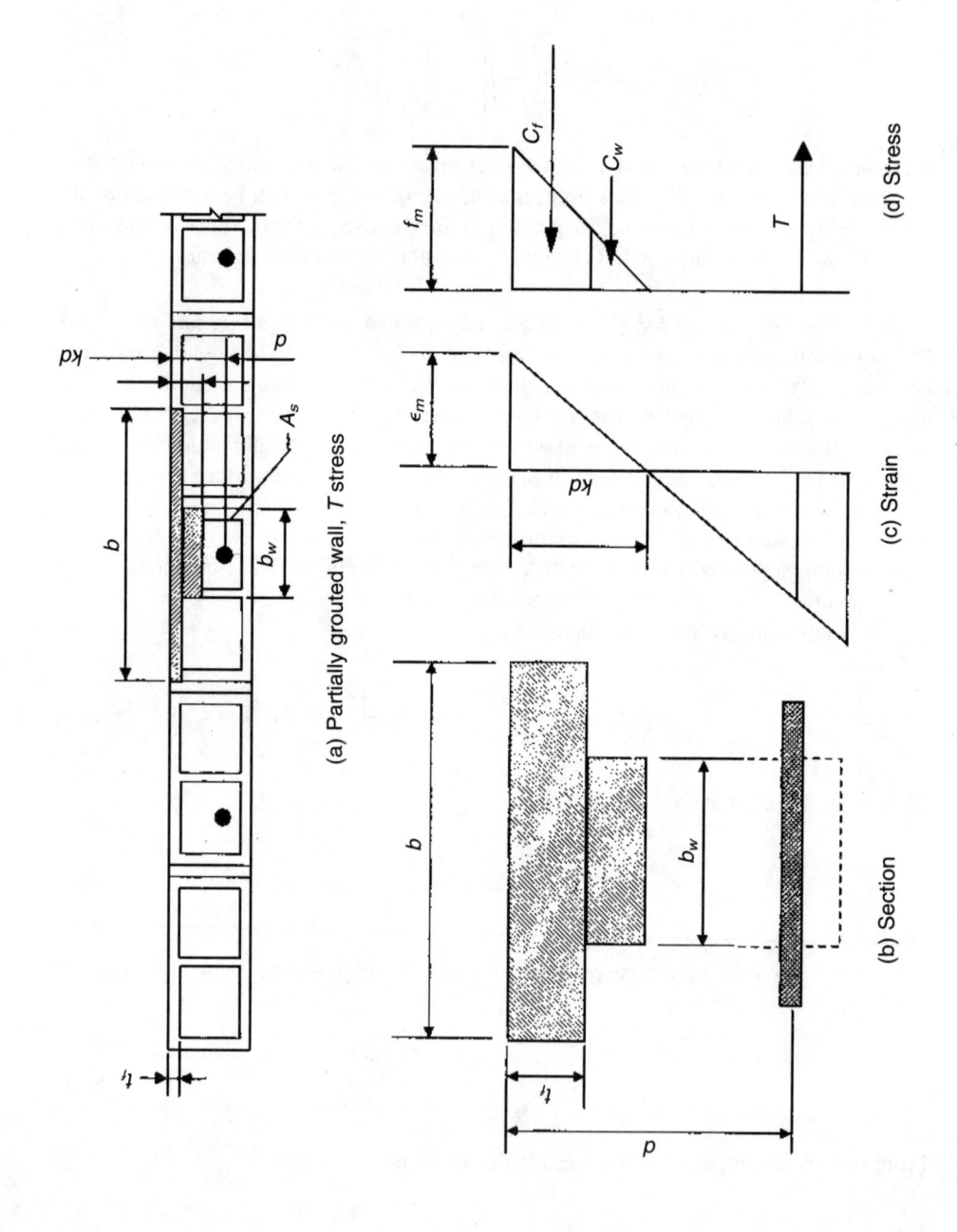

Rearranging the equation and solving for f_m gives

$$f_m = f_s \frac{k}{n(1-k)}$$

Substituting this equation of f_m into Equation 8.35 to eliminate the unit stresses, we obtain

$$k = \frac{\rho n + \left[\left(\frac{1}{2}\right)\left(\frac{t_f}{d}\right)^2\right]}{\rho n + \left(\frac{t_f}{d}\right)} \tag{8.36}$$

The distance from the top of the beam to the center of compression is

$$z = \left[\frac{3kd - 2t_f}{2kd - t_f}\right]\left[\frac{t_f}{3}\right] \quad \text{and} \quad jd = d - z$$

From the above equations, the coefficient j becomes

$$j = \frac{6 - 6\left(\frac{t_f}{d}\right) + 2\left(\frac{t_f}{d}\right)^2 + \left(\frac{t_f}{d}\right)^3\left(\frac{1}{2\rho n}\right)}{6 - 3\left(\frac{t_f}{d}\right)} \tag{8.37}$$

The resisting moments become

$$M_m = C_f jd = f_m \frac{2kd - t_f}{2kd} bt_f jd \tag{8.38}$$

$$M_s = A_s f_s jd = \rho b d f_s jd \tag{8.39}$$

The equation for the coefficient j (Eq. 8.37) is somewhat cumbersome and impractical in design. The level arm can be estimated by assuming the center of compression at the middepth of the face-shell thickness, i.e. $jd = (d - t_f/2)$.

The moment equations become

$$M_m = C_f jd = f_m \frac{2kd - t_f}{2kd} bt_f \left(d - \frac{t_f}{2}\right) \tag{8.40}$$

and

$$M_s = A_s f_s jd = \rho b d f_s \left(d - \frac{t_f}{2}\right) \tag{8.41}$$

8.3.1.9 Shear Design for Reinforced Masonry in Flexure Members and Shear Walls (UBC 2107.2.17)

Design shear f_v shall be computed based on the formula:

$$f_v = \frac{V}{bjd} \qquad (8.42)$$

For members of T or I section, b shall be substituted by the width of web b'.

When the computed f_v exceeds the allowable shear stress in masonry, F_v (Equation 8.16), web reinforcement must be provided and designed to carry the total shear force. Both vertical and horizontal shear stresses must be considered.

The area required for shear reinforcement placed perpendicular to the longitudinal reinforcement shall be computed by:

$$A_v = \frac{sV}{F_s d} \qquad (8.43)$$

where V = Design shear force, lb
 s = Spacing of stirrups or of bent bars in direction parallel to that of main reinforcement, in.
 F_s = Allowable stresses in reinforcement, psi, per UBC 2107.2.11.

8.3.2 Strength-Design Method

The design of concrete masonry structures using strength design is essentially the extension of design of reinforced concrete. The design assumptions for the strength design of masonry structures are as follows (UBC 2108.2.1.2):

1. Masonry carries no tensile stress greater than the modulus of rupture.
2. Reinforcement is completely surrounded by and bonded to masonry material, so that they work together as a homogeneous material.
3. Strain in reinforcement and masonry shall be assumed to be directly proportional to the distance from the neutral axis.
4. Stress in reinforcement below specified yield strength f_y for grade of reinforcement used shall be taken as E_s times steel strain. For strain greater than that corresponding to f_y, stress in reinforcement shall be considered independent of strain and equal to f_y.
5. Tensile strength of masonry walls shall be neglected in flexure calculations of strength, except when computing requirements for deflections.
6. Maximum usable strain ϵ_{mu} at the extreme masonry compression fiber shall:
    a. Be 0.003 for the design of beams, piers, columns, and walls.
    b. Not exceed 0.003 for moment resisting wall frames, unless lateral reinforcement as defined in UBC Section 2108.2.6.2.6 is utilized.
7. Masonry stress of $0.85 f'_m$ shall be assumed uniformly distributed over an equivalent compression zone bounded by the edges of the cross section and a straight line located parallel to the neutral axis at a distance $a = 0.85c$ from the fiber of maximum compressive strain. Distance c from fiber of maximum strain to the neutral axis shall be measured in a direction perpendicular to that axis.

8.3.2.1 Strength Requirements
The basic requirement for strength design may be expressed as follows:

$$\text{Design Strength} \geq \text{Required Strength}$$

or

$$\phi \text{ (Nominal Strength)} \geq U \tag{8.44}$$

1. Required strength: The required strength U is computed by multiplying the service loads by load factors in accordance with the factored load combinations of UBC Section 1612.2. The code gives load factors for specific combinations of loads. The factor assigned to each load is influenced by the degree of accuracy to which the load effect usually can be calculated and the variation that might be expected in the load during the lifetime of the structure. The considerations are also given to the probability of simultaneous occurrence of a specified combination of loading and the combinations of loading to determine the most critical design condition.
2. Design strength: The design strength is computed by multiplying the nominal strength by a strength reduction factor ϕ, which is always less than one. The purposes of the strength reduction factor are
    a. To account for uncertainties in design computation
    b. To reflect the importance of various types of members in the structure
    c. To allow for the probability of understrength members due to variations in material strengths, workmanship, and dimension.
    d. To reflect the degree of ductility and required reliability of the member under the load effects being considered.

    The nominal strength is computed by the code procedures, assuming that the member will have exact dimensions and material properties used in the computations.
3. Strength reduction factors (UBC 2108.1.4.1 ~ 2108.1.4.6):
    The strength reduction factors ϕ shall be as follows:
    a. Beams, piers and columns
        (1) Flexure, with or without axial load:

$$\phi = 0.8 - \frac{P_u}{A_e f'_m} \quad \text{and} \quad 0.60 \leq \phi \leq 0.80 \tag{8.45}$$

where A_e = Effective area of masonry, in^2
P_u = Factored axial load, lb
f'_m = Specified compressive strength of masonry at age of 28 days, psi.

(2) Shear: $\phi = 0.60$
b. Wall design for out-of-plane loads
  (1) Unfactor axial load $\leq 0.04 f'_m$.
      Flexure: $\phi = 0.80$
  (2) Unfactor axial load $> 0.04 f'_m$
      Axial load with or without flexure: $\phi = 0.80$
      Shear: $\phi = 0.60$

c. Wall design for in-plane loads
   (1) Axial load and axial load with flexure:

$$\phi = 0.65$$

For walls with symmetrical reinforcement in which $f_y \le 60,000$ psi, the value of ϕ may be increased linearly to 0.85 as the value of ϕP_n decreases from $0.1 f'_m A_e$ or $0.25 P_b$ (whichever is smaller) to zero.
For solid grouted walls, the value of P_b is

$$P_b = 0.85 f'_m b a_b$$

$$a_b = 0.85 d \frac{e_{mu}}{e_{mu} + \left(\dfrac{f_y}{E_s}\right)} \qquad (8.46)$$

where P_b = Nominal axial load strength of a section at balanced strain conditions, lb
f'_m = Specified compressive strength of masonry at age of 28 days, psi
b = Effective width of rectangular member or width of compression flange for T and I sections, in.
d = Distance from extreme compression fiber of flexure member to centroid of longitudinal tensile reinforcement, in.
e_{mu} = Maximum usable compressive strain of masonry
    = 0.003 for design of beams, piers, columns, and walls
    $\le$ 0.003 for moment-resisting wall frames, unless lateral reinforcement as defined in UBC Section 2108.2.6.2.6 is utilized.

   (2) Shear: $\phi = 0.60$
   $\phi = 0.80$ For shear wall that has nominal shear strength exceeds the shear corresponding to development of its nominal flexural strength for the factored-load combination.
d. Moment-resisting wall frame
   (1) Flexure, with or without axial load:

$$0.65 \le \phi = 0.85 - 2\left(\frac{P_u}{A_e f'_m}\right) \le 0.85 \qquad (8.47)$$

   (2) Shear: $\phi = 0.80$
e. Anchor bolts $\phi = 0.80$
f. Reinforcement
   (1) Development: $\phi = 0.80$
   (2) Splices: $\phi = 0.80$

8.3.2.2 Design of beams, piers, and columns

8.3.2.2.1 Nominal strength. Calculation of the nominal strength of reinforced masonry beams, piers, and columns are based on the design assumptions as mentioned in Section 8.3.2, the applicable conditions of equilibrium, and compatibility of strains.

1. **Nominal flexure strength**
   The strength capacity equations for the masonry structure are an extension of those equations for the reinforced concrete structure by replacing f'_c with f'_m, the value of f'_m being not be less than 1500 psi. For computational purpose, the value of f'_m shall not exceed 4000 psi (UBC 2108.2.3.1).

   a. *Rectangular sections with tension reinforcement only*

   The nominal moment capacity of the section is given by

$$M_n = A_s f_y \left(d - \frac{a}{2}\right) \quad \text{or} \quad M_n = 0.85 f'_m ab \left(d - \frac{a}{2}\right) \tag{8.48}$$

where

$$a = \frac{A_s f_y}{0.85 f'_m b} \tag{8.49}$$

Figure 8.4 shows the strain and stress distribution producing balanced condition on a flexure member.

Balanced strain condition exists at a cross section when the tension reinforcement reaches the strain corresponding to its specified yield strength f_y just as the masonry reaches its maximum usable strain of 0.003 at the extreme masonry compression fiber (UBC 2108.2.3.3). From similar triangles, the distance from the extreme compression fiber to the neutral axis at the balanced design is given by

$$c_b = \frac{0.003}{0.003 + \left(\dfrac{f_y}{E_s}\right)} d = \frac{87,000}{87,000 + f_y} d \tag{8.50}$$

From equilibrium, we obtain

$$A_s f_y = 0.85 f'_m b \left(0.85 d \frac{87,000}{87,000 + f_y}\right)$$

Therefore, the balanced reinforcement ratio ρ_b for rectangular sections with tension reinforcement only is given by

$$\rho_b = \frac{A_s}{b_d} = \frac{0.85(0.85) f'_m}{f_y} \left(\frac{87,000}{87,000 + f_y}\right) \tag{8.51}$$

The balanced reinforcement ratio ρ_b serves as an important index to determine the mode of failure, i.e., whether the structure will fail in tension or compression.

FIGURE 8.4 (a) Section; (b) Strain; (c) Stress

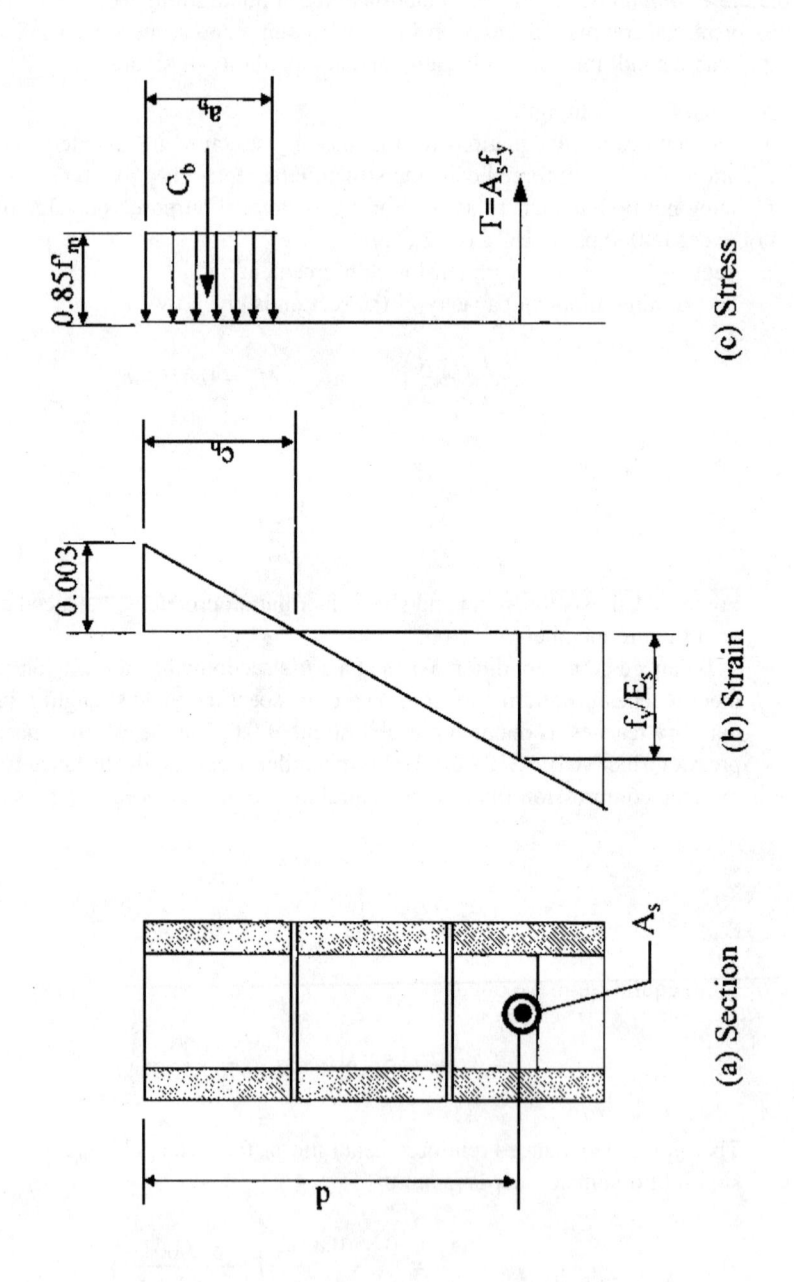

If ρ is less than ρ_b, known as under-reinforced, the tensile force is less than that at balance. In this case, the steel strain ϵ_s is greater than ϵ_y and the steel will have yield at failure. The structure will have noticeable deflection prior to the masonry reaching the crushing strain of 0.003. On the other hand, if ρ is greater than ρ_b, known as over-reinforced, the steel strain ϵ_s is less than ϵ_y when the masonry reaches the strain 0.003. The structure will fail in brittle manner with little deformation.

To ensure that brittle failures will not occur, the UBC Code requires that the ratio of reinforcement ρ provided shall not exceed $0.5\rho_b$ (UBC 2108.2.3.7), i.e.,

$$\rho_{max} = 0.50\rho_b$$

UBC Section 2108.2.3.3 also requires that all longitudinal reinforcement be included in the calculation of the balanced reinforcement ratio except the compression reinforcement contributed to resist the compression loads. The balanced reinforcement ratio is obtained by dividing the area of this reinforcement by the net area of the element. *Net area* is defined as the actual surface area of a cross section of masonry, which is the cross-sectional area minus the areas of ungrouted cores, notches, cells, and unbedded area.

b. Rectangular sections with tension and compression reinforcement
Although the use of compression steel in masonry design is very rare, occasionally the compression steel is needed to bolster the compressive capacity of masonry to avoid brittle failure.

In a rectangular section with tension reinforcement only, masonry alone takes the compression. In a double-reinforced section, the masonry and steel resist the compression together. Therefore, the bending moment is considered equal to the sum of two resisting couples, one of which is provided by the masonry compressive resistance and matching tensile resistance in the tensile steel, as in a rectangular section with tension reinforcement only. The other is provided by the compression steel and the remainder of the tension steel.

Depending on whether or not the compression steel yields, two separate cases used to determine the nominal moment capacity are discussed as follows:

Case 1. Compression steel yields
The total nominal capacity is computed as follows:

$$M_n = A'_s f_y (d - d') + (A_s - A'_s) f_y \left(d - \frac{a}{2}\right) \tag{8.52}$$

where
$$a = \frac{(A_s - A'_s) f_y}{0.85 f'_m b} \tag{8.53}$$

A'_s = Area of compression reinforcement in flexure member, in^2
d' = Distance from extreme compression face of flexure member to centroid of longitudinal compression reinforcement, in.

Based on the assumptions of balanced reinforcement ratio, the tension reinforcement reaches the strain corresponding to its specified yield strength f_y just as the masonry reaches its maximum usable strain of 0.003 at the extreme masonry compression fiber, the balance condition is defined by the following expressions:

$$(A_s - A'_s)f_y = 0.85f'_m b\left(0.85d\frac{87{,}000}{87{,}000 + f_y}\right)$$

or

$$\left(\frac{A_s - A'_s}{bd}\right) = 0.85(0.85)\left(\frac{f'_m}{f_y}\right)\left(\frac{87{,}000}{87{,}000 + f_y}\right)$$

or

$$(\rho - \rho')_b = 0.85(0.85)\left(\frac{f'_m}{f_y}\right)\left(\frac{87{,}000}{87{,}000 + f_y}\right) \qquad (8.54)$$

Case 2. Compression steel does not yield
The strain in the compression steel can be determined by the geometry of the maximum masonry strain of 0.003

$$\frac{0.003}{c} = \frac{\epsilon'_s}{c - d'} \quad \text{or} \quad \epsilon'_s = 0.003\left(\frac{c - d'}{c}\right) \qquad (8.55)$$

Assuming that the tensile steel yields, the internal forces in the member are

$$T = A_s f_y \qquad C_c = 0.85f'_m ba \qquad C_s = (E_s \epsilon'_s)A'_s \qquad (8.56)$$

From equilibrium, $C_c + C_s = T$, we have

$$0.85f'_m b(0.85c) + \left[E_s 0.003\left(\frac{c - d'}{c}\right)\right]A'_s = A_s f_y \qquad (8.57)$$

This can be reduced to the quadratic equation in c, given by

$$0.85(0.85c)f'_m bc^2 + (87{,}000A'_s - A_s f_y)c - 87{,}000A'_s d' = 0$$

Once the c distance to the neutral axis is known, the strain in the compression steel ϵ'_s can be calculated using Equation 8.55. The nominal moment capacity of the section is

$$M_n = 0.85f'_m ba\left(d - \frac{a}{2}\right) + \left[87{,}000\left(\frac{c - d'}{c}\right)\right]A'_s(d - d') \qquad (8.58)$$

The balance condition is define as

$$0.85f'_m b(0.85c_b) + A'_s f_s = A_s f_y \qquad (8.59)$$

where

$$c_b = \frac{0.003}{0.003 + \left(\frac{f_y}{E_s}\right)} d = \frac{87,000}{87,000 + f_y} d$$

and

$$f_s = E_s 0.003 \left(\frac{c_b - d'}{c_b}\right)$$

Therefore, the balanced reinforcement ratio is given by

$$\left(\frac{A_s - A'_s}{bd}\right) = 0.85(0.85)\left(\frac{f'_m}{f_y}\right)\left(\frac{87,000}{87,000 + f_y}\right) + \frac{A'_s}{bd}\left(\frac{f_s}{f_y} - 1\right)$$

or

$$(\rho - \rho')_b = 0.85(0.85)\left(\frac{f'_m}{f_y}\right)\left(\frac{87,000}{87,000 + f_y}\right) + \frac{A'_s}{bd}\left(\frac{f_s}{f_y} - 1\right) \quad (8.60)$$

2. Nominal axial compressive strength
   In accordance with UBC 2108.2.3.6.1, the maximum nominal axial compressive strength is

$$P_n = 0.80[0.85f'_m(A_e - A_s) + f_y A_s] \quad (8.61)$$

3. Nominal shear strength
   Based on conventional shear design concept, in the UBC Code, the nominal shear strength V_n includes two parts:

$$V_n = V_m + V_s$$

in which

V_m is the nominal shear strength contributed by masonry

$$V_m = C_d A_e \sqrt{f'_m} \quad 63 C_d A_e \text{ maximum}$$

C_d is the nominal shear strength coefficient as listed in UBC Table 21-K,

when $\quad \frac{M}{Vd} \leq 0.25 \quad C_d = 2.4$

when $\quad \frac{M}{Vd} \geq 1.00 \quad C_d = 1.2$

M is the maximum bending moment that occurs simultaneously with the shear load V at the section under consideration. A straight-line interpolation is permitted for the value of M/Vd between 0.25 and 1.00.

V_s is the nominal shear strength of shear reinforcement

$$V_s = A_e \rho_n f_y$$

Based on UBC Table 21-J, the maximum nominal shear strength is given by the following equations:

when $\quad \dfrac{M}{Vd} \leq 0.25 \quad (V_n)_{max} = 6.0 A_e \sqrt{f'_m} \leq 380 A_e$

when $\quad \dfrac{M}{Vd} \geq 1.00 \quad (V_n)_{max} = 4.0 A_e \sqrt{f'_m} \leq 250 A_e$

Where M is the maximum bending moment that occurs simultaneously with the shear load V at the section under consideration. A straight-line interpolation is permitted for the value of M/Vd between 0.25 and 1.00.

In the regions subjected to net tension factored loads, only shear reinforcement is assumed to provide shear resistance; V_m shall be zero. When the factored moment M_u is greater than $0.7 M_n$, the value of V_m shall be assumed to be 25 psi.

8.3.2.2.2 Code Specific Requirement. A number of provisions were included in the UBC Code, as follows:

1. Reinforcement: UBC 2108.2.3.7
2. Seismic design provision: UBC 2108.2.3.8
3. Dimensional limits for beams, piers, and columns: UBC 2108.2.3.9
4. Longitudinal and transverse reinforcement for beam: UBC 2108.2.3.10
5. Longitudinal and transverse reinforcement for piers: UBC 2108.2.3.10
6. Longitudinal and transverse reinforcement for columns: UBC 2108.2.3.10

8.3.3 Empirical Design Method

Throughout the world, masonry structures built before the 20th century are empirically designed structures. Rather than detailed analysis of loads and stresses, and calculated structural response, empirical design is based on historical precedent and rules of thumb. Design requirements are specified by the building codes; empirical requirements are simplistic in their application. Height- or length-to-thickness ratios are used in conjunction with minimum wall thickness to determine the required section of a given wall. However, empirical design methods do not incorporate reinforcing steel for load resistance, and may be used for low-rise buildings only. Buildings designed by either working design method or strength-design method are usually more efficient and economical. Therefore, empirical design may be used only for small structures and in buildings of limited height, where wind loads are low, and seismic loading is not a consideration.

8.3.3.1 Limitations of Empirical Design
UBC Section 2109 codes impose restrictions on empirical design as follows:

1. Apply only in seismic zones 0 and 1
2. Basic wind speed less than 80 miles per hour

3. Height limit: Not to exceed 35 ft in height for buildings relying on masonry walls for lateral load resistance (shear walls)
4. Lateral stability
   a. Provide shear walls parallel to the direction of the lateral forces resisted.
   b. Minimum masonry shear wall thickness = 8 in.
   c. Minimum cumulative length of walls in the each direction that shear walls are required for lateral stability is 0.4 times the long dimension of the building. The cumulative length of shear walls shall not include openings.
   d. Maximum spacing of shear walls shall not exceed the ratio listed in Table 8.9 (UBC Table 21-L).
5. Compressive stresses
   The allowable compressive stresses in masonry are given in Table 8.10 (UBC Table 21-M). Compressive stresses in masonry shall be determined by dividing the design loads, excluding wind or seismic loads, by the gross cross-sectional area based on specified rather than nominal dimension.
6. Anchor bolt shear not to exceed the allowable shear given in Table 8.11 (UBC Table 21-N).
7. Lateral support
   Masonry walls shall be laterally supported by crosswalks, pilasters, buttresses, or structural framing members horizontally, or by floors, roof, or structural framing members vertically at spaces not to exceed the requirements set forth in Table 8.12 (UBC Table 21-O).

   For cantilever walls, except parapet walls, the height-to-nominal-thickness ratio shall not exceed 6 for solid masonry or 4 for hollow masonry. For cavity walls, the value of thickness used for computing the ratio shall be the sums of the nominal thickness of inner and outer wythes of the masonry. When the walls were constructed by different classes of units and mortars, the ratio of height or length to thickness shall not exceed that allowed for the weakest of the combinations of units and mortars of which the member is composed.
8. Minimum thickness
   a. For buildings more than one story in height, the minimum nominal thickness of masonry bearing walls is set at 8 in. The thickness of unreinforced grouted brick masonry walls may be 2 in. less (6 in. minimum).

TABLE 8.9 Shear Wall Spacing Requirements for Empirical Design of Masonry

Floor or roof construction	Maximum ratio
	Shear wall spacing to shear wall length
Cast-in-place concrete	5:1
Precast concrete	4:1
Metal deck with concrete fill	3:1
Metal deck with no fill	2:1
Wood diaphragm	2:1

TABLE 8.10 Allowable Compressive Stresses for Empirical Design of Masonry

Construction: Compressive strength of unit gross area	Allowable compressive stresses*, gross cross-sectional area, psi	
	Type M or S mortar	Type N mortar
Solid masonry of brick and other solid units of clay, shale, sand-lime, or concrete brick:		
$\geq$ 8000 psi	350	300
4500 psi	225	200
2500 psi	160	140
1500 psi	115	100
Grouted masonry of clay, shale, sand-lime, or concrete:		
$\geq$ 4500 psi	275	200
2500 psi	215	140
1500 psi	175	100
Solid masonry of solid concrete masonry units:		
$\geq$ 4500 psi	225	200
2500 psi	160	140
1500 psi	115	100
Masonry of hollow load-bearing units:		
$\geq$ 2000 psi	140	120
1500 psi	115	100
1000 psi	75	70
700 psi	60	55
Hollow walls (cavity or masonry bonded)[†] solid units:		
$\geq$ 2500 psi	160	140
1500 psi	115	100
Hollow units:	75	70

* For masonry units having compressive strengths between those given in the table, linear interpolation may be used for determining allowable stresses.

[†] Where floor and roof loads are carried upon one wythe, the gross cross-sectional area is that of the wythe under load. If both wythes are loaded, the gross cross-sectional area is that of the wall minus the area of the cavity between the wythes.

    b. For solid masonry walls not over 9 ft high in one-story buildings, the minimum nominal thickness becomes 6 in. When gable construction is used, an additional 6 ft is permitted to the peak of the gable.
    c. The thickness of a parapet wall shall not be thinner than the wall below, and its height shall not exceed three times its thickness. The minimum thickness of a parapet wall is set at 8 in.
    d. Depending on the height of unbalanced fill, the minimum thicknesses of foundation walls are listed in Table 8.13.
9. Bond
    When the wall thickness consists of two or more units placed side by side, the facing and backing units shall be bonded using either masonry headers or wall ties, in accordance with UBC 2109.7.

TABLE 8.11 Allowable Shear on Bolts for Empirically Designed Masonry, Except Unburned Clay Units

Bolt diameter, in.	Embedment*, in.	Solid masonry (shear in lb)	Grouted masonry (shear in lb)
$\frac{1}{2}$	4	350	550
$\frac{5}{8}$	4	500	750
$\frac{3}{4}$	5	750	1100
$\frac{7}{8}$	6	1000	1500
1	7	1250	1850[†]
$1\frac{1}{8}$	8	1500	2250[†]

* An additional 2 in. of embedment shall be provided for anchor bolts located in the tops of columns for buildings located in seismic zones 2, 3, and 4.
[†] Permitted only with units not less than 2500 psi.

TABLE 8.12 Wall Lateral Support Requirements for Empirical Design of Masonry

Construction	Maximum l/t or h/t
Bearing walls	
Solid or solid grout	20
All other	18
Nonbearing walls	
Exterior	18
Interior	36

l = length of wall or segment, in.
h = height of wall between points of support, in.
t = effective thickness of wythe, wall or column, in.

10. Anchorage
    Masonry walls depending on other masonry walls, floor and roof diaphragms, and structural frames for lateral support shall be anchored or bonded per UBC 2109.8.

8.4 DESIGN EXAMPLES—WORKING STRESS-DESIGN METHOD

8.4.1 Lintel Design

A lintel is a horizontal beam supporting loads over an opening. It may be constructed of steel shape, reinforced concrete, precast concrete, stone, wood, and reinforced masonry. This example covers the design of reinforced masonry lintels.

TABLE 8.13 Thickness of Foundation Walls for Empirical Design of Masonry (UBC Table 21-P)

Foundation wall construction	Maximum depth of unbalanced fill, ft*	Nominal thickness, in.
Ungrouted masonry of hollow units	4	8
	5	10
	6	12
Solid masonry units	5	8
	6	10
	7	12
Fully grouted masonry of hollow or solid units	7	8
	8	10
	8	12
Partially grouted masonry of hollow units reinforced vertically with #4 bars and grout at 24 in. o.c.	7	8
Bars located not less than $4\frac{1}{2}$ in. from pressure side of wall		

*May be increased with the approval of the building official when local soil conditions warrant such an increase.

In addition to its own weight, a lintel may carry distributed loads from above, both from the wall weight and from floor or roof framing. The lintel may also carry concentrated loads from the framing members above. The shape of the loading diagram to the lintel depends upon whether arching action of the masonry above the opening can be assumed. For a lintel supporting a considerable height of wall above it, the wall tends to create an arch over the opening. When arching action occurs, the lintel supports only the masonry that is contained within a 45° triangle, with sides that begin at the supports and slope upward and inward 45° from the horizontal to converge at an apex at the center of the span. Arching action of wall over the opening may then be counted upon to support the remaining portion of the load. It is important to recognize that the horizontal thrust resulting from any arch action must be provided for by the mass of adjacent walls or by properly designed tension devices. Therefore, arching action should not be considered when the end of the lintel is located near a wall corner, near a control or building expansion joint, or in stacked bond walls.

Where uniform floor or roof loads are applied to the wall above the apex of the triangle, it may be assumed that arching action will carry these loads around the opening without placing load on the lintel. When uniform floor or roof loads are applied below the apex of the triangle, arching action cannot take place; so it is assumed these loads will be carried and applied uniformly on the lintel. Also, when a uniform floor or roof load is applied below the apex of the triangle, it is assumed that all the weight of the masonry above the lintel is uniformly supported by the lintel. An illustration of uniform load distribution on lintels is shown in Figure 8.5.

Concentrated loads from beams or trusses framing into the masonry wall above an opening may be considered to transfer downward from the apex of a triangle located at the point of load application. The sides of the triangle make an angle of 60° with the horizontal. The

MASONRY DESIGN **8.33**

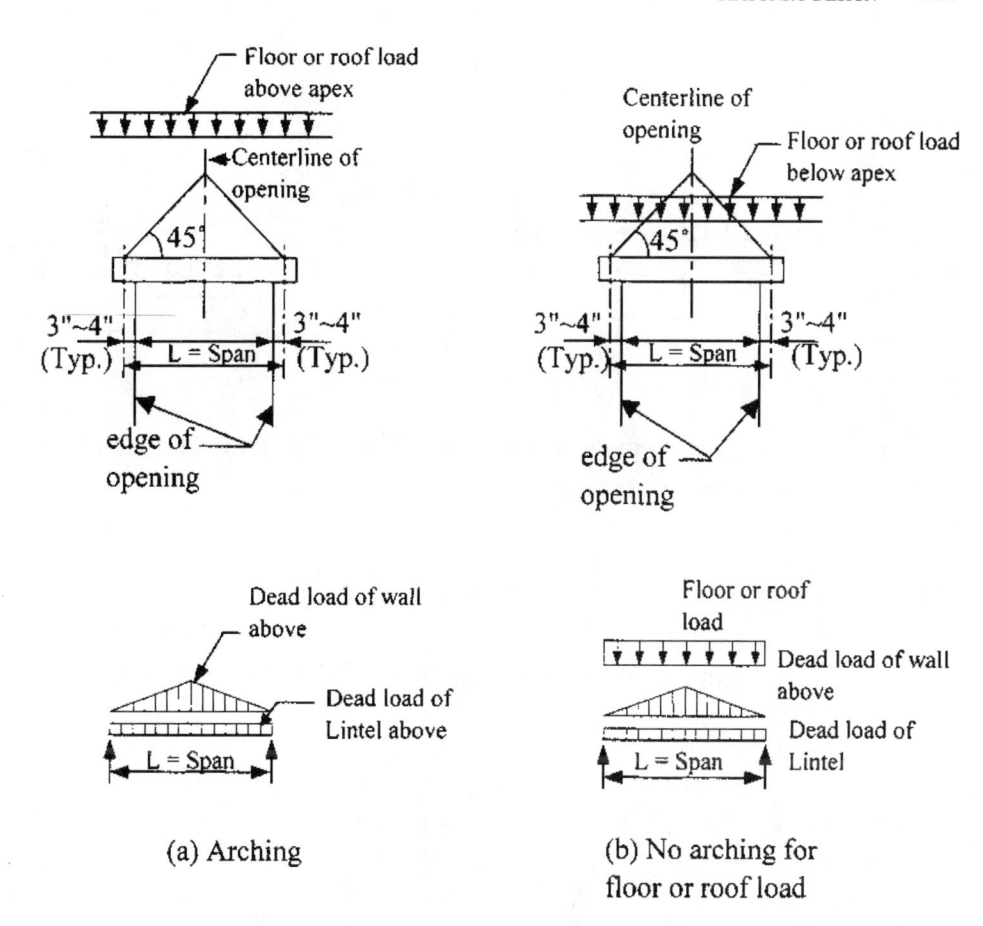

Approximate Uniform Load Distribution on Lintel

FIGURE 8.5

(a) Arching

(b) No arching for floor or roof load

load is transferred as a uniform load over the base of the triangle. This uniform load may extend over only a portion of the lintel, as shown in Figure 8.6.

EXAMPLE 8.1

A door opening 3-ft 4-in. wide by 7-ft 4-in. high is located in the wall, as shown in Figure 8.7a. Wall units are 8-in. non-load-bearing CMU hollow concrete blocks. Wall height = 12 ft. Assume Type S mortar, f'_m = 1500 psi, an 8-in. × 8-in. CMU lintel, and Grade 60 reinforcing steels. Use half-stresses for non-continuous inspection. Determine the reinforcement required using working stress-design method.

Solution

Due to the location of the lintel within the wall panel, the confinng end thrust necessary to provide arching action may be assumed. Therefore, the lintel must support its own weight plus the weight of the triangle of masonry above the door and below the arch.

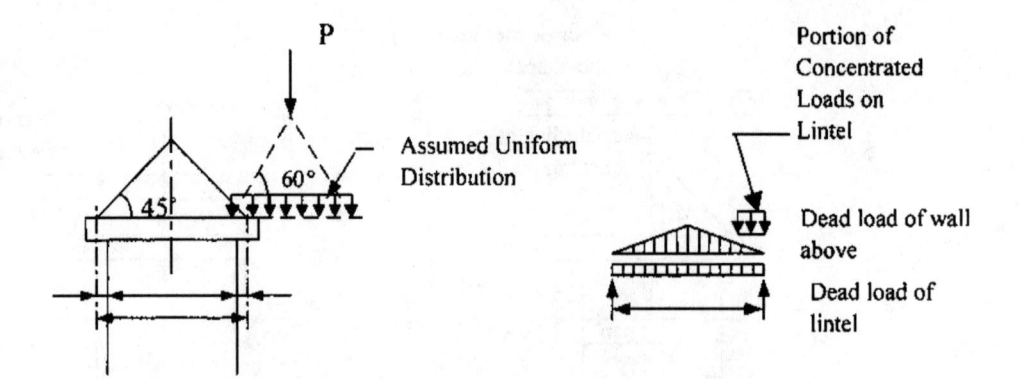

Approximate Concentrated Load Distribution on Lintel
FIGURE 8.6

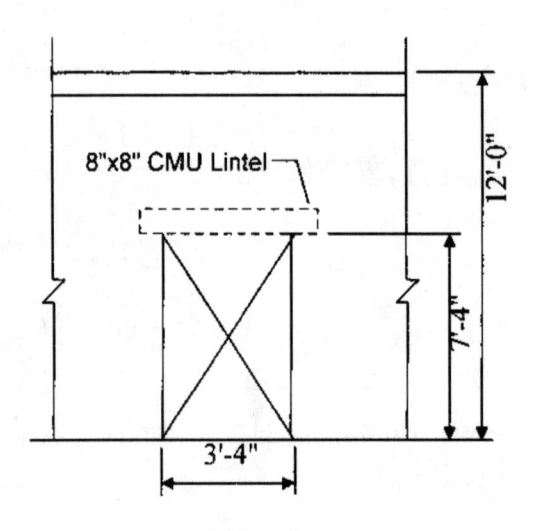

(a) Elevation

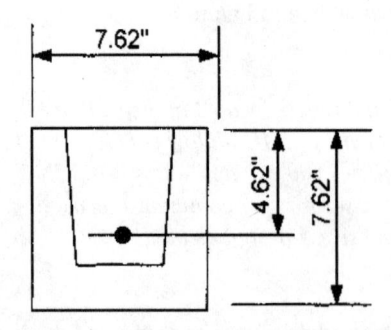

(b) Lintel cross section

FIGURE 8.7

8.34

1. Flexure design
   a. Determine the maximum moment due to the loading M_{max} as follows:

   $$M_{max} = \frac{wL^2}{8} + \frac{w'L^3}{24}$$

   where w = The lintel weight = 62 lb/ft (Table 8.4, solid grouted)
   w' = The weight of the masonry triangle (assume no reinforced filled cells) = 50 psf (Table 8.4, normal weight)
   L = Span length = (3 ft 4 in.) + 2(4 in.) = 4 ft-0 in.

   $$M_{max} = \frac{(62)(4.0)^2}{8} + \frac{(50)(4.0)^3}{24} = 257 \text{ ft-lb}$$

   b. Determine the area of reinforcement
   The balance-steel ratio in the working stress design is determined by the equation

   $$\rho_b = \frac{n}{2\left(\dfrac{f_s}{f_m}\right)\left[n + \left(\dfrac{f_s}{f_m}\right)\right]}$$

   where E_m = Modulus of elasticity of concrete unit masonry (UBC 2106.2.12.1)
   E_s = Modulus of elasticity of steel (UBC 2106.2.12.2)
   $E_m = 750 f'_m$ reduce by one-half since no special inspection provided

   $$n = \frac{E_s}{E_m} = \frac{29{,}000{,}000}{\left(\frac{1}{2}\right)750(1500)} = 51.56$$

   $f_s = 24{,}000$ psi (UBC 2107.2.11)
   $f_m = 0.33 f'_m = 0.33 \times 1500 = 495$ psi < 2000 psi   (UBC 2107.2.6)

   Reduce by one-half, $f_m = 0.5(495) = 247.5$ psi

   $$\rho_b = \frac{n}{2\left(\dfrac{f_s}{f_m}\right)\left[n + \left(\dfrac{f_s}{f_m}\right)\right]} = \frac{51.56}{2\left(\dfrac{24{,}000}{247.5}\right)\left(51.56 + \dfrac{24{,}000}{247.5}\right)} = 0.0018$$

   b = The lintel width = $7\frac{5}{8}$ in.
   d = The effective depth of lintel = 4.62 in.
   $A_{sb} = 0.0018(7.62)(4.62) = 0.06$ in^2

   The minimum horizontal reinforcement required above any wall opening is one #4 bar, $A_s = 0.20$ in$^2 > 0.06$ in^2. Therefore, the design section is over-reinforced and the compressive stresses in masonry will control over the tensile stress in the reinforcement.
   c. Calculate the masonry moment

8.36 CHAPTER 8

The masonry resisting moment M_m is determined as follows:

$$M_m = Cjd = \left(\frac{1}{2}\right) f_m kjbd^2$$

where $k = \sqrt{2\rho n + (\rho n)^2} - \rho n$

and $\rho = \dfrac{A_s}{bd} = \dfrac{0.2}{(7.62)(4.62)} = 0.00568$

$\rho n = (0.00568)(51.56) = 0.2929$

$j = 1 - \left(\dfrac{k}{3}\right) = 1 - \left(\dfrac{0.527}{3}\right) = 0.824$

$k = \sqrt{2(0.2929) + (0.2929)^2} - 0.2929 = 0.527$

$M_m = \left(\dfrac{1}{2}\right)\dfrac{(247.5)(0.527)(0.824)(7.62)(4.62)^2}{(12\text{ in./ft})} = 728 \text{ ft-lb}$

$M_{max} = 257$ ft-lb < 728 ft-lb   OK.

2. Shear check
   a. Determine the maximum shear V as follows:

   $$V = \frac{wL}{2} + \frac{w'L^2}{8} = \frac{(62)(4)}{2} + \frac{(50)(4)^2}{8} = 224 \text{ lb}$$

   b. Determine the shear stress f_v in the lintel due to the maximum shear V_{max} as follows:

   $$f_v = \frac{V}{bjd} = \frac{224}{(7.62)(0.824)(4.62)} = 7.72 \text{ psi}$$

   c. Determine the allowable shear stress F_v as follows:

   $F_v = 1.0\sqrt{f'_m} = 1.0\sqrt{1500} = 38.72$ psi   50 psi maximum

   Reduce by half-stresses for non-continuous inspection.

   $F_v = \left(\dfrac{1}{2}\right)(38.72) = 19.36$ psi > 7.72 psi   OK.

3. Check bearing
   a. Determine the maximum bearing stress f_{br}, assuming a triangular stress distribution as follows:

   $$f_{bg} = \frac{V_{max}}{A_{brg}}$$

   where $V_{max} = 224$ lb.
   A_{brg} = Bearing area of the lintel = 8 in. × 7.62 in. = 61 in^2
   $f_{br} = \dfrac{(2)(224)}{61} = 7.34$ psi

b. Determine the allowable bearing stress F_{br} as follows:

$$F_{br} = 0.26 f'_m = (0.26)(1500) = 390 \text{ psi}$$

Reduce by half-stresses for non-continuous inspection.

$$F_{br} = \left(\frac{1}{2}\right)(390) = 195 \text{ psi} > 7.34 \text{ psi} \quad \text{OK.}$$

EXAMPLE 8.2
A door opening 12 ft wide by 10 ft high is located in a load-bearing wall, as shown in Figure 8.8. Assume the wall above the lintel is reinforced vertically at 32 in. o.c. There is a continuous 8-in. bond beam at the top of wall. The lintel carries a uniform dead load of 100 plf and a uniform live load of 250 plf. There is also a concentrated live load, P is located 7 ft to the right of the door's centerline. These loads are applied at the top of the wall. Wall units are 8-in. load-bearing CMU hollow concrete block. Wall height = 14 ft. Assume Type S mortar, f'_m = 1500 psi, Grade 60 reinforcing steels. Assume that special inspection is provided. Design the lintel to support the given dead and live loads using working stress-design method.

Solution
Since the loads are applied below the apex of the triangle, arching action cannot be assumed. The lintel must be designed for the full applied dead and live loading above.

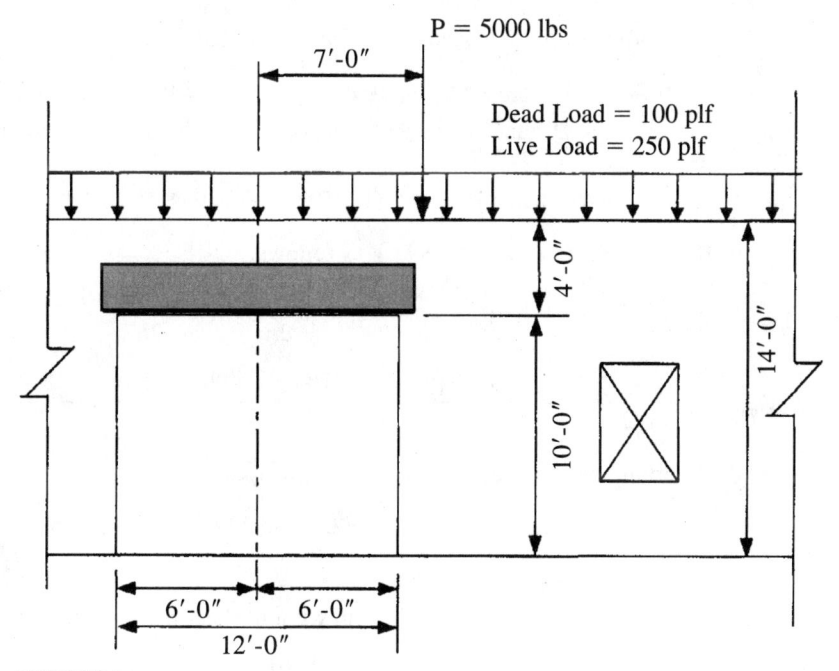

FIGURE 8.8

1. Shear design
   a. Determine the lintel loads as follows:
      The lintel depth will be determined so that shear reinforcement is not required. To establish the dead load, assume a lintel depth of 24 in.
      Unit weight of 8-in. solid grouted wall = 92 psf; partially grouted at 32 in. o.c. = 65 psf (Table 8.4)
      Uniform dead loads include:

      The weight of lintel = (2 ft)(92 psf) = 184 plf (solid grouted)
      The wall above = $\left(\frac{8}{12} \text{ ft}\right)(92 \text{ psf}) + \left(\frac{16}{12}\right)(65) = 148$ plf (includes 8-in. bond beam and 16-in. partially grouted wall)
      Uniform applied dead loads = 100 plf
      Total uniform dead load $W_{DL} = 184 + 148 + 100 = 432$ plf
      Uniform applied live loads $W_{LL} = 250$ plf

      Uniform distribution of the concentrated live load on the lintel W_P is determined as follows:

      Location of concentrated load above lintel $h' = (4 \text{ ft}) - (2 \text{ ft}) = 2$ ft
      Base of triangle = $(2)h' \tan 30° = (2)\tan 30° h' = 1.155h' = (1.155)(2) = 2.31$ ft

      $$W_P = \frac{P}{2h' \tan 30°} = \frac{5000}{2.31} = 2165 \text{ plf}$$

      Portion of lintel loaded, $a = h' \tan 30° + 0.5L - x'$

      where L = Design span length of the lintel = $12 \text{ ft} + \frac{8}{12} \text{ ft} = 12.67$ ft
      x' = Distance from centerline of span to the point of load application = 7 ft
      $a = (2)\tan 30° + (0.5)(12.67) - 7.0 = 0.49$ ft

   b. Determine the maximum shear V as follows:

      $$V_{max} = \frac{W_{DL}L}{2} + \frac{W_{LL}L}{2} + \frac{W_P a(2L - a)}{2L}$$

      $$V_{max} = \frac{432 \times 12.67}{2} + \frac{250 \times 12.67}{2} + \frac{2165 \times 0.49[(2 \times 12.67) - 0.49]}{2 \times 12.67} = 5361 \text{ lbs}$$

   c. Determine the allowable shear stress F_v as follows:

      $$F_v = 1.0\sqrt{f'_m} = 1.0\sqrt{1500} = 38.72 \text{ psi} \qquad 50 \text{ psi maximum}$$

      $\therefore F_v = 38.72$ psi

   d. Determine the minimum lintel depth without shear reinforcement as follows:

      $$f_v = \frac{V}{bjd} = \frac{5361}{(7.62)(0.9)(d_{reqd})} = 38.72 \text{ psi}$$

Assume $j = 0.90$ (check with actual $j = 0.899$, no revision required)

$$d_{reqd} = \frac{(5361)}{(7.62)(0.9)(38.72)} = 20.19 \text{ in.}$$

For a 24-in. deep lintel, the actual effective beam depth,

$$d_{act} = 20.62 \text{ in.} > 20.19 \text{ in.} \quad \text{The 24-in. lintel depth OK.}$$

2. Flexural design
   a. Determine the maximum moment due to the loading M_{max} at midspan as follows:

   $$M_{max} = \frac{W_{DL}L^2}{8} + \frac{W_{LL}L^2}{8} + \frac{W_P a^2}{4}$$

   $$M_{max} = \frac{(432)(12.67)^2}{8} + \frac{(250)(12.67)^2}{8} + \frac{(2165)(0.49)^2}{4} = 13{,}815 \text{ ft-lb}$$

   b. Determine the area of reinforcement
   The balance steel ratio in the working stress design is determined by the equation

   $$\rho_b = \frac{n}{2\left(\dfrac{f_s}{f_m}\right)\left[n + \left(\dfrac{f_s}{f_m}\right)\right]}$$

   where E_m = Modulus of elasticity of concrete unit masonry (UBC 2106.2.12.1)
   E_s = Modulus of elasticity of steel (UBC 2106.2.12.2)
   $E_m = 750 f_{m'}$ Assume that special inspection is provided.

   $$n = \frac{E_s}{E_m} = \frac{29{,}000{,}000}{750(1500)} = 25.77$$

   $f_s = 24{,}000$ psi (UBC 2107.2.11)
   $f_m = 0.33 f'_m = 0.33(1500) = 495$ psi < 2000 psi (UBC 2107.2.6)

   $$\rho_b = \frac{n}{2\left(\dfrac{f_s}{f_m}\right)\left[n + \left(\dfrac{f_s}{f_m}\right)\right]} = \frac{25.77}{2\left(\dfrac{24{,}000}{495}\right)\left(25.77 + \dfrac{24{,}000}{495}\right)} = 0.0036$$

   b = The lintel width = $7\frac{5}{8}$ in.
   d = The effective depth of lintel = 20.62 in.
   $A_{sb} = 0.0036(7.62)(20.62) = 0.57 \text{ in}^2$

   The minimum horizontal reinforcement required above any wall opening is one #4 bar, $A_s = 0.20$ in.^{2} < 0.57 in.^{2} Try two #4 bars, $A_s = 0.40$ in.^{2}

   c. Calculate the masonry moment
   The masonry resisting moment M_m is determined as follows:

   $$M_m = Cjd = \left(\frac{1}{2}\right) f_m k j b d^2$$

where $k = \sqrt{2\rho n + (\rho n)^2} - \rho n$

and $\rho = \dfrac{A_s}{bd} = \dfrac{0.40}{(7.62)(20.62)} = 0.00255$

$\rho n = (0.00255)(25.77) = 0.0656$

$j = 1 - \left(\dfrac{k}{3}\right) = 1 - \left(\dfrac{0.3025}{3}\right) = 0.8992$

$M_m = \left(\dfrac{1}{2}\right)\dfrac{(495)(0.3025)(0.8992)(7.62)(20.62)^2}{(12\ \text{in./ft})} = 18{,}176\ \text{ft-lb}$

$M_{\max} = 13{,}815\ \text{ft-lb} < 18{,}176\ \text{ft-lb}\quad\text{OK.}$

d. Calculate the reinforcing steel moment

The reinforcing resisting moment M is determined as follows:

$$M = Tjd = A_s f_s(jd) = \dfrac{(0.40)(24{,}000)(0.8992)(20.62)}{12} = 14{,}8333\ \text{ft-lb}$$

Note that the reinforcing moment governs the design.

3. Check bearing

a. Determine the maximum bearing stress f_{br}, assuming a triangular stress distribution as follows:

$$f_{bg} = \dfrac{V_{\max}}{A_{br}}$$

where $V_{\max} = 5361\ \text{lb}$

A_{brg} = The bearing area of the lintel = 8 in. × 7.62 in. = 61 in.^{2}

$f_{br} = \dfrac{(2)(5361)}{61} = 176\ \text{psi}$

b. Determine the allowable bearing stress F_{br} as follows:

$F_{br} = 0.26 f'_m = (0.26)(1500) = 390\ \text{psi}$

$F_{br} = 390\ \text{psi} > 176\ \text{psi}\quad\text{OK.}$

EXAMPLE 8.3

Two door openings 12-ft wide by 10-ft high are located in a load-bearing wall, as shown in Figure 8.9. The three masonry courses above the openings will be solid grouted and used as the lintel. The lintel carries a uniform roof dead load of 600 plf and a uniform roof live load of 150 plf. Wall units are 8-in. load-bearing CMU hollow concrete blocks. Wall height = 12 ft. Wall panel length is 30 ft. Assume Type S mortar, $f'_m = 1500$ psi, Grade 60 reinforcing steels. Assume that special inspection is provided. Design the lintel over the door using working stress-design method.

Solution

Due to the size of openings and the length of the wall panel, the walls adjacent to the openings cannot provide the necessary confining end thrust for the arching action. Therefore,

FIGURE 8.9

Dead Load = 600 plf
Live Load = 150 plf

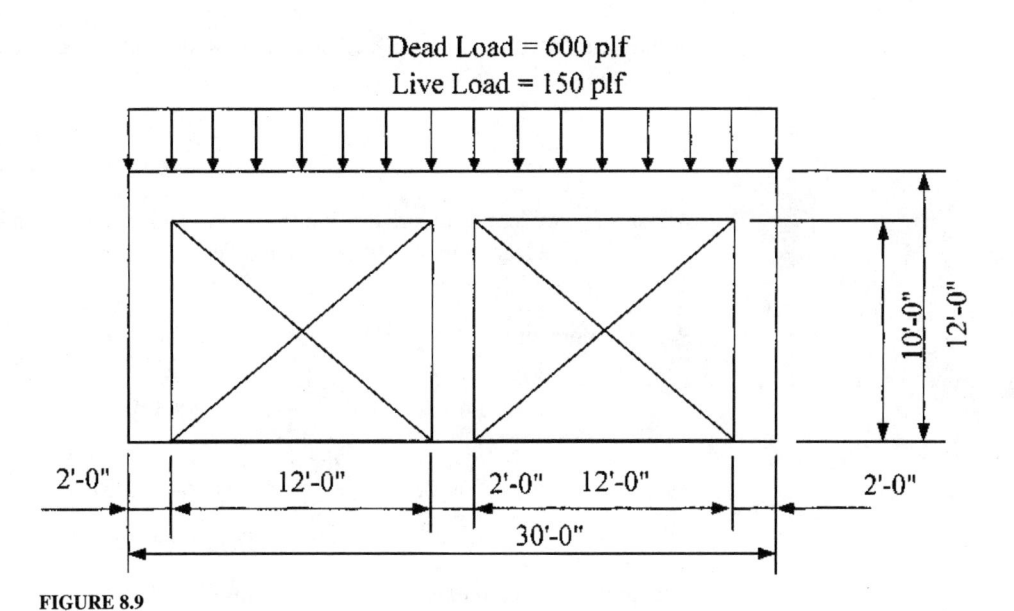

(Dimensions: 2'-0", 12'-0", 2'-0", 12'-0", 2'-0"; overall 30'-0"; height 10'-0" opening, 12'-0" overall)

the lintel must be designed to support its own weight plus the full applied dead and live loads above.

1. Flexure design
   a. Determine the positive and negative moments due to the loading:
   The masonry lintel will be analyzed as a braced frame member in lieu of a moment analysis, since the masonry frame meets all requirements for the approximate method mentioned in ACI-8.3.3. The ACI moment coefficients will be used to determine the approximate design moments and shears.
   Loads:

   Lintel weight = (92 psf)(2 ft) = 184 plf (Table 8.4, solid grouted)

   Roof dead load = 600 plf

   Total dead load, W_{DL} = 184 + 600 = 784 plf

   Roof live load, W_{LL} = 150 plf

   ACI moment coefficient:

   Positive moment, discontinuous end is integral with the support,

   $$\text{Pos. } M = \frac{WL_n^2}{14}$$

   Negative moment, at exterior face of first interior support (two spans),

   $$\text{Neg. } M = \frac{WL_n^2}{9}$$

Negative moment, at interior face of exterior supports for members built where support is a column,

$$\text{Neg. } M = \frac{WL_n^2}{16}$$

where L_n = Clear span length for positive moment and average of adjacent clear spans for negative moment
= 12 ft-0 in.

Positive moment,

$$\text{Pos. } M_{DL} = \frac{(784)(12)^2}{14} = 8064 \text{ ft-lb}$$

$$\text{Pos. } M_{LL} = \frac{(150)(12)^2}{14} = 1543 \text{ ft-lb}$$

$$\text{Pos. } M_{TOT} = 8064 + 1543 = 9607 \text{ ft-lb}$$

Negative moment, at exterior face of first interior support,

$$\text{Neg. } M_{DL} = \frac{(784)(12)^2}{9} = 12{,}544 \text{ ft-lb}$$

$$\text{Neg. } M_{LL} = \frac{(150)(12)^2}{9} = 2400 \text{ ft-lb}$$

$$\text{Neg. } M_{TOT} = 12{,}544 + 2400 = 14{,}944 \text{ ft-lb}$$

The negative moment at exterior face of first interior support governs the flexure design. Conservatively, it also provides same steel area for positive moment.

b. Determine the area of reinforcement.

The balance-steel ratio in the working stress design is determined by the equation

$$\rho_b = \frac{n}{2\left(\dfrac{f_s}{f_m}\right)\left[n + \left(\dfrac{f_s}{f_m}\right)\right]}$$

where E_m = Modulus of elasticity of concrete unit masonry (UBC 2106.2.12.1)
E_s = Modulus of elasticity of steel (UBC 2106.2.12.2)
$E_m = 750 f'_m$ no reduction required since special inspection is provided

$$n = \frac{E_s}{E_m} = \frac{29{,}000{,}000}{750(1500)} = 25.78$$

$f_s = 24{,}000$ psi (UBC 2107.2.11)
$f_m = 0.33 f'_m = 0.33(1500) = 495$ psi < 2000 psi (UBC 2107.2.6)

$$\rho_b = \frac{n}{2\left(\dfrac{f_s}{f_m}\right)\left[n + \left(\dfrac{f_s}{f_m}\right)\right]} = \frac{25.78}{2\left(\dfrac{24{,}000}{495}\right)\left(25.78 + \dfrac{24{,}000}{495}\right)} = 0.0036$$

b = The lintel width = $7\frac{5}{8}$ in.
d = The effective depth of lintel = 20.62 in.
$A_{sb} = 0.0036(7.62)(20.62) = 0.57$ in^2

The minimum horizontal reinforcement required above any wall opening is one #4 bar, $A_s = 0.20$ in$^2 < 0.57$ in^2 Try two #4 bars, $A_s = 0.40$ in^2

c. Calculate the masonry moment
The masonry resisting moment M_m is determined as follows:

$$M_m = Cjd = \left(\frac{1}{2}\right) f_m kjbd^2$$

where $\quad k = \sqrt{2\rho n + (\rho n)^2} - \rho n$

and, $\quad \rho = \dfrac{A_s}{bd} = \dfrac{0.40}{(7.62)(20.62)} = 0.00255$

$\rho n = (0.00255)(25.77) = 0.0656$

$k = \sqrt{2(0.0656) + (0.0656)^2} - 0.0656 = 0.3025$

$j = 1 - \left(\dfrac{k}{3}\right) = 1 - \left(\dfrac{0.3025}{3}\right) = 0.8992$

$$M_m = \left(\frac{1}{2}\right) \dfrac{(495)(0.3025)(0.8992)(7.62)(20.62)^2}{(12 \text{ in./ft})} = 18{,}176 \text{ ft-lb}$$

$M_{NEG} = 14{,}944$ ft-lb $< 18{,}176$ ft-lb   OK.

d. Calculate the reinforcing steel moment
The reinforcing resisting moment M is determined as follows:

$$M = Tjd = A_s f_s (jd) = \dfrac{(0.40)(24{,}000)(0.8992)(20.62)}{12} = 14{,}833 \text{ ft-lb}$$

$M_{max} = 13{,}815$ ft-lb $< 14{,}833$ ft-lb   OK.

Note that the reinforcing moment governs the design.

2. Shear check

a. Determine the maximum shear V:
Using ACI shear coefficient, shear in end members at face of first interior support;

$$V = \dfrac{1.15 W L_n}{2}$$

$$V = \dfrac{1.15 W_{DL} L_n}{2} + \dfrac{1.15 W_{LL} L_n}{2} = \dfrac{(1.15)(784)(12)}{2} + \dfrac{(1.15)(150)(12)}{2} = 6445 \text{ lbs}$$

Shear at face of all other supports

$$V = \dfrac{W L_n}{2}$$

$$V = \dfrac{W_{DL} L_n}{2} + \dfrac{W_{LL} L_n}{2} = \dfrac{(784)(12)}{2} + \dfrac{(150)(12)}{2} = 5604 \text{ lbs}$$

b. Determine the shear stress f_v in the lintel due to the maximum shear V_{max} as follows:

$$f_v = \frac{V}{bjd} = \frac{6445}{(7.62)(0.8992)(20.62)} = 45.62 \text{ psi}$$

c. Determine the allowable shear stress F_v as follows (UBC 2107.2.8):

$$F_v = 1.0\sqrt{f'_m} = 1.0\sqrt{1500} = 38.72 \text{ psi} \quad 50 \text{ psi maximum}$$

$$F_v = 38.72 \text{ psi} < f_v = 45.62 \text{ psi}$$

Therefore, shear reinforcement is required in the beam.

d. Determine the area of shear reinforcement required A_v as follows (UBC 2107.2.17): The area required for shear reinforcement placed perpendicular to the longitudinal reinforcement shall be computed by:

$$A_v = \frac{sV}{F_s d}$$

where V = Design shear force, lb
S = Spacing of stirrups or of bent bars in direction parallel to that of main reinforcement, in.
F_s = Allowable stresses in reinforcement, psi, per UBC 2107.2.11
= 24,000 psi

Although there is no code mentioning the maximum spacing for shear reinforcement in UBC Working Stress Design of Masonry, UBC 2108.2.3.7 (Strength Design of Masonry) requires that the maximum spacing of shear steel shall not exceed one-half the depth of member, nor 48 in.
Use $s = 8$ in.

$$A_v = \frac{(8)(6445)}{(24,000)(20.62)} = 0.104 \text{ in}^2$$

Provide one #3 bar ($A_v = 0.11 \text{ in}^2$) spaced at 8 in. o.c.

e. Determine the allowable shear stress F_v where shear reinforcement is designed to take entire shear force (UBC 2107.2.8)

$$F_v = 3.0\sqrt{f'_m} = 3.0\sqrt{1500} = 116.19 \text{ psi} \quad 150 \text{ psi maximum}$$

$$F_v = 116.19 \text{ psi} > f_v = 45.62 \text{ psi} \quad \text{OK.}$$

8.4.2 Reinforced Masonry Column and Pilaster Design

A column is a vertical structural member primarily designed to support axial loads and whose width does not exceed three times the thickness. In a reinforced column, both the reinforcement and masonry resist compression. Columns build integrally with the wall but projecting from either or both faces of the wall are called pilasters. In addition to supporting vertical loads, the pilaster serves as a stiffening element for the adjoining wall panels and also carries the wall panel reactions caused by the wind or seismic effects. When a wall is designed and reinforced to span horizontally between the pilasters, it is usually assumed to act as a one-way slab, spanning in that direction. The pilasters will carry the entire lateral loads. Pilasters are designed similar to columns except that the pilasters are laterally supported in the direction of the wall, while columns are typically unsupported in both directions. Both columns and pilasters must be designed to sustain the combined effects of axial compressive stresses and flexural stresses.

In general, to design masonry columns or pilasters, the load and moment combinations at top, mid-height, and bottom of the columns should be examined. At top and bottom of a column or pilaster, the combination of the axial load P plus the eccentric moment Pe should be considered. At the mid-height of column, the combination of the axial load P and moments caused by lateral loading and the eccentricity moment $Pe/2$ should be considered. Normally, masonry columns or pilasters will be given lateral support at top by roof or floor system members and at bottom by the footing with dowel bars. For this condition, the conservative assumption is that the column spans vertically as a simple span. When other support fixity conditions exist, moment and axial load shall be calculated in accordance with accepted principles of mechanics.

UBC specifies provisions and limitations for masonry columns and pilasters as follows:

1. In seismic zones 0 and 1. There is no special design and construction provision.
2. In other seismic zones. The area of vertical reinforcement shall not be less than 0.5 percent and not more than 4 percent of the effective area of the column. The effective area of solid grouted masonry is usually equal to gross area Ag (UBC 2106.2.5). At least four #3 bars must be provided. The minimum clear distance between parallel bars in columns shall be $2\frac{1}{2}$ times the bar diameter (UBC 2107.2.13). Lateral ties to the longitudinal bars and additional ties around anchor bolts shall be provided in accordance with the provisions of UBC 2106.3.6 and UBC 2106.3.7. The lateral ties are used to hold the longitudinal bars in position and to provide lateral support to restrain the longitudinal bars from buckling.

    The ties shall be arranged so that every corner and alternate longitudinal bars have lateral support provided by the corner of a complete tie having an included angle of not more than 135° or by a standard hook at the end of a tie. No bar shall be farther than 6 in. on either side from such a laterally supported longitudinal bar. The spacing of the ties shall not exceed 16 longitudinal bar diameters, 48 tie diameters or the least dimension of the column but not more than 18 in. The minimum tie size is a $\frac{1}{4}$-inch diameter bar for longitudinal bars up to #7, and a #3 bar for larger longitudinal bars. Ties smaller than #3 may be used for longitudinal bars larger than #7, provided the total cross-sectional of such smaller ties crossing a longitudinal plane is equal to that of the larger ties at their required spacing.
3. In seismic zones 3 and 4, in addition to the provisions listed above, the masonry columns and pilasters shall be designed and constructed with the following additional requirements and limitations:
    a. The maximum tie spacing shall not exceed 8 in. for the full height of column when the column is stressed by tensile or compressive axial overturning forces from seismic loading. In all other columns, ties shall be spaced at 8-in. maximum in the tops and bottoms of columns for a distance of one-sixth the clear column height, 18 in., or the maximum column cross-sectional dimension, which is

greater. For the remaining height of column, tie spacing shall not shall exceed 16 longitudinal bar diameters, 48 tie diameters, or the least dimension of the column but not more than 18 in. Column ties shall terminate with a minimum 135°-hook with extensions not less than six bar diameters or 4 in. Such extensions shall engage the longitudinal column reinforcement and project into the interior of the column.

b. The least nominal dimension of a reinforced masonry column should not be less than 12 in., except when the allowable stresses are reduced by one-half for working stress design, this value may be reduced to 8 in. (UBC 2197.1.3.2).

EXAMPLE 8.4
A reinforced masonry pilaster, 16-ft high, supports truss dead load end reaction of 25 kips and live load reaction of 15 kips. The eccentricity of truss reaction e is 2 in. The spacing of pilasters is 25 ft. The wall between pilasters is subjected to a wind pressure of 20 psf. The wind load may act both inward and outward. $f'_m = 1500$ psi, Type S mortar. Assume that special inspection is provided. Determine the pilaster size and vertical reinforcement, neglecting the weight of the pilaster.

Solution
1. Design assumptions:
   Assume that the wall spans horizontally between pilasters and all lateral loads on the wall are transmitted to the pilasters. The pilasters are simply supported by roof trusses and spread footings.
2. Check the minimum and maximum reinforcement requirements.
   Try 16 in. × 16 in. pilaster with six #9 bars. ($A_s = 6.0$ in.²)
   The minimum reinforcement $A_{s\ (min)}$ and the maximum reinforcement $A_{s\ (max)}$ are determined as follows:

$$A_{s\ (min)} = 0.005A_e \quad \text{(UBC 2107.2.13)}$$
$$A_{s\ (max)} = 0.04A_e$$

   where A_e is the effective area of column. For a solid grouted masonry, A_e is equal to the gross area A_g:

$$A_g = bt = (\text{actual width})(\text{actual depth}) = (15.62)(15.62) = 244 \text{ in}^2$$
$$A_{s\ (min)} = (0.005)(244)$$
$$= 1.22 \text{ in}^2 < 6.0 \text{ in}^2 \quad \text{OK.}$$
$$A_{s\ (max)} = (0.04)(244)$$
$$= 9.76 \text{ in}^2 > 6.0 \text{ in}^2 \quad \text{OK.}$$

3. Check assumed pilaster for the given loads at the top.
   a. Check axial load
   The allowable axial compressive force P_a is determined as follows:

$$P_a = [0.25f'_m A_g + 0.65A_s F_{sc}]R$$

At top of column, no $P\Delta$ effect exist, $R = 1.0$

$$F_{sc} = 0.4f_y = 24{,}000 \text{ psi} \quad (f_y = 60 \text{ ksi}) \quad \text{(UBC 2107.2.11)}$$

$$P_a = \frac{[(0.25)(1500)(15.62)(15.62) + (0.65)(6.0)(24{,}000)](1.0)}{1000} = 185 \text{ kips}$$

$$P_a > P = 15 + 25 = 40 \text{ kips} \quad \text{OK.}$$

b. Check bending moment
  The eccentric moment at top of column is

$$M = Pe = (40.0)(2.0) = 80.0 \text{ kip-in.}$$

The area of the reinforcement that is in tension = 3.0 in.² (three #9)
(1) Calculate the masonry moment
  The masonry resisting moment M_m is determined as follows:

$$M_m = Cjd = \left(\frac{1}{2}\right)f_m kjbd^2$$

where $f_m = 0.33f'_m = 0.33(1500) = 495 \text{ psi} < 2000 \text{ psi}$ (UBC 2107.2.6)
  $b = 15.62$ in.
  $d = 15.62 - 3.5 = 12.12$ in.
  $n = \dfrac{E_s}{E_m} = \dfrac{29{,}000{,}000}{750(1500)} = 25.78$ (Assume full inspection)
  $k = \sqrt{2\rho n + (\rho n)^2} - \rho n$

and $\rho = \dfrac{A_s}{bd} = \dfrac{3.0}{(15.62)(12.12)} = 0.01585$
  $\rho n = (0.01585)(25.77) = 0.4084$
  $k = \sqrt{2(0.4084) + (0.4084)^2} - 0.4084 = 0.5834$

$$j = 1 - \left(\frac{k}{3}\right) = 1 - \left(\frac{0.5834}{3}\right) = 0.8055$$

$$M_m = \left(\frac{1}{2}\right)\frac{(495)(0.5834)(0.8055)(15.62)(12.12)^2}{(1000 \text{ lb/kips})} = 266.87 \text{ kip-in}$$

$$M = 80.0 \text{ kip-in.} < 266.87 \text{ kip-in.} \quad \text{OK}$$

(2) Calculate the reinforcing steel moment
  The reinforcing resisting moment M is determined as follows:

$$M_s = Tjd = A_s f_s (jd) = \frac{(3.0)(24{,}000)(0.8055)(12.12)}{1000} = 702.91 \text{ kip-in.}$$

$$M = 80.0 \text{ kip-in.} < 702.91 \text{ kip-in.} \quad \text{OK.}$$

Moment capacity M_a = smaller of (M_m and M_s) = 266.87 kip-in.

c. Check combine axial and bending moment, unity formula
   The unit formula can be rewritten in the following format:

$$\frac{P}{P_a} + \frac{M}{M_a} \leq 1.0$$

$$\frac{40}{185} + \frac{80}{267} = 0.52 < 1.0 \quad \text{OK.}$$

4. Check assumed pilaster for the given loads at mid-height.
   a. Check axial load
      The allowable axial compressive force P_a is determined as follows:

$$P_a = [0.25 f'_m A_g + 0.65 A_s F_{sc}] R$$

At mid-height, the $P\Delta$ effect is maximum, reduce the allowable axial load by the load-reduction factor, R

$h' = (16)(12) = 192$ in. (The effective height)

$r = (0.289)(15.62) = 4.51$ in. (Approximate radius of gyration for a rectangular section, UBC Table 21-H-1)

$$\frac{h'}{r} = \frac{192}{4.51} = 42.57 \leq 99$$

Therefore,

$$R = 1 - \left(\frac{h'}{140r}\right)^2 = 1 - \left(\frac{192}{(140)(4.51)}\right)^2 = 0.908$$

$F_{sc} = 0.4 f_y = 24,000$ psi ($f_y = 60$ ksi) (UBC 2107.2.11)

$$P_a = \frac{[(0.25)(1500)(15.62)(15.62) + (0.65)(6.0)(24,000)](0.908)}{1000} = 168 \text{ kips}$$

$P_a > P = 40$ kips OK.

b. Check bending moment
   The eccentric moment at mid-height is

$$M = \frac{Pe}{2} = \frac{(40.0)(2.0)}{2} = 40.0 \text{ kip-in.}$$

The wind load moment at mid-height is determined as follows:

Tributary width = the spacing of pilasters = 25 ft
Wind pressure = 20 psf
Wind load = (20)(25) = 500 plf

Wind load moment

$$M_{wind} = \frac{W_{wind} L^2}{8} = \frac{(500)(192)^2 (1 \text{ ft}/12 \text{ in.})}{8(1000 \text{ lb/kips})} = 192 \text{ kip-in.}$$

The total moment at mid-height = 40 + 192 = 232 kip-in.
Moment capacity at mid-height, M_a, = 266.87 kip-in. > 232 kip-in. OK.

c. Check combine axial and bending moment, unity formula
The unit formula can be rewritten in the following format:

$$\frac{P}{P_a} + \frac{M}{M_a} \leq 1.0$$

$$\frac{40}{168} + \frac{232}{267} = 1.11 < 1.33 \quad \text{OK} \quad \text{(UBC 1612.3.2)}$$

8.4.3 Reinforced Masonry Wall Design for Out-of-plane Loads

Depending on the structural requirements of the design, masonry walls may be single-wythe or multi-wythe, solid or hollow, grouted or ungrouted, reinforced or unreinforced, load-bearing or non-load-bearing. Most masonry walls are designed to span vertically and transfer the lateral loads due to wind or earthquake to the roof, floor, or foundation. However, under certain circumstances such as in a wall-pilaster system, the wall may be designed to span horizontally between pilasters, which in turn span vertically to transfer the lateral loads to the horizontal support elements above and below. The masonry walls also axially transfer the loads from roofs, floors, or beams to foundations.

EXAMPLE 8.5
A building is to be constructed using 12-in.-thick CMU blocks. The total wall height is 24 ft. Determine the reinforcing bar size required resisting lateral wind force of 25 lb/ft², axial dead load of 1000 lb/ft and axial live load of 500 lb/ft. The eccentricity of axial load e is 4 in. The moment due to lateral wind load and to axial eccentricity are additive. $f'_m = 1500$ psi. Use full inspection.

Solution
1. Determine the maximum applied moment that must be resisted by the wall.
   Assume that the wall acts as a simply supported vertical beam.
   The total axial load P = 1000 + 500 = 1500 lb/ft
   The horizontal reaction at the bottom of the wall R_a is

$$R_a = \frac{Pe}{h(12)} + \frac{wh}{2} = \frac{(1500 \text{ lb/ft})(4 \text{ in.})}{(24 \text{ ft})(12 \text{ in./ft})} + \frac{(25 \text{ lb/ft}^2)(24 \text{ ft})}{2} = 321 \text{ lb/ft of wall}$$

Locate where maximum moment occurs, x distance from the bottom of the wall:

$$M_x = R_a x - \frac{wx^2}{2} = (321)x - \frac{(25)(x^2)}{2}$$

Maximum moment is found at the location where shear is zero.

$$V = R_a - wx = 321 - (25)x = 0$$

Solving for x:

$$x = \frac{321}{25} = 12.84 \text{ ft from bottom of wall}$$

Maximum moment in the wall M_{max} is

$$M_{max} = (321)(12.84) - \frac{(25)(12.84)^2}{2} = 2061 \text{ ft-lb/ft of wall}$$

Assume the wall is partially grouted at the vertical reinforcing steel bars, which are spaced at 24 in. o.c. Design moment in the wall per bar is as follows:

$$\text{Moment per bar} = \frac{(2061 \text{ ft-lb/ft})(24 \text{ in. o.c.})}{(12 \text{ in./ft})} = 4122 \text{ ft-lb/per bar}$$

2. Determine the required steel area.
   To estimate the steel required, assume that the allowable steel stress governs and the flexural compression area is rectangular.

   $M = Tjd = A_s f_s (jd)$

   $j = 0.90$ (assumed),

   $d = \dfrac{t}{2} = \dfrac{11.62}{2} = 5.81$ in. (steel at center of 12-in. nominal block)

   $f_s = (1.33)(24{,}000 \text{ psi})$
   $= 31{,}920 \text{ psi}$ (allow for one-third increase due to wind load)

   Try $A_s = \dfrac{M}{f_s jd} = \dfrac{(4122)(12)}{(31{,}920)(0.9)(5.81)} = 0.30 \text{ in}^2/24 \text{ in.}$

   Use # 6 bars at 24 in. o.c. ($A_s\ 0.44 = \text{in}^2$)

3. Check moment in the wall
   Determine masonry stress block kd distance

$$k = \frac{pn + \left(\dfrac{1}{2}\right)\left(\dfrac{t_f}{d}\right)^2}{pn + \left(\dfrac{t_f}{d}\right)}$$

where $\rho = \dfrac{A_s}{bd} = \dfrac{0.44}{(24)(5.81)} = 0.00316$

$n = \dfrac{E_s}{E_m} = \dfrac{29{,}000{,}000}{750(1500)}$ (Assume full inspection)

$pn = (0.00316)(25.77) = 0.0814$

Face-shell thickness, $t_f = 1.5$ in. (12 in. CMU)

$$\left(\frac{t_f}{d}\right) = \frac{1.50}{5.81} = 0.258$$

$$k = \frac{0.0814 + 0.5(0.258)^2}{0.0814 + (0.258)} = 0.338$$

$$kd = (0.338)(5.81) = 1.96 > t_f = 1.50 \text{ in.}$$

Therefore, the stress block is a T section.

$$j = \frac{6 - 6\left(\frac{t_f}{d}\right) + 2\left(\frac{t_f}{d}\right)^2 + \left(\frac{t_f}{d}\right)^3 \left(\frac{1}{2\rho n}\right)}{6 - 3\left(\frac{t_f}{d}\right)}$$

$$j = \frac{6 - 6(0.258) + 2(0.258)^2 + (0.258)^3 \left[\frac{1}{2(0.0814)}\right]}{6 - 3(0.258)} = 0.898$$

The allowable flexural compressive stress f_m is (UBC 2107.2.6).

$$F_b = 0.33 f'_m \quad 2000 \text{ psi maximum}$$
$$F_b = (0.33)(1500) = 495 \text{ psi} < 2000 \text{ psi}$$

The resisting moment for the masonry is

$$M_m = C_f jd = f_m \frac{2kd - t_f}{2kd} b t_f jd$$

$$M_m = (495) \left[\frac{2(1.96) - 1.5}{2(1.96)}\right] \left[\frac{(24)(1.5)(0.898)(5.81)}{(12 \text{ in./ft})}\right] = 4783 \text{ ft-lb}$$

UBC 1612.3.2 permits a one-third increase in allowable stresses for all combinations including W (wind) or E (earthquake). Therefore,

$$M_m = \left[1 + \left(\frac{1}{3}\right)\right](4783) = 6377 \text{ ft-lb} > 4122 \text{ ft-lb/per bar} \quad \text{OK.}$$

The reinforcing steel resisting moment is

$$M_s = A_s f_s jd = \frac{(0.44)(24{,}000)(0.898)(5.81)}{(12 \text{ in./ft})} \left[1 + \left(\frac{1}{3}\right)\right] = 6122 > 4122 \quad \text{OK.}$$

The steel resisting moment controls the design. $M_r = 6122$ ft-lb

4. Check axial load in the wall.
   a. Calculate axial stress at maximum moment.
      Wall load:

      The weight of wall for the 12-in. CMU wall with reinforcing steel at 24 in. o.c. is 102 lb/ft² (Table 8.4).

      Wall load = (102 lb/ft²)(24 − 12.84) = 1138 lb per ft width of wall

      Axial load at top of wall = 1500 lb per ft

      Total axial load = 1138 + 1500 = 2638 lb per ft width of wall

   For the 12-in. CMU wall with reinforcing steel at 24 in. o.c., the equivalent wall thickness for computing compression, EST = 7.5 in. (Table 8.5)
   Axial stress,

   $$f_a = \frac{P}{bt} = \frac{2638}{(12)(7.5)} = 29.3 \text{ psi}$$

   b. Calculate allowable axial compressive stress (UBC 2107.2.5).

   $$F_a = 0.25 f'_m R$$

   where

   $$R = 1 - \left(\frac{h'}{140r}\right)^2 \quad \text{for } \frac{h'}{r} \leq 99$$

   $$R = \left(\frac{70r}{h'}\right)^2 \quad \text{for } \frac{h'}{r} > 99$$

   For the 12-inch CMU wall with reinforcing steel at 24 inches o.c., the radius of gyration = 3.82 in. (Table 8.6)

   $h' = (24)(12) = 288$ in.   (effective height, pin-pin support)

   $$\frac{h'}{r} = \frac{288}{3.82} = 75.39 < 99$$

   $$R = 1 - \left(\frac{h'}{140r}\right)^2 = 1 - \left(\frac{288}{140(3.82)}\right)^2 = 0.71$$

   $F_a = 0.25 f'_m R = (0.25)(1500)(0.71) = 266 \text{ psi} > 29.3 \text{ psi}$ OK.

5. Check combine load

   The unit equation will be used to check the combined stress condition. Since wind load is part of combination, UBC 1612.3.2 permits a one-third increase in allowable stresses for all combinations including W (wind) or E (earthquake). Therefore,

   $$F_a = 266 \text{ psi} \times \left[1 + \left(\frac{1}{3}\right)\right] = 355 \text{ psi}$$

   $f_a = 29.3$ psi
   $M = 4122$ ft-lb
   $M_r = 6122$ ft-lb

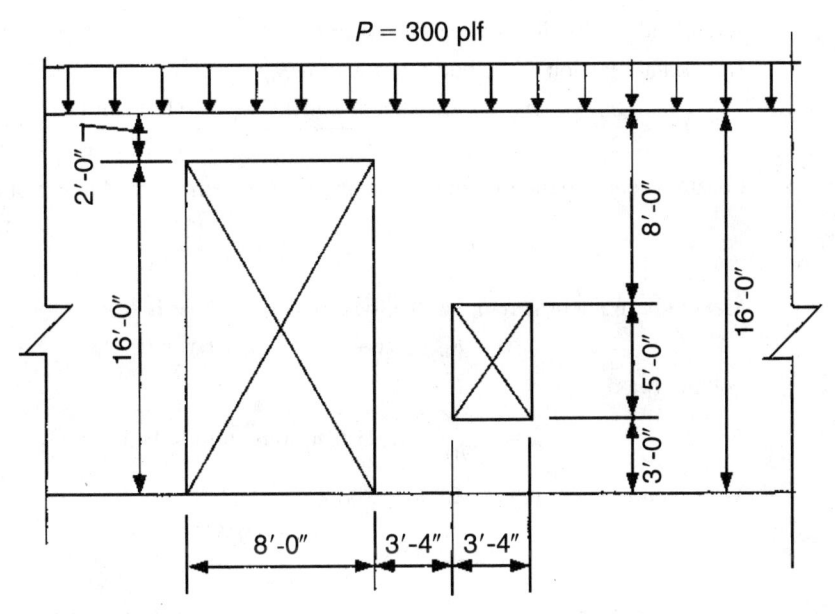

FIGURE 8.10

Unit equation $\dfrac{f_a}{F_a} + \dfrac{M}{M_r} \leq 1.0$

$$\dfrac{29.3}{355} + \dfrac{4122}{6122} = 0.08 + 0.67$$

$$= 0.75 \leq 1.0 \quad \text{OK.}$$

6. Summary: One #6 bar per cell @ 24 in. o.c. is adequate.

EXAMPLE 8.6

Design the section of wall between the 8-ft-0-in. door and 3-ft-4-in. window openings shown in the front elevation in Figure 8.10. Wall is to be constructed using 12-in.-thick CMU blocks. At the windows, the wall above and below the opening generally must span to the sides of the opening, then vertically to the roof diaphragm and to the floor. Determine the reinforcing bar size required resisting lateral wind force of 30 lb/ft² and axial load of 300 lb/ft. The eccentricity of axial load e is 1 in. The wind load is positive (inward) on the exterior face of the wall. The axial load is applied on the interior side of wall centerline. Thus, the moments due to lateral wind load and to axial eccentricity are not additive. $f'_m = 1500$ psi. Assume special inspection is provided.

Solution

1. Assume that the portion of wall between the openings will resist the applied axial loads and the wind loads between the middle of the door and the middle of the window.

$$\text{Tributary width} = \dfrac{8 \text{ ft}}{2} + 3.33 \text{ ft} + \dfrac{3.33 \text{ ft}}{2} = 9 \text{ ft}$$

Wind load = (30 lb/ft²)(9 ft) = 270 lb/ft

Axial load = (300 lb/ft)(9 ft) = 2700 lb

Horizontal reaction at the bottom of wall is R_a:

$$R_a = \frac{wh}{2} - \frac{Pe}{h(12)} = \frac{(270 \text{ lb/ft})(16 \text{ ft})}{2} - \frac{(2700 \text{ lb})(1 \text{ in.})}{(16 \text{ ft})(12 \text{ in./ft})} = 2146 \text{ lb}$$

Locate where maximum moment occurs, x distance from the bottom of the wall:

$$M_x = R_a x - \frac{wx^2}{2} = (2146)x - \frac{(270)(x^2)}{2}$$

Maximum moment is found at the location where shear is zero

$$V = R_a - wx = 2146 - (270)x = 0$$

Solving for x

$$x = \frac{2146}{270} = 7.95 \text{ ft from bottom of wall}$$

Maximum moment in the wall M_{max} is

$$M_{max} = (2146)(7.95) - \frac{(270)(7.95)^2}{2} = 8528 \text{ ft-lb}$$

2. Check axial loads

   Assume the lintel over the door is fully grouted from the top of the opening to the top of the wall.

   The weight of lintel, solid grouted 12-in. CMU = 140 lb/ft^2

   The weight of 12-in. CMU = 69 lb/ft^2

   Wall load at maximum moment

   $$= \left(\frac{1}{2}\right)(140 \text{ lbs/ft}^2)(2 \text{ ft})(8 \text{ ft}) + (140 \text{ lbs/ft}^2)(3 \text{ ft-4 in.})(7.95 \text{ ft})$$

   $$+ \left(\frac{1}{2}\right)(69 \text{ lbs/ft}^2)(3 \text{ ft-4 in.})(8 \text{ ft}) = 5750 \text{ lb}$$

   Total axial load = 2700 + 5750 = 8450 lb

   For the 12-in. CMU wall fully grouted, the equivalent wall thickness for computing compression, EST = 11.6 in. (Table 8.5)

   Axial stress,

   $$f_a = \frac{P}{bt} = \frac{8450}{(40)(11.6)} = 18.2 \text{ psi}$$

   Calculate allowable axial compressive stress (UBC 2107.2.5)

   $$F_a = 0.25 f'_m R$$

   where $R = 1 - \left(\dfrac{h'}{140r}\right)^2$ for $\dfrac{h'}{r} \leq 99$

   $R = \left(\dfrac{70r}{h'}\right)^2$ for $\dfrac{h'}{r} > 99$

For the 12-in. CMU wall fully grouted, the radius of gyration = 3.34 in. (Table 8.6)

$h' = (16)(12) = 192$ in.   (effective height, pin-pin support)

$$\frac{h'}{r} = \frac{192}{3.34} = 57.59 < 99$$

$$R = 1 - \left(\frac{h'}{140r}\right)^2 = 1 - \left(\frac{192}{140(3.34)}\right)^2 = 0.83$$

$F_a = 0.25 f'_m R = (0.25)(1500)(0.83) = 311$ psi > 18.2 psi   OK.

3. Determine the required steel area.

To estimate the steel required, assume that the allowable steel stress governs and the flexural compression area is rectangular.

$$M = Tjd = A_s f_s (jd)$$

$j = 0.90$ (assumed),

$d = \dfrac{t}{2} = \dfrac{11.62}{2} = 5.81$ in.   (steel at center of 12-in. nominal block)

$f_s = (1.33)(24,000$ psi$)$
$ = 31,920$ psi (allow for one-third increase due to wind load)

Try $A_s = \dfrac{M}{f_s jd} = \dfrac{(8528)(12)}{(31,920)(0.9)(5.81)} = \dfrac{0.61 \text{ in}^2}{40 \text{ in.}}$

Use five #4 bars ($A_s = 1.00$ in^2)

4. Check moment in the wall.

Determine masonry stress block kd distance

$$k = \frac{\rho n + \left(\frac{1}{2}\right)\left(\frac{t_f}{d}\right)^2}{\rho n + \left(\frac{t_f}{d}\right)}$$

where $\rho = \dfrac{A_s}{bd} = \dfrac{1.00}{(40)(5.81)} = 0.00430$

$n = \dfrac{E_s}{E_m} = \dfrac{29,000,000}{750(1500)} = 25.78$ (Assume full inspection)

$\rho n = (0.00430)(25.78) = 0.1108$

Face-shell thickness, $t_f = 1.50$ in. (12 in. CMU, Table 8.3)

$\left(\dfrac{t_f}{d}\right) = \dfrac{1.50}{5.81} = 0.258$

$k = \dfrac{0.1108 + 0.5(0.258)^2}{0.1108 + (0.258)} = 0.391$

$kd = (0.391)(5.81) = 2.27 > t_f = 1.50$ in.

Therefore, the stress block is a T section.

$$j = \cfrac{6 - 6\left(\cfrac{t_f}{d}\right) + 2\left(\cfrac{t_f}{d}\right)^2 + \left(\cfrac{t_f}{d}\right)^3\left(\cfrac{1}{2\rho n}\right)}{6 - 3\left(\cfrac{t_f}{d}\right)}$$

$$j = \cfrac{6 - 6(0.258) + 2(0.258)^2 + (0.258)^3\left[\cfrac{1}{2(0.1108)}\right]}{6 - 3(0.258)} = 0.892$$

The allowable flexural compressive stress f_m is (UBC 2107.2.6)

$$F_b = 0.33 f'_m \quad 2000 \text{ psi maximum}$$
$$F_b = (0.33)(1500) = 495 \text{ psi} < 2000 \text{ psi}$$

The resisting moment for the masonry is

$$M_m = C_f j d = f_m \frac{2kd - t_f}{2kd} b t_f j d$$

$$M_m = (495)\left[\frac{2(2.27) - 1.50}{2(2.27)}\right]\left[\frac{(40)(1.50)(0.892)(5.81)}{(12 \text{ in./ft})}\right] = 8589 \text{ ft-lb}$$

UBC 1612.3.2 permitted a one-third increase in allowable stresses for all combinations including W (wind) or E (earthquake). Therefore,

$$M_m = \left[1 + \left(\frac{1}{3}\right)\right](8589) = 11{,}452 \text{ ft-lb} > 8528 \text{ ft-lb/per bar} \quad \text{OK.}$$

The reinforcing steel resisting moment is

$$M_s = A_s f_s j d = \frac{(1.0)(24{,}000)(0.892)(5.81)}{(12 \text{ in./ft})}\left[1 + \left(\frac{1}{3}\right)\right] = 13{,}820 > 8589 \quad \text{OK.}$$

The masonry resisting moment controls the design. $M_r = 11{,}452$ ft-lb

5. **Check combine load.**
   The unit equation will be used to check the combined stress condition. Since wind load is part of combination, UBC 1612.3.2 permits a one-third increase in allowable stresses for all combinations including W (wind) or E (earthquake). Therefore,

$$F_a = 311 \text{ psi} \times \left[1 \times \left(\frac{1}{3}\right)\right] = 415 \text{ psi}$$

$f_a = 18.2$ psi
$M = 8528$ ft-lb
$M_r = 11{,}452$ ft-lb

Unit equation $\dfrac{f_a}{F_a} + \dfrac{M}{M_r} \leq 1.0$

$$\dfrac{18.2}{415} + \dfrac{8528}{11{,}452} = 0.04 + 0.74$$
$$= 0.78 \leq 1.0 \quad \text{OK.}$$

6. Summary: One #4 bar per cell, total five #4 bars in solid-grouted 12-in. CMU is adequate.

8.4.4 Reinforced Masonry Wall Design for In-plane Loads (Shear Wall Design)

A shear wall is a wall designed principally to resist horizontal shear forces such as seismic forces and wind forces acting in the plane of the wall. These vertical members resist the horizontal load from the diaphragm at the top of the wall and transfer it to the foundation at the base of the wall. A diaphragm may be considered as analogous to a large horizontal plate girder loaded in its plane to distribute lateral force to shear walls. The roof or floor slab constitutes the web and the walls or bond beams act as flanges. The stiffness of the diaphragm affects the distribution of lateral forces to the lateral force-resisting system. No diaphragm is infinitely rigid, and no diaphragm capable of carrying load is infinitely flexible. For the purpose of analysis, diaphragms are classified into three groups: rigid, semirigid or semiflexible, or flexible.

Diaphragms constructed of cast-in-place concrete slab or heavy metal decking are considered rigid. Rigid diaphragm is not expected to deflect appreciably, and will cause each vertical element to deflect the same amount. The amount of force that it takes to cause that deflection depends upon the rigidity of the element in question. Therefore, the distribution of the lateral force is in direct proportion to the relative rigidities of those vertical elements. Rigid diaphragms are also considered capable of transferring torsional shear deflections and forces, and therefore must be designed for shears resulting from both direct shear and torsional moment. The torsional moment takes place when the center of mass and the center of rigidity of the lateral force-resisting system do not coincide. The torsional moment is computed as the diagram load times the eccentricity, the distance from the center of mass and the center of rigidity. The center of mass tends to rotate about the center of rigidity.

On the other hand, diaphragms made up of distinct units, such as wood joists with wood decking or precast concrete planks, will undergo large deflections. These diaphragms are classified as flexible diaphragms. Flexible diaphragm is analogous to a series of simple beams spanning between very rigid supports (i.e., the vertical resisting elements). The lateral force is distributed to the vertical elements on a tributary area basis independent of their stiffness. It should be noted that a flexible diaphragm is not considered capable of distributing torsional stresses, resulting from the condition where the center of mass and the center of rigidity do not coincide.

Semirigid, or semiflexible, diaphragms exhibit significant deflection under load, but also have sufficient stiffness to distribute a portion of their load to the vertical elements in direct proportion to the rigidity of those elements. The action is analogous to a continuous beam system.

UBC Section 1630.6 specifies the criteria to determine the type of diaphragm for the purpose of distribution of story shear and torsional moment. Diaphragms are considered to be flexible when the maximum lateral deformation of the diaphragm is more than two times the average story drift of the associated story. This may be determined by comparing the computed midpoint in-plane deflection of the diaphragm itself under lateral load with the story drift of adjoining vertical-resisting elements under equivalent tributary lateral load.

8.4.4.1 Wall Rigidities. The first step in designing the shear wall is to determine the relative rigidities or stiffness of the shear wall elements. In masonry structure, the rigidity of a wall element is dependent on its dimension, the modulus of elasticity E_m, the shear modulus E_v, and the condition of supports at top and the bottom of the wall. Because most masonry piers act as short, deep beams, when a force is applied at the top of the wall, both flexure and shear contributions to the displacement must be considered. The total deflection Δ, equals the sum of the deflection due to bending moment Δ_m, and the deflection due to shear Δ_v. For walls fixed at top and bottom, the total deflection Δ_f is

$$\Delta_f = \Delta_m + \Delta_v = \frac{Ph^3}{3E_m I} + \frac{1.2Ph}{AE_v} \qquad (8.62)$$

where P = Lateral force on wall, lb
h = Height of wall, in.
E_m = Modulus of elasticity in bending, psi
E_v = Shear modulus, psi $E_v = 0.4E_m$
I = Cross-sectional moment of inertia of wall in the direction of bending, in^4
A = Cross-sectional area of wall, in^2

For walls fixed at bottom only, cantilever from the foundation, the total deflection, Δ_c, is

$$\Delta_c = \Delta_m + \Delta_v = \frac{Ph^3}{12E_m I} + \frac{1.2Ph}{AE_v} \qquad (8.63)$$

By substituting $0.4E_m$ for E_v, $\frac{td^3}{12}$ for I, and td for A, the total deflections can be written as

$$\Delta_f = \frac{P}{E_m t}\left[\left(\frac{h}{d}\right)^3 + 3\left(\frac{h}{d}\right)\right] \qquad (8.64)$$

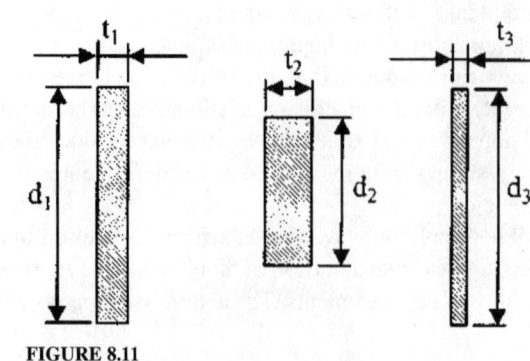

FIGURE 8.11

and

$$\Delta_c = \frac{P}{E_m t}\left[4\left(\frac{h}{d}\right)^3 + 3\left(\frac{h}{d}\right)\right] \tag{8.65}$$

where t = Wall thickness, in.
 d = Effective length of wall, in.

The rigidity of the wall k is defined as the inverse of the total deflection of wall under unit horizontal load, as stated in the following equations:

Rigidity of fixed wall $k_f = \dfrac{1}{\Delta_f}$ (8.66)

Rigidity of cantilever wall $k_c = \dfrac{1}{\Delta_c}$ (8.67)

Consider a rigid diaphragm supported by a system of parallel shear walls as shown in Figure 8.11.

Assuming that the center of mass coincides with the center of wall rigidity, since the diaphragm is rigid, the walls must deflect equally. The diaphragm shear force P will be distributed to the walls in proportion to the their relative stiffness. Thus, each wall i will carry a portion of diaphragm shear P_i equal to

$$P_i = \left(\frac{k_i}{\sum_{i=1}^{n} k_i}\right) P = R_i P \tag{8.68}$$

where R_i is the relative rigidity for any wall panel i.

$$R_i = \frac{k_i}{\sum_{i=1}^{n} k_i} \tag{8.69}$$

Since the distribution of the lateral force is in direct proportion to the relative rigidities of the vertical elements, not the actual rigidity, only the relative values are required, and any

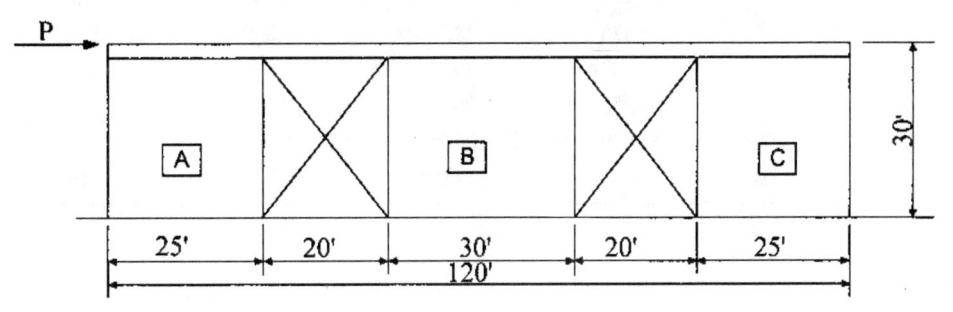

FIGURE 8.12

TABLE 8.14 Deflection and Rigidity of Walls or Piers for Distribution of Lateral Forces

$\dfrac{h}{d}$	Δ_f fixed ends $\left(\dfrac{h}{d}\right)^3 + 3\left(\dfrac{h}{d}\right)$	Δ_f cantilever $4\left(\dfrac{h}{d}\right)^3 + 3\left(\dfrac{h}{d}\right)$	k_f fixed ends $\dfrac{1}{\Delta_f}$	k_c cantilever $\dfrac{1}{\Delta_c}$
0.1	0.301	0.304	3.322	3.289
0.2	0.608	0.632	1.645	1.582
0.3	0.927	1.008	1.079	0.992
0.4	1.264	1.456	0.791	0.687
0.5	1.625	2.000	0.615	0.500
0.6	2.016	2.664	0.496	0.375
0.7	2.443	3.472	0.409	0.288
0.8	2.912	4.448	0.343	0.225
0.9	3.429	5.616	0.292	0.178
1.0	4.000	7.000	0.250	0.143
1.1	4.631	8.624	0.216	0.116
1.2	5.328	10.512	0.188	0.095
1.3	6.097	12.688	0.164	0.079
1.4	6.944	15.176	0.144	0.066
1.5	7.875	18.000	0.127	0.056
1.6	8.896	21.184	0.112	0.047
1.7	10.013	24.752	0.100	0.040
1.8	11.232	28.728	0.089	0.035
1.9	12.559	33.136	0.080	0.030
2.0	14.000	38.000	0.071	0.026
2.1	15.561	43.344	0.064	0.023
2.2	17.248	49.192	0.058	0.020
2.3	19.067	55.568	0.052	0.018
2.4	21.024	62.496	0.048	0.016
2.5	23.125	70.000	0.043	0.014
2.6	25.376	78.104	0.039	0.013
2.7	27.783	86.832	0.036	0.012
2.8	30.352	96.208	0.033	0.010
2.9	33.089	106.256	0.030	0.009
3.0	36.000	117.000	0.028	0.009

value for P, E_m, and t could be used. Assuming the convenient but arbitrary values of $P = 1.0 \times 10^6$ lbs., $t = 1$ in., and $E_m = 1.0 \times 10^6$ psi gives $P/(E_m t) = 1.0$, Equations 8.64 and 8.65 can be rewritten in the following form:

$$\Delta_f = \left(\frac{h}{d}\right)^3 + 3\left(\frac{h}{d}\right) \tag{8.70}$$

$$\Delta_c = 4\left(\frac{h}{d}\right)^3 + 3\left(\frac{h}{d}\right) \tag{8.71}$$

Values of Δ_f, Δ_c, k_f, and k_c are provided in tabular form in Table 8.14.

When walls of different moduli of elasticity E_m or thicknesses t are being compared, the tabulated values can be factored by the ratios of $E_m t$.

In the case of a solid wall with no openings or with full-height openings, the computations of the deflection are quite simple. For instance, consider a masonry shear wall with two full-height openings, as shown in Figure 8.12. The diaphragm shear P is assumed to be shared by the three independent wall panels, A, B, and C. These wall panels are assumed to be tied together by a rigid diaphragm and deflect equally. Each wall panel is assumed to behave like a vertical cantilever. The shear P will be distributed to the wall panels in proportion to their relative stiffness.

Where the shear wall has openings for doors, windows, etc., however, the computations for deflection and rigidity are much more complex. Several methods are being used to determine the relative wall deflections and rigidities of a wall consisting of several connected piers. One of the methods suggested in a Concrete Masonry Association of California publication is described in the following procedure:

Step 1. Obtain the deflection of the solid cantilever wall.

Step 2. Deduct the deflection of a cantilever strip having a height equal to that of the highest opening in the wall.

Step 3. Compute the deflection of all composite piers with openings lying within that strip.

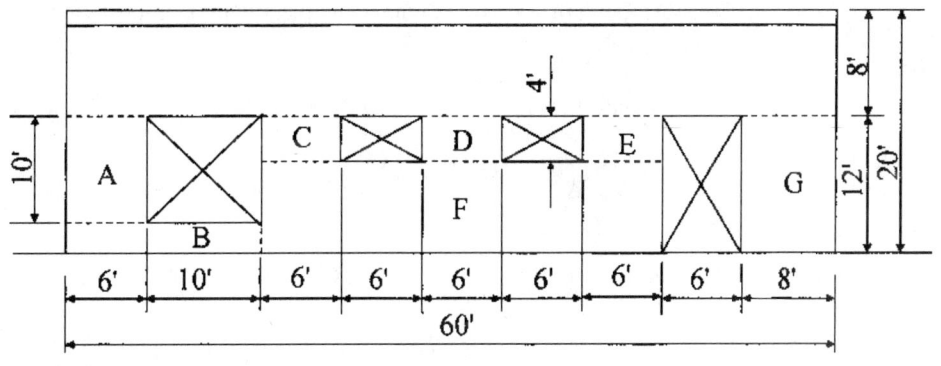

FIGURE 8.13

8.62 CHAPTER 8

Step 4. Add these deflections of the individual piers to the modified wall deflection obtained in Step 2 to obtain the final wall deflection.

Step 5. Take the reciprocal of the final wall deflection calculated in Step 4 to obtain the relative rigidity of the wall.

EXAMPLE 8.7

Determine the shear carried by the individual elements of the shear wall shown in Figure 8.13. Assuming a rigid diaphragm transmits a total seismic force of 150 kips to these 12-in. CMU masonry shear walls. These walls are designed assuming $f'_m = 1500$ psi, and no special inspection will be provided.

Solution

1. Determine the deflection of the solid cantilever wall.

   $h = 20$ ft $d = 60$ ft $\dfrac{h}{d} = 0.33$

   $$\Delta_c = 4\left(\dfrac{h}{d}\right)^3 + 3\left(\dfrac{h}{d}\right) = (4)(0.33)^3 + (3)(0.33) = 1.134$$

   $$k_c = \dfrac{1}{\Delta_c} = \dfrac{1}{1.134} = 0.882$$

2. Determine the deflection of the cantilever strip having a height equal to the highest opening in the wall.

   $h = 12$ ft $d = 60$ ft $\dfrac{h}{d} = 0.20$

   $$\Delta_c = 4\left(\dfrac{h}{d}\right)^3 + 3\left(\dfrac{h}{d}\right) = (4)(0.20)^3 + (3)(0.20) = 0.632$$

   Deduct deflection of the strip with openings.

   $$\Delta = 1.134 - 0.632 = 0.502$$

3. Compute the deflection of all composite piers with openings lying within that strip.
   a. Piers A-B

   Solid wall: $h = 12$ ft $d = 16$ ft $\dfrac{h}{d} = 0.75$

   $$\Delta_f = \left(\dfrac{h}{d}\right)^3 + 3\left(\dfrac{h}{d}\right) = (0.75)^3 + (3)(0.75) = 2.672$$

   Opening strip: $h = 10$ ft $d = 16$ ft $\dfrac{h}{d} = 0.625$

   $$\Delta_f = \left(\dfrac{h}{d}\right)^3 + 3\left(\dfrac{h}{d}\right) = (0.625)^3 + (3)(0.625) = 2.119$$

Pier A: $h = 10$ ft $\quad d = 6$ ft $\quad \dfrac{h}{d} = 1.67$

$$\Delta_f = \left(\dfrac{h}{d}\right)^3 + 3\left(\dfrac{h}{d}\right) = (1.67)^3 + (3)(1.67) = 9.667$$

$$\Delta = 2.672 - 2.119 + 9.667 = 10.22$$

$$k_f = \dfrac{1}{\Delta} = \dfrac{1}{10.22} = 0.098$$

b. Piers C-D-E-F

Solid wall: $h = 12$ ft $\quad d = 30$ ft $\quad \dfrac{h}{d} = 0.40$

$$\Delta_f = \left(\dfrac{h}{d}\right)^3 + 3\left(\dfrac{h}{d}\right) = (0.40)^3 + (3)(0.40) = 1.264$$

Opening strip: $h = 4$ ft $\quad d = 30$ ft $\quad \dfrac{h}{d} = 0.13$

$$\Delta_f = \left(\dfrac{h}{d}\right)^3 + 3\left(\dfrac{h}{d}\right) = (0.13)^3 + (3)(0.13) = 0.392$$

Piers C-D-E
Piers C, D, and E have the same height and depth.

$h = 4$ ft $\quad d = 6$ ft $\quad \dfrac{h}{d} = 0.67$

$$\Delta_f = \left(\dfrac{h}{d}\right)^3 + 3\left(\dfrac{h}{d}\right) = (0.67)^3 + (3)(0.67) = 2.31$$

$$k_f = \dfrac{1}{\Delta} = \dfrac{1}{2.31} = 0.433$$

$$\Sigma k_i = 0.433 \times 3 = 1.299 \qquad \Delta = \dfrac{1}{\Sigma k_i} = \dfrac{1}{1.299} = 0.770$$

For Piers C-D-E-F

$$\Delta = 1.264 - 0.392 + 0.770 = 1.642$$

$$k_f = \dfrac{1}{\Delta} = \dfrac{1}{1.642} = 0.609$$

c. Pier G

$h = 12$ ft $\quad d = 8$ ft $\quad \dfrac{h}{d} = 1.50$

$$\Delta_f = \left(\dfrac{h}{d}\right)^3 + 3\left(\dfrac{h}{d}\right) = (1.5)^3 + (3)(1.5) = 7.875$$

$$k_f = \dfrac{1}{\Delta} = \dfrac{1}{7.875} = 0.127$$

For entire strip with openings:

$$\Sigma k_i = 0.098 + 0.609 + 0.127 = 0.834 \qquad \Delta = \frac{1}{\Sigma k_i} = \frac{1}{0.834} = 1.199$$

4. Final wall deflection

$$\Delta = 0.502 + 1.199 = 1.701$$

5. Rigidity, $k_c = \dfrac{1}{\Delta} = \dfrac{1}{1.701} = 0.588$

6. Force distribution:
   The portion of diaphragm shear carried by each individual wall element is in proportion to its relative rigidity.
   Total rigidity: $\Sigma k_i = 0.098 + 0.609 + 0.127 = 0.834$

   Piers A-B: $\dfrac{k}{\Sigma k_i} = \dfrac{0.098}{0.834} = 11.8\% \qquad P = (11.8\%)(150) = 17.7$ kips

   Piers C-D-E-F: $\dfrac{k}{\Sigma k_i} = \dfrac{0.609}{0.834} = 73.0\% \qquad P = (73.0\%)(150) = 109.5$ kips

   Pier G: $\dfrac{k}{\Sigma k_i} = \dfrac{0.127}{0.834} = 15.2\% \qquad P = (15.2\%)(150) = 22.8$ kips

   For Piers C, D, and E take the force assigned to Piers C-D-E-F and distribute it to Piers C, D, and E. Since Piers C, D, and E have the same rigidity, they will share the force equally.

$$P_C = P_D = P_E = \left(\frac{1}{3}\right)(109.5) = 36.5 \text{ kips}$$

8.4.4.2 Center of Mass and Center of Rigidity.
The lateral force caused by an earthquake acts at the center of mass. The location of the center of mass is determined by the summation of the first moment of the respective wall weight and floor (and/or roof) weights taken about the centerline of any wall and divided by the summation of the weights. The equations for calculating the center of mass are:

$$x_{cm} = \frac{\sum_i w_i x_i}{\sum_i w_i} \qquad \text{and} \qquad y_{cm} = \frac{\sum_i w_i y_i}{\sum_i w_i} \qquad (8.72)$$

where w_i represents the weight of elements with centroidal coordinates x_i and y_i.

The center of rigidity is the point at which horizontal reaction is assumed to be concentrated. Similar to the center of mass, the location of center of rigidity is obtained by dividing the summation of first moment of the relative lateral rigidity about the centerline of any wall by the summation of the relative lateral rigidity. Hence, the center of rigidity is located as follows:

$$x_r = \frac{\sum_i R_{yi} x_i}{\sum_i R_{yi}} = \frac{\sum_i k_{yi} x_i}{\sum_i k_{yi}} \qquad \text{and} \qquad y_r = \frac{\sum_i R_{xi} y_i}{\sum_i R_{xi}} = \frac{\sum_i k_{xi} y_i}{\sum_i k_{xi}} \qquad (8.73)$$

where R_{yi} = Relative rigidity of a particular wall with axis in the y direction
   R_{xi} = Relative rigidity of a particular wall with axis in the x direction.
   $\sum_i R_{yi} = 1$ and $\sum_i R_{yi} = 1$
   k_{yi} = Rigidity of a particular wall with axis in the y direction. For such a wall, it is assumed that $k_{xi} = 0$.
   k_{xi} = Rigidity of a particular wall with axis in the y direction. For such a wall, it is assumed that $k_{yi} = 0$.

EXAMPLE 8.8
Locate the center of mass and center of rigidity for a building as shown in Figure 8.14. All walls are a total of 20-ft high; 16 ft-8 in. between supports with a 3 ft-4 in. parapet. Assume the roof diaphragm is a rigid diaphragm and weight to be 75 psf. All walls are 8-in. CMU partially grouted at 24-in o.c. without opening.

Solution
1. Locate center of mass
   a. Roof weight:
      Roof weight = (75 psf)(60 ft)(40 ft) = 180,000 lb = 180 kips
   b. Wall weights:
      Unit weight of 8-in. CMU partially grouted at 24 in o.c. = 69 psf (Table 8.4)

   $$\text{Wall weight} = 69 \text{ psf} \left[\left(\frac{h}{2} + \text{parapet} \right)(\text{wall length}) - (\text{opening area}) \right]$$

   $$\text{West wall} = (69 \text{ psf}) \left[\left(\frac{16.67}{2} + 3.33 \right)(40) - 0 \right] = 32{,}195 \text{ lb} \approx 32.20 \text{ kips}$$

   $$\text{East wall} = (69 \text{ psf}) \left[\left(\frac{16.67}{2} + 3.33 \right)(40) - 0 \right] = 32{,}195 \text{ lb} \approx 32.20 \text{ kips}$$

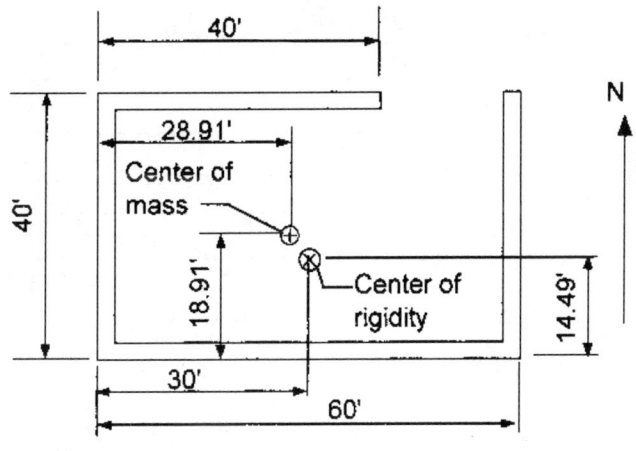

FIGURE 8.14

North wall = (69 psf)$\left[\left(\dfrac{16.67}{2} + 3.33\right)(40) - 0\right]$ = 32,195 lb ≈ 32.20 kips

South wall = (69 psf)$\left[\left(\dfrac{16.67}{2} + 3.33\right)(60) - 0\right]$ = 48,293 lb ≈ 48.30 kips

Item	w, weight	x*	y*	wx	wy
Roof	150 kips	30.00	20.00	4500	3000
West wall	32.20 kips	0.00	20.00	0	644
East wall	32.20 kips	60.00	20.00	1932	644
North wall	32.20 kips	20.00	40.00	644	1288
South wall	48.30 kips	30.00	0.00	1449	0

* Neglecting the wall thickness

By taking first moment about centerlines of west wall and south wall,

$$\Sigma w = 294.9 \text{ kips} \quad \Sigma wx = 8525 \quad \Sigma wy = 5576$$

$$x_{cm} = \dfrac{\sum_i w_i x_i}{\sum_i w_i} = \dfrac{8525}{294.9} = 28.91 \text{ ft east of west wall}$$

$$y_{cm} = \dfrac{\sum_i w_i y_i}{\sum_i w_i} = \dfrac{5576}{294.9} = 18.91 \text{ ft north of south wall}$$

2. Locate center of rigidity

The rigidities of all walls can be calculated using Equations 8.71 and 8.67.

$$\Delta_c = 4\left(\dfrac{h}{d}\right)^3 + 3\left(\dfrac{h}{d}\right) \quad \text{and} \quad \text{Rigidity } k_c = \dfrac{1}{\Delta_c}$$

Wall	x, ft	y, ft	h, ft	d, ft	$\dfrac{h}{d}$	N-S direction Δ_c	k_c	R_i	E-W direction Δ_c	k_c	R_i
West	0	—	20	40	0.50	2.0	0.50	0.50	—	—	—
East	60	—	20	40	0.50	2.0	0.50	0.50	—	—	—
North	—	40	20	40	0.50	—	—	—	2.00	0.50	0.36
South	—	0	20	60	0.33	—	—	—	1.13	0.88	0.64
					Σ	—	1.0	1.0	—	1.38	1.0

From Equation 8.73, the center of rigidity is at

$$x_r = \frac{\sum_i k_{yi} x_i}{\sum_i k_{yi}} = \frac{(0.5)(0) + (0.50)(60)}{(0.50 + 0.50)} = \frac{(0.5)(0) + (0.50)(60)}{1.0}$$

$= 30.0$ ft east of west wall

and

$$y_r = \frac{\sum_i k_{xi} y_i}{\sum_i k_{xi}} = \frac{(0.88)(0) + (0.50)(40)}{(0.88 + 0.50)} = \frac{(0.64)(0) + (0.36)(40)}{1.0}$$

$= 14.49$ ft north of south wall

8.4.4.3 Torsional Shear Force.
A torsional moment will be generated when the center of mass does not coincide with the center of rigidity of the lateral force-resisting system. UBC Section 1630.6 recommends the calculated mass center at each level in each direction be displaced a distance equal to 5 percent of the building dimension at that level perpendicular to the direction of the force under consideration. These additional eccentricities are intended to account for the torsional moment induced by the torsional moment between center of mass and rigidity in other stories and the effects of the rotational component of the ground motion, unbalanced dead and live load, openings in the floors, etc. Thus, the torsional design moment may be expressed as

$$\begin{aligned} \text{Torsional moment} &= P_x e_y & e_y &= e_y \text{ (calculated)} + 0.05L \\ \text{Torsional moment} &= P_y e_x & e_x &= e_x \text{ (calculated)} + 0.05W \end{aligned} \quad (8.74)$$

where P_x = Applied horizontal force applied in the x direction
P_y = Applied horizontal force applied in the y direction
L = Dimension of building in the y direction
W = Dimension of building in the x direction.

The value of torsional shear force resisted by a particular wall element with an axis parallel to the y axis, due to the applied horizontal load P_y is given by the formula

$$P_{yi} = \frac{R_{yi} \bar{x}_i}{J_r} P_y e_x \quad (8.75)$$

$$J_r = \Sigma(R_{xi} \bar{y}_i^2 + R_{yi} \bar{x}_i^2) \quad (8.76)$$

where J_r = Polar moment of inertia (rotational stiffness) of all walls in a story
$\bar{x}_i, \bar{y}_i$ = Perpendicular distance from the center of rigidity to the axis of a particular wall
R_{xi} = Relative wall rigidity of a particular wall with axis in the x direction. For such a wall, it is assumed that $R_{yi} = 0$
R_{yi} = Relative wall rigidity of a particular wall with axis in the y direction. For such a wall, it is assumed that $R_{xi} = 0$.

Similarly, for a wall element with an axis parallel to the x axis, due to applied horizontal load P_x, is given by the formula

$$P_{xi} = \frac{R_{xi}\bar{y}_i}{J_r} P_x e_y \qquad (8.77)$$

Therefore, the total horizontal shear force P_{yi}, resisted by each wall i, with an axis parallel to the y axis, due to the applied horizontal load P_y, is equal to

$$P_{yi} = \frac{k_{yi}}{\sum_{i=1}^{n} k_{yi}} P_y + \frac{R_{yi}\bar{x}_i}{J_r} P_y e_x = R_{yi}P_y + \frac{R_{yi}\bar{x}_i}{J_r} P_y e_x \qquad (8.78)$$

The torsional shear force P_{xi}, resisted by each wall i, with an axis parallel to the x axis, due to the applied horizontal load P_y, equal to

$$P_{xi} = \frac{R_{xi}\bar{y}_i}{J_r} P_y e_x \qquad (8.79)$$

Similarly, the total horizontal shear force P_{xi}, resisted by each wall i, with an axis parallel to the x axis, due to the applied horizontal load P_x, is equal to

$$P_{xi} = \frac{k_{xi}}{\sum_{i=1}^{n} k_{xi}} P_x + \frac{R_{xi}\bar{y}_i}{J_r} P_x e_y = R_{xi}P_x + \frac{R_{xi}\bar{y}_i}{J_r} P_x e_y \qquad (8.80)$$

The torsional shear force P_{yi}, resisted by each wall i, with an axis parallel to the y axis, due to the applied horizontal load P_x, is equal to

$$P_{yi} = \frac{R_{yi}\bar{x}_i}{J_r} P_x e_y \qquad (8.81)$$

EXAMPLE 8.9
The plan of a 40-ft × 60-ft building, as shown in Figure 8.14 of Example 8.8, subjected to a seismic lateral force of 100 kips in the north-south direction. Determine the design lateral force carried by the west and east walls and design the shear wall to withstand the shear force applied. All walls are 8-in. CMU partially grouted at 24 in o.c. without opening. Assume $f'_m = 1500$ psi and special inspection will be provided.

Solution
1. Locate center of mass
   The location of center of mass has been determined in Example 8.8 as follows:
   $$x_{cm} = 28.91 \text{ ft east of west wall}$$
   $$y_{cm} = 18.91 \text{ ft north of south wall}$$

2. Locate center of rigidity
   The location of center of rigidity has been determined in Example 8.8 as follows:
   $$x_r = 30.00 \text{ ft east of west wall}$$
   $$y_r = 14.49 \text{ ft north of south wall}$$
3. Calculate the polar moment of inertia

Wall	R_{yi}	x_i	R_{xi}	y_i	$\bar{x}_i$	$\bar{y}_i$	$R_{xi}\bar{y}_i^2$	$R_{yi}\bar{x}_i^2$
West	0.50	0.00	—	—	30.00	—	—	450
East	0.50	60.00	—	—	30.00	—	—	450
North	—	—	0.36	0.00	—	25.51	234	—
South	—	—	0.64	40.00	—	14.49	134	—

$$J_r = \Sigma(R_{xi}\bar{y}_i^2 + R_{yi}\bar{x}_i^2) = 234 + 134 + 450 + 450 = 1268 \text{ ft}^2$$

4. Calculate torsional eccentricity and torsional moment
   The calculated eccentricity in x direction is
   $$e_x \text{ (calculated)} = |x_r - x_{cm}| = 30.00 - 28.91 = 1.09 \text{ ft}$$
   $$e_x = e_x \text{ (calculated)} + 0.05W = 1.09 + (0.05)(60.0) = 4.09 \text{ ft}$$
   $$\text{Torsional moment} = P_y e_x = (100)(4.09) = 409 \text{ ft-kip}$$

   The total shear carried by the west wall is
   $$P_{yi} = R_{yi}P_y + \frac{R_{yi}\bar{x}_i}{J_r}P_y e_x = (0.50)(100) + \frac{(0.50)(30)}{1268}(100)(4.09) = 54.83 \text{ kips}$$

   The total shear carried by the east wall is
   $$P_{yi} = R_{yi}P_y + \frac{R_{yi}\bar{x}_i}{J_r}P_y e_x = (0.50)(100) + \frac{(0.50)(30)}{1268}(100)(4.09) = 54.83 \text{ kips}$$

   Since the seismic lateral force could be in either direction, only the positive torsional shear is considered.
   For north wall, the torsional shear is
   $$P_{xi} = \frac{R_{xi}\bar{y}_i}{J_r}P_y e_x = \frac{(0.36)(25.51)}{1268}(100)(4.09) = 2.96 \text{ kips}$$

   For south wall, the torsional shear is
   $$P_{xi} = \frac{R_{xi}\bar{y}_i}{J_r}P_y e_x = \frac{(0.64)(14.49)}{1268}(100)(4.09) = 2.99 \text{ kips}$$

5. Design east and west wall
   a. Shear check
      When checking the shear capacity of the masonry, UBC Section 2107.1.7 calls for the actual design force to be 1.5 times the computed total shear stresses.

Shear stress in east and west walls are determined as follows:
For 8-in. CMU partially grouted at 24 in o.c. without opening, the in-plane shear area = 50.5 in²/ft

$$f_v = \frac{1.5V}{td} = \frac{(1.5)(54.83)(1000)}{(50.5)(40 \text{ ft})} = 40.7 \text{ psi}$$

The allowable shear stress is calculated based on the M/Vd ratio.

For a cantilever pier, $M = Vh$ and $\dfrac{M}{Vd} = \dfrac{h}{d}$

For a fixed end pier, $M = \dfrac{Vh}{2}$ and $\dfrac{M}{Vd} = \dfrac{h}{2d}$

$h = 16 \text{ ft } 9 \text{ in.}$ $d = 40 \text{ ft } 0 \text{ in.}$ $\dfrac{h}{d} = \dfrac{201}{480} = 0.42 < 1$

$$F_v = \frac{1}{3}\left(4 - \frac{M}{Vd}\right)\sqrt{f'_m} = \left(\frac{1}{3}\right)(4 - 0.42)\sqrt{f'_m} = 1.19\sqrt{f'_m} = 1.19\sqrt{1500} = 46 \text{ psi}$$

$$\left(80 - 45\frac{M}{Vd}\right)_{\text{max}} = [80 - (45)(0.42)] = 61.1 \text{ psi maximum}$$

Therefore, no shear reinforcement is required.

b. Flexure check
Maximum moment occurs at the bottom of wall.

$$M = Vh = (54.83 \text{ kips})(16 \text{ ft } 8 \text{ in.}) = 10{,}966 \text{ kip-in.}$$

Determine area of reinforcing steel required.
To estimate the steel required, assume that the allowable steel stress governs and the flexural compression area is rectangular.

$$M = Tjd = A_s f_s (jd)$$
$$j = 0.90 \text{ (assumed)}, \quad d \approx 40 \text{ ft} = 480 \text{ in.}$$

$f_s = (1.33)(24{,}000 \text{ psi})$
    $= 31{,}920 \text{ psi}$ (allow for one-third increase due to wind load)

Try $A_s = \dfrac{M}{f_s jd} = \dfrac{(10{,}966)(1000)}{(31{,}920)(0.9)(480)} = 0.80 \text{ in}^2$

Use two #6 bars. ($A_s = 0.88 \text{ in}^2$)
Check compression stress of masonry
Flexural compressive stress in masonry is determined as follows:

$$f_b = \frac{2M}{bd^2 kj}$$

where $b = 7.62$ in.
$d \approx 40$ ft $= 480$ in.
$$n = \frac{E_s}{E_m} = \frac{29{,}000{,}000}{750(1500)} = 25.78 \quad \text{(Special inspection will be provided)}$$
$$k = \sqrt{2\rho n + (\rho n)^2} - \rho n$$

and $\rho = \dfrac{A_s}{bd} = \dfrac{0.88}{(7.62)(480)} = 0.00024$

$\rho n = (0.00024)(25.77) = 0.0062$

$k = \sqrt{2(0.0062) + (0.0062)^2} - 0.0062 = 0.1053$

$j = 1 - \left(\dfrac{k}{3}\right) = 1 - \left(\dfrac{0.1053}{3}\right) = 0.9649$

$$f_b = \frac{(2)(10{,}966)(1{,}000)}{(7.62)(480)^2(0.1053)(0.9646)} = 123 \text{ psi} < 660 \text{ psi} \quad \text{OK.}$$

Allowable flexural compressive stress is (UBC 2107.2.6)

$$f_m = \left[1 + \left(\frac{1}{3}\right)\right]0.33 f'_m = \left[1 + \left(\frac{1}{3}\right)\right]0.33(1500) = 660 \text{ psi} < 2000 \text{ psi}$$

Check flexural tensile stress in reinforcement:

$$f_s = \frac{M}{A_s j d} = \frac{(10{,}966)(1000)}{(0.88)(0.9646)(480)} = 26{,}914 \text{ psi} < 31{,}920 \text{ psi} \quad \text{OK.}$$

8.5 DESIGN EXAMPLES—STRENGTH-DESIGN METHOD

8.5.1 Load Factors Using Strength Design

Service load or actual loads are generally used for allowable load design. However, for strength-design procedure, the actual load or code-specified loads are increased by load factors. UBC Section 1612.2.2 presents a series of load factors and combinations of factored loads:

$U = 1.4D$ (8.82)

$U = 1.2D + 1.6L + 0.5 (L_r \text{ or } S)$ (8.83)

$U = 1.2D + 1.6(L_r \text{ or } S) + (f_1 L \text{ or } 0.8W)$ (8.84)

$U = 1.2D + 1.3W + f_1 L + 0.5(L_r \text{ or } S)$ (8.85)

$U = 1.2D + 1.0E + (f_1 L + f_2 S)$ (8.86)

$U = 0.9D \pm (1.0E \text{ or } 1.3W)$ (8.87)

where D = Dead load
           L = Live load, except roof live load, including any permitted live-load reduction
           L_r = Roof live load, including any permitted live-load reduction
           S = Snow load

W = Load due to wind pressure
E = Earthquake load
W = Wind load
f_1 = 1.0 for floors in places of public assembly, for live loads in excess of 100 psf, and for garage live load
 = 0.5 for other live loads
f_2 = 0.7 for roof configurations (such as saw tooth) that do not shed snow off the structure
 = 0.2 for other roof configurations.

8.5.2 Lintel Design

EXAMPLE 8.10

A door opening 3-ft 4 in. wide by 7-ft 4 in. high is located in the wall as shown in Figure 8.15a. Wall units are 8-in. non-load-bearing CMU hollow concrete blocks. Wall height = 12 ft. Assume Type S mortar, f'_m = 1500 psi, an 8-in. × 8-in. CMU lintel, and Grade 60 reinforcing steels. Determine the reinforcement required using strength-design method. (It should be noted that the strength design method always calls for special inspection during construction as set forth in UBC Section 1701.5, Item 7).

Solution
This same problem was solved previously in Example 8.1, using allowable stress design method. Loads distributed to lintels are discussed in Section 8.4.1 and Example 8.1.

1. Flexure design
   Since the factored axial compressive force on lintel does not exceed $0.05 A_e f'_m$, design lintel as a beam.

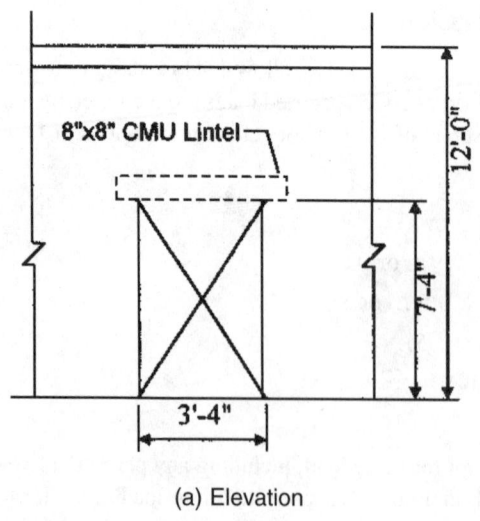

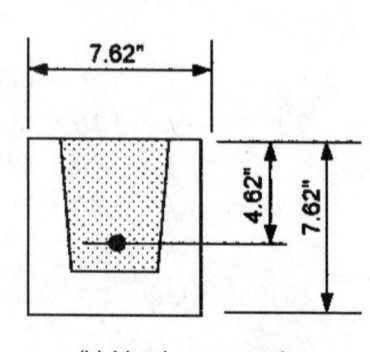

(a) Elevation   (b) Lintel cross section

FIGURE 8.15

a. Determine the factored applied moment as follows:
Maximum dead load moment due to the applied loads on lintel is

$$M_{DL} = \frac{wL^2}{8} + \frac{w'L^3}{24}$$

where w = Lintel weight = 62 lb/ft. (Table 8.4, solid grouted)
w' = Weight of the masonry triangle (assume no reinforced filled cells) = 50 psf (Table 8.4, normal weight)
L = Span length = (3 ft 4 in.) + 2(4 in.) = 4 ft 0 in.

$$M_{DL} = \frac{(62)(4.0)^2}{8} + \frac{(50)(4.0)^3}{24} = 257 \text{ ft-lb}$$

Applied factored load and factored moment are
Factored load: $U = 1.4D$
Factored moment: $M_u = (1.4)M_{DL} = (1.4)(257) = 360$ ft-lb

b. Determine the balanced reinforcement ratio and maximum steel ratio permitted.

$$\rho_b = \frac{0.85(0.85)f'_m}{f_y}\left(\frac{87{,}000}{87{,}000 + f_y}\right) = \frac{0.85(0.85)(1500)}{60{,}000}\left(\frac{87{,}000}{87{,}000 + 60{,}000}\right)$$
$$= 0.0107$$

Maximum steel ratio permitted is

$$\rho_{max} = 0.50\rho_b = (0.50)(0.0107) = 0.0053$$

Thus, the maximum reinforcement area is

$$A_{s\,max} = \rho_{max}bd$$

where b = Lintel width = $7\frac{5}{8}$ in.
d = Effective depth of lintel = 4.62 in.
$A_{s(max)} = 0.0053(7.62)(4.62) = 0.19$ in.^{2}

c. Select reinforcement; evaluate the capacity of the singly reinforced beam and the applied factored moment with the design ultimate capacity.
Try one #4 bar, $A_s = 0.20$ in$^2 \approx 0.19$ in^2

$$a = \frac{A_s f_y}{0.85 f'_m b} = \frac{(0.20)(60{,}000)}{(0.85)(1500)(7.62)} = 1.24 \text{ in.}$$

The nominal moment capacity is

$$M_n = A_s f_y \left(d - \frac{a}{2}\right) = (0.20)(60{,}000)\left(4.62 - \frac{1.24}{2}\right)\left(\frac{1}{12}\right) = 4{,}000 \text{ ft-lb}$$

Design ultimate capacity is

$$\phi M_n = (0.80)(4000) = 3200 \text{ ft-lb}$$

Since $M_u = 360$ ft-lb $< \phi M_n = 3200$ ft-lb, the moment capacity is adequate.
d. Check the cracking moment strength of the beam.
The cracking moment is

$$M_{cr} = Sf_r = \frac{(73.7)(235)}{12} = 1443 \text{ ft-lb}$$

where $\quad S = \dfrac{bt^2}{6} = \dfrac{(7.62)(7.62)^2}{6} = 73.7 \text{ in}^2 \quad$ and $\quad f_r = 235 \text{ psi}$

$$1.3M_{cr} = (1.3)(1443)$$
$$= 1876 \text{ ft-lb}$$
$$< M_n = 4000 \text{ ft-lb}$$

2. Shear check
   a. Determine the factored applied V_u:
   Maximum dead load shear due to the applied loads on lintel is

$$V_{DL} = \frac{wL}{2} + \frac{w'L^2}{8} = \frac{(62)(4)}{2} + \frac{(50)(4)^2}{8} = 224 \text{ lb}$$

   Applied factored load and factored shear are
   Factored load: $U = 1.4D$,
   Factored shear: $V_u = (1.4)V_{DL} = (1.4)(224) = 314$ lb
   b. Determine the nominal shear strength of masonry V_m:

$$V_m = C_d A_e \sqrt{f'_m} \qquad 63 C_d A_e \quad \text{maximum}$$

   C_d is the nominal shear strength coefficient as listed in UBC Table 21-K.
   For fully grouted simply supported lintel,

$$C_d = 2.4, \text{ and } A_e = bt = (7.62)(7.62) = 58 \text{ in}^2$$

   Therefore, $V_m = (2.4)(58)\sqrt{1500} = 5391$ lb $< 63(2.4)(58) = 8770$ lb maximum
   Since, $V_u < V_m$, no shear reinforcement is required.

8.5.3 Reinforced Masonry Column and Pilaster Design

EXAMPLE 8.11
A reinforced masonry pilaster, 16 ft high, supports truss dead-load end reaction of 25 kips and live-load reaction of 15 kips. The eccentricity of truss reaction e is 2 in. The spacing of pilasters is 25 ft. The wall between pilasters is subjected to a wind pressure of 20 psf. The wind load may act both inward and outward. $f'_m = 1500$ psi, Type S mortar. Determine the pilaster size and vertical reinforcement, neglecting the weight of the pilaster. (Note that this same problem was solved previously in Example 8.4 using allowable stress-design method.)

Solution

1. Design assumptions
   Assume that the wall spans horizontally between pilasters and all lateral loads on the wall are transmitted to the pilasters. The pilasters are simply supported by roof trusses and spread footings.

   Try 16-in. × 16-in. pilaster with six #9 bars. ($A_s = 6.0$ in^2)

2. Check the minimum and maximum reinforcement requirements.
   The minimum reinforcement $A_{s(\min)}$ and the maximum reinforcement $A_{s(\max)}$ are determined as follows:

   $A_{s(\max)} = 0.03 A_e$
   $A_{s(\min)} = 0.005 A_e$ (UBC 2108.2.3.12.2)

   where A_e is the effective area of column.
   Note that UBC Section 2108.23.12.4 requires all columns to be solid grouted. For a solid grouted masonry, A_e is equal to the gross area A_g;

   $A_g = bt = $ (actual width)(actual depth) $= (15.62)(15.62) = 244$ in^2
   $A_{s(\max)} = (0.03)(244) = 7.32$ in$^2 > 6.0$ in^2 OK.
   $A_{s(\min)} = (0.005)(244) = 1.22$ in$^2 < 6.0$ in^2 OK.

3. Calculate all of the service loads and factored loads.
   a. The service loads are

      Roof dead load = 25 kips
      Roof live load = 15 kips
      Wind Load
         Tributary width = the spacing of pilasters = 25 ft
         Wind pressure = 20 psf
         Wind load = (20)(25) = 500 plf

   b. The factored loads are
      Since only roof dead load, roof live load, and wind load are considered in the design, the code requires the following factored-load combinations (UBC 1612.2.1):

      $U = 1.4D$
      $U = 1.2D + 0.5L_r$
      $U = 1.2D + 1.6L_r + 0.8W$
      $U = 1.2D + 1.3W + 0.5L_r$
      $U = 0.9D \pm 1.3W$

4. Check assumed pilaster for the axial loads at the top
   The required design axial load is the larger value of the following load combinations:

   $1.4D = 1.4(25) = 35$ kips
   $1.2D + 0.5L_r = 1.2(25) + 0.5(15) = 37.5$ kips

$$D + 1.6L_r + 0.8W = 1.2(25) + 1.6(15) + 0.8(0) = 54 \text{ kips}$$
$$D + 1.3W + 0.5L_r = 1.2(25) + 1.3(0) + 0.5(15) = 37.5 \text{ kips}$$
$$0.9D \pm 1.3W = 0.9(25) \pm 1.3(0) = 22.5 \text{ kips}$$

Thus, the design axial load is
$$P_u = 54 \text{ kips}$$

The maximum nominal axial compressive strength is determined in accordance with the following formula:
$$P_n = 0.80[0.85f'_m(A_e - A_s) + f_y A_s]$$
$$= 0.80 \frac{[(0.85)(1500)(244 - 6.0) + (60,000)(6.0)]}{1000} = 530.8 \text{ kips}$$

The strength-reduction factor ϕ is determined from
$$0.65 \leq \phi = 0.85 - 2\left(\frac{P_u}{A_e f'_m}\right) \leq 0.85$$
$$\phi = 0.85 - 2\left(\frac{54(1000)}{(244)(1500)}\right) = 0.55 < 0.65, \quad \text{Use } \phi = 0.65$$
$$\phi P_n = (0.65)(530.8) = 345.0 \text{ kips} > 54 \text{ kips} \quad \text{OK.}$$

5. Check assumed pilaster for the given loads at mid-height
   The area of the reinforcement that is in tension $= 3.0 \text{ in}^3$ (three #9 bars)
   $d = 15.62 - 3.5 = 12.12 \text{ in.}$
   a. The factored moment M_u at the midheight of the column is determined by the following formula:
   $$M_u = \frac{w_u h^2}{8} + P_{uf}\frac{e}{2}$$
   where P_{uf} is the factored load from tributary roof loads.
   The eccentric moments at mid-height are
   $$\text{Roof dead load, } M_{DL} = P_{DL}\frac{e}{2} = \frac{(25.0)(2.0)}{2} = 25.0 \text{ kip-in.}$$
   $$\text{Roof live load, } M_{LR} = P_{LR}\frac{e}{2} = \frac{(15.0)(2.0)}{2} = 15.0 \text{ kip-in.}$$
   $$\text{Wind load moment, } M_{wind} = \frac{W_{wind} L^2}{8} = \frac{(500)(192)^2(1 \text{ ft}/12 \text{ in.})}{8}$$
   $$= 192 \text{ kip-in.}$$

   b. The required design moment is the larger value of the following load combinations:
   $1.4D = 1.4(25) = 35 \text{ kip-in.}$
   $1.2D + 0.5L_r = 1.2(25) + 0.5(15) = 37.5 \text{ kip-in.}$
   $D + 1.6L_r + 0.8W = 1.2(25) + 1.6(15) + 0.8(192) = 207.6 \text{ kip-in.}$

$$D + 1.3W + 0.5L_r = 1.2(25) + 1.3(192) + 0.5(15) = 287.1 \text{ kip-in.}$$
$$0.9D \pm 1.3W = 0.9(25) \pm 1.3(192) = 272.1 \text{ kip-in.}$$

Thus the design axial loads include the following two cases:

Case 1: Maximum moment and the corresponding axial load
$$P_u = 37.5 \text{ kips}$$
$$M_u = 287.1 \text{ kip-in.}$$

Case 2: Minimum axial load and the corresponding bending moment
$$P_u = 22.5 \text{ kips}$$
$$M_u = 272.1 \text{ kip-in.}$$

Case 1: Maximum moment and the corresponding axial load
Depth of rectangular stress block a

$$a = \frac{P_u + A_s f_y}{0.85 f'_m b} = \frac{(37.5)(1000) + (3.0)(60,000)}{(0.85)(1500)(15.62)} = 10.92 \text{ in.}$$

The nominal moment capacity is

$$M_n = A_s f_y \left(d - \frac{a}{2}\right) = (3.0)(60,000)\left(12.12 - \frac{10.92}{2}\right)\left(\frac{1}{1000}\right) = 1199 \text{ kip-in}$$

Design ultimate moment capacity is

$$\phi M_n = (0.65)(1199) = 779 \text{ kip-in.} > 287.1 \text{ kip-in.} \quad \text{OK.}$$

Case 2: Minimum axial load and the corresponding bending moment
Depth of rectangular stress block a

$$a = \frac{P_u + A_s f_y}{0.85 f'_m b} = \frac{(22.5)(1000) + (3.0)(60,000)}{(0.85)(1500)(15.62)} = 10.17 \text{ in.}$$

The nominal moment capacity is

$$M_n = A_s f_y \left(d - \frac{a}{2}\right) = (3.0)(60,000)\left(12.12 - \frac{10.17}{2}\right)\left(\frac{1}{1000}\right) = 1266 \text{ kip-in}$$

Design ultimate moment capacity is

$$\phi M_n = (0.65)(1266) = 823 \text{ kip-in.} > 272.1 \text{ kip-in.} \quad \text{OK.}$$

6. Therefore, the section is adequate for strength.

CHAPTER 9
INTRODUCTION TO SEISMIC DESIGN

9.1 GENERAL

Strong earthquakes, such as those in Loma Prieta, Calif., 1989; Northridge, Calif., 1994; Kobe, Japan, 1995; Turkey 1999; and Taiwan 1999, have taken thousands of lives, caused billions of dollars in damages, and incurred other indirect costs as a result of damage to buildings, highways, and other public facilities. The structural engineering community is intensifying efforts to minimize the loss of lives, property, and commerce due to structural failures in future earthquakes. Structural engineers are not good at predicting the nature—intensity, duration, location—of earthquakes. There are no scientific methods that can predict earthquakes accurately. However, engineers can use the lessons learned from past quakes and apply state-of-the-art engineering knowledge to design, build, and retrofit bridges and buildings that will perform well during a given level of ground shaking.

Building and bridge codes have added and updated earthquake-resistance provisions to minimize the hazards caused by ground shaking. Structures designed and built using modern code provisions (mid-1970s or later) performed reasonably well in recent earthquakes. This clearly points to the importance of incorporating modern requirements into the design and construction of new structures, and seismic retrofitting of existing structures. The development of seismic codes is based on the following principles for structural performance:

1. Use realistic seismic ground motion intensities and forces in the design.
2. Design to resist small-to-moderate earthquakes within the elastic range of the structural components without significant damage, but possibly with some non-structural damage.
3. Design to resist large earthquakes without collapse of all or part of the structure, but possibly with some structural as well as non-structural damage.
4. Where feasible, confine structural damage to locations that are easily detected and accessible for inspection and repair.

Many pre-1934 unreinforced masonry buildings did not perform well at epicentric regions of major earthquakes. Failures resulted from inadequate anchorage of the masonry walls to the roof and the floor diaphragms, and from inadequate strength and ductility in the basic building materials. Many major cities in high seismicity regions impose rehabilitation standards on unreinforced masonry commercial buildings constructed prior to 1934. Most rehabilitation schemes use through-bolts and face plates to connect the wood floor and roof diaphragms to the masonry walls. These bolts and plates are visible on the exterior of a building, helping to identify rehabilitated structures. Rehabilitation also includes structural strengthening by adding interior wood shear-wall partitions, new posts, or other supplemental load-resisting elements; infilling openings; or increasing wall thickness.

Reinforced concrete structures designed prior to the mid-1970s do not have adequate reinforcing steel to ensure good performance during an earthquake. These concrete structures are termed *non-ductile concrete structures* and have collapsed in past quakes. Modern reinforced concrete structures, designed and constructed in accordance with the seismic provisions of current building codes, performed well, as did steel bridges and framed buildings. Some steel-framed buildings, however, experienced brittle failures at welded connections.

Structural engineers must continue to learn and upgrade seismic-design criteria and construction practices. Building and bridge codes have been developed and updated based on better understanding of seismic risk through lessons from past earthquakes and research in structural, geotechnical, and earthquake engineering. There are far fewer structural failures when modern building and bridge codes are followed and enforced.

9.2 SEISMIC HAZARD

The causes of earthquakes are not well understood, and experts do not fully agree how available knowledge should be interpreted to specify ground motions for use in design. Predicting the ground motions at a site is a complex task in which one must combine the uncertainties of the influence of the source mechanism of the earthquake; magnitude, distance, and duration of the quake; geology of the travel path; and nature of the underlying soil.

The use of probabilistic concepts has allowed the uncertainties to be considered in the evaluation of ground-motion characteristics. Probabilistic seismic hazard analysis allows uncertainties to be identified, quantified, and combined in a mathematical way. The most probable ground motions for a location then can be determined to design structures to help prevent collapse and protect the public.

The U.S. Geological Survey has produced national maps of seismic hazards since 1948. These maps are revised as new studies improve the understanding of seismic hazard, and they provide information essential to creating and updating the seismic-design provisions of building and bridge codes used in the United States. The maps can be obtained from USGS, Central Region, Geologic Hazards Team, Golden, Colorado, or from the USGS Web site: http://quake.wr.usgs.gov/QUAKES/FactSheets/RiskMaps.

A typical seismic hazard map could be titled, "Ground motions having 10 percent probability of being exceeded in 50 years." The 10 percent is an *exceedance probability* and the 50 years is an *exposure time*. This means that the ground motions shown in this map have a 10 percent probability of being exceeded in 50 years. An event with this probability has a return period of about 475 years. This means that these ground motions have an annual probability of occurrence of $\frac{1}{475}$ per year. This is commonly used in building and bridge codes as the *design earthquake*, and the ground motions associated with the design earthquake are termed *design ground motions*. The ground motions are generally expressed as *peak ground accelerations* (PGA). The probability is based on the relationship

$$P_0 = \left(1 - e^{-t/T}\right) \tag{9.1}$$

where P_0 = Probability of exceedance
t = Exposure time in years
T = Return period in years

9.3 SEISMIC ZONES

Building and bridge codes traditionally divide a seismic hazard map into zones, in which common levels of seismic design are applied. For example, in the 1997 Uniform Building Code, the seismic hazard map of the United States is divided into six zones of different levels of peak ground accelerations (PGA), as shown in Figure 16.2 of the code:

Seismic Zone	Peak Ground Acceleration
0	< 0.075
1	0.075
2A	0.15
2B	0.20
3	0.30
4	0.40

Building codes have been replacing maps that use numbered zones with maps showing contours of design ground motion. The 1997 Uniform Building Code is the only building code that still uses seismic zones. And it is anticipated that the new edition of the UBC will drop the zones and adopt the contours of design ground motion. The new format will avoid the need to revise zone boundaries by petition from various states.

9.4 SITE CHARACTERISTICS

Earthquakes occur on faults. A *fault* is a thin zone of crushed rock between two tectonic plates. It also can be a fracture within a tectonic plate or in the crust of the earth where rocks have moved relative to one another. A fault can be any length—from inches to thousands of miles. Active faults move at average of a fraction of an inch to about 4 inches per year. For example, the Juan de Fuca plate is subducting beneath the North American plate along the Cascadia Subduction Zone off the coast of Washington state at a rate of 1.2 to 1.6 inches per year. When the rock on one side of a fault suddenly slips with respect to the other, energy is released abruptly, causing ground motions that rattle buildings and bridges. The larger slips correspond to larger energy release and larger ground motions. As expected, larger rupture length results in larger earthquake magnitude. The well-known San Andreas Fault in California has a length of more than 650 miles, extending to a depth of more than 10 miles, and it has been the source of many large earthquakes, including the famous 1906 San Francisco earthquake, which had a magnitude 8.3 on the Richter Scale. The following table gives an approximate relationship between earthquake magnitude and length of fault that has slipped.

Magnitude	Length of Slipped Fault (miles)
8.0	190
7.0	25
6.0	5
5.0	2.1
4.0	0.83

The 1989 Loma Prieta earthquake, magnitude 7.1 on the Richter Scale, was reported to have a rupture length of 25 to 30 miles.

Studies on the influence of earthquake magnitude on the duration and acceleration of strong motion suggest the following approximate relationships:

Magnitude	Maximum Acceleration (%g)	Duration (second)
8.0	50	34
7.5	45	30
7.0	37	24
6.5	29	18
6.0	22	12
5.5	15	6
5.0	9	2

An earthquake has one magnitude, but many intensities are distributed over the region. The intensity, or ground shaking, of a specific site depends on the above factors, as well as the distance from the earthquake source, the geology of the travel path, and the nature of the underlying soil. The intensity generally decreases with distance from the earthquake source. Local amplification, however, can occur as the earthquake waves pass from bedrock into softer geologic materials and soil layers. It is very difficult to establish good attenuation relationships, because there are so many possible combinations and permutations of the thickness and stiffness of geologic materials and soil layers in the travel path. When the soil layers of a site are known, several methods are available to estimate the amplification or deamplification of the ground shaking or acceleration. For example, the one-dimensional, linear computer program SHAKE has been used to estimate the amplification or deamplification for a specific site. Generic site amplification factors have been developed by empirical and analytical studies and are used in building and bridge codes for general soil categories.

The *hypocenter* is the point deep down in the fault where the fault begins the slip that causes the earthquake. The *epicenter* is the corresponding point on the surface above the hypocenter. The rupture of a fault initiates at the hypocenter and propagates upward/downward and along one or both directions of the fault. Sometimes the rupture breaks through to the surface of the earth, showing evidence of rupture. The upward or downward rupture can result in ground uplift or subsidence in excess of 15 feet. Seismic waves are generated along the entire length of the fault. The direction in which the rupture propagates interests structural engineers. This *directivity effect* causes significant amplification of shaking and velocity impulse to structures in the near-fault region, which may be within a few miles of the major fault zones. The directivity, or focusing of energy along the fault in the direction of rupture, is a significant factor for most large earthquakes. The directivity effect was observed in the 1989 Loma Prieta and the 1995 Kobe earthquakes. Shaking intensity attenuates at a much faster rate in the direction perpendicular to the fault rupture plane than along the fault axis. In the Loma Prieta earthquake, San Francisco and Oakland, which are in line with the fault axis, felt stronger shaking than would be expected. San Jose, which is perpendicular to the fault, experienced weaker shaking.

9.5 EARTHQUAKE RISK MITIGATION

The principal ways in which earthquakes cause damage are by strong ground shaking; by the secondary effects of ground failures, such as surface rupture, ground cracking, landslides, liquefaction, uplift, and subsidence; or by tsunamis and seiches. Most building and bridge damage is caused by ground shaking—it causes 99 percent of the earthquake damage to homes in California. Therefore, the focus of earthquake risk mitigation should be to develop plans and take action to mitigate the damage. It takes the cooperative efforts of citizens, and city, county, and state governments to safeguard against major structural failures and loss of lives and property, and to maintain commerce.

Structural designers play an important role in the chain of earthquake risk mitigation. Drawing from observations of structural performance in past major earthquakes, they can develop design and construction criteria to supplement building and bridge codes that protect life and lessen property loss, and maintain post-earthquake functionality.

The two most important lessons from recent major earthquakes are:

1. Thousands of non-ductile buildings and bridges were damaged or collapsed.
2. Surface faulting ruptured lifelines, buildings, bridges, and other critical facilities constructed over or across a fault.

Non-ductile structures, such as unreinforced masonry buildings, inadequately reinforced concrete buildings, and bridges are prone to catastrophic failures. We have a large inventory of non-ductile structures in high seismicity regions in the United States. Surface faulting has caused some spectacular rupturing or fracturing of buildings, bridges, and dams, as evidenced in recent earthquakes in Turkey and Taiwan. Structural designers must work with governmental agencies to identify these failure-prone structures and take necessary actions to mitigate the risk.

In California, the Alquist-Priolo Earthquake Fault Zoning Act was passed in 1972 to mitigate the hazard of surface faulting to structures for human occupancy. This state law was a direct result of the 1971 San Fernando earthquake associated with extensive surface fault ruptures that damaged numerous homes, commercial buildings, and other structures. The main purpose of the act is to prevent construction of structures used for human occupancy on the surface traces of active faults. Surface rupture is the most easily avoided seismic hazard for new construction. For existing structures, the only mitigation is to relocate the structures. In 1990, the California Legislature passed the Seismic Hazards Mapping Act to address non-surface fault rupture earthquake hazards, including liquefaction and seismically induced landslides. California approved Proposition 192 in March 1996 to provide $2 billion in bonds to retrofit seven of the state's toll bridges and more than 1000 highway bridges identified as lacking strength and/or ductility.

Utah is situated on the 240-mile Wasatch Fault, which has the potential to produce large earthquakes above magnitude 7.5 on the Richter Scale. The highly populated areas of Salt Lake City, Ogden, and Provo are on soft lake sediments that will shake especially violently during large earthquakes. The Wasatch Fault has not caused a powerful earthquake for the past 150 years, but the people of Utah are aware of the possibility. With the help of public agencies, communities have acted to reduce loss of life and property in future earthquakes. For example, the century-old Salt Lake City and County Building has been made safer by the installation of base isolation devices beneath its unreinforced masonry structure. Utah has made major improvements in the public infrastructure to reduce seismic risk. At least 10 fire stations and four major hospitals in Salt Lake City have been strengthened or replaced with new earthquake-resistant structures. More than 400 public and private school buildings in the region have been evaluated for seismic resistance. Three high schools and one grade school have been strengthened or replaced.

Utah and California set the examples on what can and should be done to mitigate earthquake risk.

9.6 1997 UNIFORM BUILDING CODE (UBC) EARTHQUAKE PROVISIONS

9.6.1 General Provisions

Earthquake design requirements are in Division IV of the UBC. The requirements are meant to safeguard against major structural failures and loss of life, not to limit damage or maintain function. The code provides minimum requirements that reflect the need to design and build seismic-resistant structures. It incorporates the lessons learned from recent earthquakes, notably the 1994 Northridge earthquake, and it points out the need to meet seismic detailing requirements and limitations even when other design forces, such as wind, control the design.

Some key aspects of the UBC earthquake provisions are discussed in the subsections that follow.

9.6.2 Occupancy Categories

The UBC design ground motion is based on a 10 percent probability of being exceeded in 50 years, which is an earthquake having a return period of 475 years. Buildings designed in accordance with UBC are expected to perform without major structural failures and loss of life. However, for essential facilities, such as hospitals, fire and police stations, emergency response centers (ERC), structures housing equipment for ERC, and facilities housing toxic or explosive substances, the UBC assigns higher seismic importance factors I and I_p to provide higher seismic resistance. UBC Table 16-K contains the definitions for the occupancy categories and the assignments of seismic importance factors.

9.6.3 Soil Profile Type

Lessons from past earthquakes show that ground shaking is stronger in soft soil than in hard rock. There is amplification or deamplification in different soil types. UBC defines six soil profile types, S_A, S_B, S_C, S_D, and S_E, in Table 16-J. Soil Profile Type S_F is defined as soils requiring site-specific evaluation as follows:

1. Soils vulnerable to potential failure or collapse under seismic loading, such as liquefiable soils, quick and highly sensitive clays, and collapsible, weakly cemented soils
2. Peats and/or highly organic clays, where the thickness of peat or highly organic clay exceeds 10 feet
3. Very high plasticity clays with a plasticity index, $PI > 75$, where the depth of clay exceeds 25 feet
4. Very thick soft/medium stiff clays, where the depth of clay exceeds 120 feet

To account for the site effects of the soil profile types on structures, seismic response coefficients are used by UBC to amplify seismic zone factors Z. The seismic response coefficients to be assigned to each structure are listed in Table 16-Q for C_a and Table 16-R for C_v. There is deamplification for Soil Profile Type S_A, which is hard rock, and no amplification for S_B, which is rock. There is significant amplification for types S_C, S_D, S_E, and S_F.

9.6.4 Near-source Factor

In Section 9.4 we noted that the directivity effect causes significant amplification of ground shaking to structures in the near-fault or near-source region. To account for the near-source effect, UBC defines three seismic source types A, B, C and assigns Near-Source Factors N_a and N_v to each seismic source type, depending on the distance to the known seismic source. Subduction sources are not included in these definitions and should be evaluated on a site-specific basis. Near-source factors are given in Tables 16-S and 16-T. Seismic source types are defined in Table 16-U, which is reproduced below:

Seismic Source Type	Seismic Source Description	Seismic Source Definition	
		Max. Moment Magnitude M	Slip Rate SR (in./year)
A	Faults that are capable of producing large magnitude events and that have a high rate of seismic activity	$M \geq 7.0$	$SR \geq 0.20$
B	All faults other than Types A and C	$M \geq 7.0$ $M < 7.0$ $M \leq 7.0$	$SR < 0.20$ $SR > 0.08$ $SR < 0.08$
C	Faults that are not capable of producing large magnitude earthquakes and that have a relatively low rate of seismic activity	$M < 6.5$	$SR \leq 0.08$

9.6.5 Seismic Factors

The UBC uses the Seismic Force Overstrength Factor Ω_o to ensure that the structures have minimum design strength over and above the seismic force determined by analysis. This may be considered as a seismic force amplification factor to obtain structural overstrength against earthquake forces higher than those anticipated in the design.

The UBC introduces the Response Modification Factor R to recognize the ductility of a structure. *Ductility* is the ability of a structure to undergo inelastic deformation without collapse. It is not economical to design a structure to resist large earthquakes elastically. The Response Modification Factor approach takes advantage of the inherent energy dissipation capacity of a structural system as it undergoes inelastic deformation in the components or connections. This approach demands stringent detailing requirements to ensure ductile behavior of the structural system.

The values of the Seismic Force Overstrength Factor Ω_o and the Response Modification Factor R are given in Table 16-N of the UBC.

9.6.6 Redundancy Factor

The UBC introduces a Redundancy Factor ρ to recognize the importance of providing multiple-load paths in a structural system. Engineers consider it good practice to build in as much redundancy as feasible in a structural system. The building code provisions began addressing redundancy after the 1994 Northridge earthquake. The AASHTO LRFD bridge design specifications stipulate that multiple-load-path and continuous structures should be used unless there are compelling reasons not to use them. More stringent design provisions are imposed on nonredundant structures than on redundant structures.

9.6.7 Design and Analysis

The UBC identifies design requirements that must be followed to ensure adequate strength to withstand lateral displacements induced by the design basis ground motion, considering the inelastic response of the structure and the inherent redundancy, overstrength, and ductility of the lateral-force resisting system. The design and analysis procedures and methods are outlined in the UBC. Diligently following the provisions in UBC and updating with new research findings and experience, structural designers will achieve seismic-resistant structures consistent with the level of performance desired.

9.7 IMPORTANCE OF PROPER DETAILING

Proper detailing is paramount in the design and construction of seismic-resistant structures. This fact has been confirmed by every recent major earthquake. The 1994 Northridge earthquake showed that new bridges designed and built to current design criteria and construction standards performed well, as did existing bridges retrofitted to current retrofit standards.

With each major earthquake, structural designers continue to learn and modify the criteria and construction practices to ensure that new and retrofitted structures will perform well. Building and bridge codes have been undergoing progressive improvement based on research, experience, and costly lessons from recent major earthquakes. Codes traditionally have focused on life safety. Modern codes, such as the UBC and AASHTO, now pay attention to structural performance beyond the issues of life safety, including more stringent, performance-based criteria. This is a direct reflection on the costly disruption of building use, commerce, and communications. In June 1990, the governor of California instituted the first requirements for highway bridges to perform to levels above the provisions of life safety. His order called for the preparation of an action plan to ensure that "all transportation structures maintained by the state are safe from collapse in the event of an earthquake and that vital transportation links are designed to maintain their function following an earthquake." With the shift of emphasis to seismic performance criteria, structural designers need to pay greater attention to proper detailing to ensure adequate redundancy and ductility in the structures to meet the performance levels expected. For example, in a low-level earthquake there should be only minimal damage, and in a significant earthquake, collapse should be prevented, but significant damage might occur. For critical structures, only repairable damage would be expected. The facility should be functional within a few days after the earthquake.

Proper detailing includes but is not limited to the following:

1. Consider structural system reliability. A structure is a system of members, components, and connections of the structure, including the foundation. Each structural element contributes to the integrity and safety of the bridge. Every member, component, and connection must serve its function to resist and transmit seismic forces, and to

accommodate displacements as expected in the design. Additionally, the structural elements should be designed to have reserve strength and ductility to absorb and dissipate energy of higher magnitude without fracture or collapse.

2. Provide at least one continuous viable load path to transmit inertial loads to the foundation. All members, components, and connections along the load path must be capable of resisting the imposed load effects. Experience in past earthquakes has shown that when one or more of the members, components, or connections behaved in a ductile manner, damage was much reduced.

3. Avoid irregularities in the structures as much as is practicable. Irregularities include geometric and stiffness irregularities, discontinuities in lateral-force path, capacity and diaphragm, and large skews.

4. Consider commercially available and tested base isolation devices to limit the damaging seismic forces on the structure, and to maintain post-earthquake serviceability of existing and new critical structures.

5. Provide adequate anchors between building and foundation—properly designed and detailed anchorage systems can have good ductility and absorb considerable energy without breaking.

6. Provide adequate reinforcing steel and confinement reinforcement in concrete and masonry members to ensure ductile behavior under high seismic forces.

7. Design and detail steel members and connections to avoid local and global buckling, rupturing of welds and brittle fracture.

Good detailing practices are covered in the Uniform Building Code, the provisions of building code requirements for Reinforced Concrete (ACI 318-95) and commentary, the AISC LRFD Design Manual, and the FHWA Seismic Retrofit Manual for Highway Bridges.

APPENDIX A
REFERENCES

GENERAL

1. International Conference of Building Officials. *1997 Uniform Building Code, Volume 2 Structural Engineering Design Provisions*. Whittier, Calif.
2. American Association of State Highway and Transportation Officials. *AASHTO LRFD Bridge Design Specifications*, Second Edition 1998 and Interims. Washington, D.C.
3. Merritt, F.S., Ricketts, J.T., and Loftin, M.K. 1995. *Standard Handbook for Civil Engineers*, Fourth Edition. New York: McGraw-Hill.
4. Hicks, T.G. 1999. *Handbook of Civil Engineering Calculations*. New York: McGraw-Hill.
5. Gaylord, E.H., Gaylord C.H., and Stallmeyer, J.E. 1997. *Structural Engineering Handbook*. New York: McGraw-Hill.
6. Chen, W.F., and Duan, L. 1999. *Bridge Engineering Handbook*. CRC Press.
7. Young, W.C. 1990. *Roark's Formulas for Stress and Strain*. New York: McGraw-Hill.
8. Lindeburg, M.R. 1999. *Practice Problems for the Civil Engineering PE Exam*, Seventh Edition, Professional Publications.
9. Newnan, D.G. 1995. *Civil Engineering—License Problems and Solutions*, Twelfth Edition. Engineering Press Bookstore.

CHAPTER 1: PROPERTIES OF MATERIALS

1. Popov, E.P., Nagarajan, S., and Lu, Z.A. *Mechanics of Materials*, 2nd Edition. Prentice-Hall.
2. Timoshenko, S. *Strength of Materials, Part I Elementary Theory and Problems*, 2nd Edition, 14th Printing. D. Van Nostrand Company.
3. Halperin, D.A. *Statics and Strength of Materials for Technology*. New York: John Wiley & Sons.
4. American Society for Testing and Materials, *Annual Book of ASTM Standards, Section 3 Metals Test Methods and Analytical Procedures*. West Conshohocken, Pennsylvania.

CHAPTER 2: PROPERTIES OF SECTIONS

1. Meriam, J.L. *Mechanics—Part 1 Statics*, 2nd Edition. New York: John Wiley & Sons.
2. Meriam, J.L. and Kraige, L.G. 1997. *Statics, Volume 1*. New York: John Wiley & Sons.
3. Nash, W.A. *Statics and Mechanics of Materials*, Schaum's Outline Series. New York: McGraw-Hill.

CHAPTER 3: STRENGTH OF MATERIALS

1. Beedle, L.S. 1958. *Plastic Design of Steel Frames*. New York: John Wiley & Sons.
2. Chajes A. 1974. *Principles of Structural Stability Theory*. Prentice-Hall.
3. Gere, J.M., and Timoshenko, S.P. 1977. *Mechanics of Materials*, Fourth Edition, PWS Publishing Company.
4. Jensen A. 1962. *Statics and Strength of Materials*. New York: McGraw-Hill.
5. Nash W. *Strength of Materials*, Fourth Edition, Schaum's Outline Series. New York: McGraw-Hill.
6. Pytel, A. and Singer, F.L. 1987. *Strength of Materials*, Fourth Edition. New York:Harper & Row Publishers.
7. Beer, F.P., Johnston, E.R. Jr. and DeWolf, J.T. 1992. *Mechanics of Materials*, Second Edition. New York: McGraw-Hill.
8. AISC. *Manual of Steel Construction—Allowable Stress Design*. American Institute of Steel Construction, Inc. Chicago, Illinois

CHAPTER 4: PRINCIPLES OF STATICS

1. Meriam, J.L. *Mechanics—Part 1 Statics*, Second Edition. New York: John Wiley & Sons.
2. Meriam, J.L., and Kraige, L.G. 1997. *Statics, Volume 1*, April. New York: John Wiley & Sons.
3. Nash, W.A. *Statics and Mechanics of Materials*, Schaum's Outline Series. New York: McGraw-Hill.
4. Gere, J.M., and Timoshenko, S.P. 1997. *Mechanics of Materials*, Fourth Edition. PWS Publishing Company.
5. Wilbur, J.B., and Norris, C.H. 1948. *Elementary Structural Analysis*. March. New York: McGraw-Hill.
6. Timoshenko, S.P. and Young, D.H. *Theory of Structures*, Second Edition, New York: McGraw-Hill.
7. Huang, T.C. 1967. *Engineering Mechanics, Volume I Statics*. Addison-Wesley.

CHAPTER 5: INTRODUCTION TO DESIGN AND ANALYSIS

1. Levy, M. and Salvadori, M.1992. *Why Buildings Fall Down—How Structures Fail*. (paperback 1994). W.W. Norton & Company.
2. Canfield, D.T. and Bowman, J.H. 1954. Business, *Legal and Ethical Phases of Engineering*, Second Edition. New York: McGraw-Hill.

3. ASCE. 1993. *Minimum Design Loads for Buildings and Other Structures*. (formerly ANSI A58.1-1982. ASCE. New York.

CHAPTER 6: CONCRETE DESIGN

1. International Conference of Building Officials. *1997 Uniform Building Code, Volume 2, Structural Engineering Design Provisions*" Whittier, California.
2. American Concrete Institute. 1995. *ACI 318-95 Building Code Requirements and Commentary for Reinforced Concrete*. Farmington Hills, Michigan.
3. Leet, Kenneth M. and Bernal, D. 1997. *Reinforced Concrete Design*, Third Edition. New York: McGraw-Hill.
4. Collins, M.P. and Mitchell, D. 1997. *Prestressed Concrete Structures*, Second Printing. Response Publications, Canada.
5. CRSI. 1999. *CRSI Handbook*.
6. Ghosh, S.K., Fanella, D.A., and Rabbat, B.G. 1996. *Note on ACI318-95*, Sixth Edition. Portland Cement Association.
7. Aitcin, P.C. 1998. *High-Performance Concrete*. E & FN Spon.
8. ACI Committee 315. *ACI Detailing Manual—1994*, Publication SP-66(94). American Concrete Institute, Detroit, Michigan.
9. Lin, T.Y. and Burns, N.H. 1981. *Design of Prestressed Concrete Structures*, Third Edition. New York: John Wiley and Sons.
10. Wang, C.K. and Salmon, C.G. 1985. *Reinforced Concrete Design*, Fourth Edition. Harper & Row.
11. MacGregor, J.G. 1992. *Reinforced Concrete Design*, Second Edition. Prentice Hall.
12. Fintel, M. 1985. *Handbook of Concrete Engineering*, Second Edition. Van Nostrand Reinhold.
13. American Concrete Institute. 1999. *ACI Manual of Concrete Practice*. Farmington Hills, Michigan, .
14. Prestressed Concrete Association. 1992. *PCACOL-Strength Design of Reinforced Concrete Column Sections* (a computer program. Portland Cement Association. Skokie, Illinois.
15. Prestressed Concrete Institute. 1971. *PCI Design Handbook*, Fourth Edition. Chicago, Illinois.

CHAPTER 7: STEEL DESIGN

1. International Conference of Building Officials. *1997 Uniform Building Code, Volume 2 Structural Engineering Design Provisions*. Whittier, California.
2. American Association of State Highway and Transportation Officials. *AASHTO LRFD Bridge Design Specifications*, Second Edition 1998 and Interims. Washington, D.C.
3. American Institute of Steel Construction. 1994. *Load & Resistance Factor Design—Volumes I and II, Manual of Steel Construction*, Second Edition. Chicago, Illinois.
4. McCormac, J.C. 1995. *Structural Steel Design—LRFD Method*, Second Edition. Harper Collins College Publishers.
5. Tamboli, A.R. 1997. *Steel Design Handbook—LRFD Method*. New York: McGraw-Hill.
6. Trahair, N.S. 1997. *The Behaviour and Design of Steel Structures*.
7. Beedle, L.S. 1958. *Plastic Design of Steel Frames*. New York: John Wiley and Sons.
8. Galambos, T.V. ed. 1988. *SSRC—Guide to Stability Criteria for Metal Structures*, Fourth Edition. New York: John Wiley and Sons.

CHAPTER 8: MASONRY DESIGN

1. International Conference of Building Officials. *1997 Uniform Building Code, Volume 2 Structural Engineering Design Provisions*. Whittier, California.
2. American Concrete Institute. 1995. *ACI 530-95 Building Code Requirements for Masonry Structures*. Farmington Hills, Michigan.
3. Derecho, A.T., Schultz, D.M. and Fintel, M. 1974. *Analysis and Design of Small Reinforced Concrete Buildings for Earthquake Forces*, Engineering Bulletin. Portland Cement Association.
4. Amrhein, J.E. 1992. *Reinforced Masonry Handbook*, Fifth Edition. Masonry Institute of America.
5. Joint Department of the Army, Navy, and Air Force. *TM 5-809-3/NAVFAC DM-2.9/AFM 88-3*, Chapter 3, "Masonry Structural Design for Buildings."
6. Schneider, R.R. and Dickey, W.L. 1987. *Reinforced Masonry Design*, Second Edition., Prentice Hall.

CHAPTER 9: INTRODUCTION TO SEISMIC DESIGN

1. International Conference of Building Officials. *1997 Uniform Building Code, Volume 2 Structural Engineering Design Provisions*. Whittier, California.
2. American Association of State Highway and Transportation Officials. *AASHTO LRFD Bridge Design Specifications*, Second Edition 1998 and Interims. Washington, D.C.
3. Frangopol, D.M. and Nakib, R. 1991. "Redundancy in Highway Bridges," *Engineering Journal*, AISC, Vol. 28, No. 1.
4. ATC-3. 1978. *Tentative Provisions for the Development of Seismic Regulations for Buildings* (amended 1982). Applied Technology Council, Redwood City, California.
5. ATC-6-2. 1983. *Seismic Retrofitting Guidelines for Highway Bridges*, December. Applied Technology Council, Redwood City, California.
6. ATC-32. 1996. *Improved Seismic Design Criteria for California Bridges: Provisional Recommendations*, June. Applied Technology Council, Redwood City.
7. Williams, A. 1998. *Seismic Design of Buildings and Bridges*, Second Edition. Engineering Press.
8. Priestley, M.J.N., Seible, F. and Calvi, G.M. 1996. *Seismic Design and Retrofit of Bridges*. New York: John Wiley & Sons.
9. BSSC. 1997. *NEHRP Recommended Provisions for Seismic Regulations for New Buildings and Other Structures*, Part 1: "Provisions" and Part 2: "Commentary." Federal Emergency Management Agency (Report Nos. FEMA 302 and 303), Washington, D.C.
10. BSSC. 1997. *NEHRP Guidelines for the Seismic Rehabilitation of Buildings*. Federal Emergency Management Agency (Report Nos. FEMA 273 and 274), Washington, D.C.
11. Ambrose, J. and Vergun, D. 1995. *Simplified Building Design for Wind and Earthquake Forces*, Third Edition. New York: John Wiley and Sons.

INDEX

A

Admixture, 6.7
Aesthetics, 5.2
Aggregates, 6.7
Allowable stress design, 5.4
Analysis, 5.1, 9.9
Axial force, 7.26
Axial loading, design for, 6.16

B

Beam design charts, 7.22
Beam deflection, 7.25
Methods for determining, 3.26-3.35
Beams, 7.17
    Composite, 7.51
Curved, 3.38
Deflection of, 3.25
Loads acting on, 3.19
Plastic deformations of, 3.41
Statically determinate, 3.36
Statically indeterminate beams, 3.36
Stresses in, 3.18
Beams with shear connectors, strength of, 7.52
Bearing, 6.31
Bearing plates, 7.56
Bearing strength at bolt holes, 7.38

Bending, 7.26
Bending moment, 3.13
Bolts, full-tensioned, 7.34
    Snug-tight, 7.34
Breaking strength
Brittle fracture, 1.12

C

Catenary cables, 4.27
Centroid of an area, 2.1
Centroid of a line, 2.4
Centroid of a volume, 2.4
Collapse mechanism, 3.42
Column formulas, 7.14
Columns, 3.42
    Beam, 3.44
    Critical buckling load of, 3.42
    Effective length of, 3.43
    Long, 6.21
Combined axial and flexural loading, 6.17
Combined stresses, 3.46
Combined tension, 7.37
Compression, 3.1
Compression members, 7.12
    Strain energy in, 3.7
Compressive strength, 6.2

Concrete, mechanical properties of, 6.2
    Prestressed, 6.36
Concrete design, 6.1
Concrete quality, 6.6
Connections, bolted, 7.33
    Minimum strength of, 7.36
    Slip-critical, 7.41
    Types of, 7.35
    Welded, 7.45
Constructibility, 5.2
Crack width limitation, 6.15
Creep, 1.12, 6.4
Curing, 6.6, 6.8

D

Design, 5.1, 9.9
Design codes, 5.8
Design data, 8.5
Design examples, 8.31, 8.71
Design loads, 5.6
Design methods, 5.3
    Working stress, 8.7
Design requirements, general, 8.6
Design rupture strength, 7.42
Design shear strength, 7.24, 7.53
Design specifications, 5.8
Design tensile strength, 7.2
Design tension, 7.37
Detailing, 6.16, 6.20, 9.9
Determinate force system, 3.3
Determinacy, 4.15
Development length, 6.5
Dowels, 6.31
Ductility, 1.8

E

Earthquake provisions, 9.6
Earthquake risk mitigation, 9.5
Economy, 5.3
Edge distance, 7.36

Effective width, 7.52
Elastic analysis, 3.5
Elasticity, modulus of, 1.4, 6.3, 8.5
Empirical design method, 8.28
Equilibrium, 4.10

F

Fatigue life, 1.8
Flexible cables, 4.23
Flexural loading, design for, 6.8, 6.16
Flexure, design for, 7.17
Flexure check, 6.30
Forces, composition of, 4.4
    Resolution of, 4.4
Footings, design of, 6.30
    Sizing, 6.30
Fracture toughness, 1.11
Free-body diagram, 4.11

G

Grades, 6.4
Gross area, 7.4
Grout, 8.4

H

Hardness, 1.10
Hooke's law, 1.4, 1.13

I

Inclined axes, 2.17
Indeterminate force system, 3.3

J

Joints, methods of, 4.18

L

Lengths, 6.5
Lintel design, 8.31, 8.72
Load, 3.15
Load and resistance factor design, 5.5

M

Masonry units, 8.2
Materials, properties of, 1.1
    Strength of, 3.1
    Testing, 1.13
Maximum spacing, 7.36
Minimum spacing, 7.36
Mohr's circle, 2.20
Moment and couple, 4.7
Moment of inertia, 2.6
    Methods for determining, 2.10-2.15
Moment relationships, 3.15
Mortar, 8.2

N

Near-source factor, 9.7
Net area, 7.4
    Effective, 7.6
Neutral axis, 3.20
Neutral axis in concrete slab, 7.54
Newton's laws, 4.3

O

Occupancy categories, 9.7

P

Pappus-Guldinus, theorems of, 2.4
Parabolic cables, 4.24
Percentage elongation, 1.6
Percentage reduction in area, 1.7
Pilaster design, 8.44, 8.74
Placing, 6.6, 6.8
Plastic analysis, 3.5
Plastic hinge, 3.41
Plastic moment, 3.22
Plastic section modulus, 3.25
Poisson's ratio, 1.7
Principal stresses and planes, 3.47
    Determining, 3.55
Product of inertia, 2.16
    Transfer of axes for, 2.17
Proportional limit, 1.4
Proportioning, 6.6, 6.7

R

Radius of gyration, 2.20
Rectangular doubly reinforced beam, 6.12
Rectangular singly reinforced beam, 6.9
Redundancy factor, 9.8
Reinforced masonry column design, 8.44, 8.74
Reinforced masonry wall design, 8.49, 8.57
Reinforcing accessories, 8.4
Reinforcement, 6.4
    Concrete protection of, 6.6
Relaxation, 1.13
Resilience, modulus of, 1.9

S

Safety, 5.2
Scalar quantities, 4.2
Secant modulus, 1.7
Section modulus, 3.20
Sections, properties of, 2.1
    Method of, 4.21
Seismic design, 9.1
Seismic factors, 9.8
Seismic hazard, 9.2
Seismic zones, 9.3
Shear, 3.15
    Design for, 6.21
Shear and bending moment diagram, 3.16
Shear center, 3.36
Shear check, 6.30
Shear connectors, 7.53
    Placement and spacing of, 7.54
    Required number of, 7.53
Shear deformation, 3.9
Shear friction, 6.24
Shear in bearing-type connections, 7.37

Shear modulus, 3.9
Shearing force, 3.13
Shear strength, 7.37
Sizes, 6.5
Shrinkage, 1.12, 6.4
Site characteristics, 9.3
Slenderness ratio, 3.43
Soil profile type, 9.7
Splices, 6.5
Static friction, 4.10
Statics, principles of, 4.1
Steel design, 7.1
Steel sections, classification of, 7.12
Strain, 1.1
    Normal, 1.3
    Shear, 3.8
Strain energy in pure shear, 3.9
Strain hardening, 1.5
Strength design method, 5.4, 8.20
Strength during construction, 7.53
Strength requirements, 8.21
Stress, 1.1
    Normal, 1.2
    Sheer, 3.7
    Working, 1.7
Stress-strain diagrams, 1.3
Stress-strain relationship, 6.3
Structural designer, responsibilities of, 5.1
Structural steels, attributes of, 7.2
Structures, 4.14
Superposition, method of, 4.23
System of forces, 4.3

T

Tangent modulus, 1.7
Tensile strength, 6.2
Tension, 3.1

Tension members, 7.2
    Design of, 7.10
Strain energy in, 3.7
Test specimens, 1.2
Thermal coefficient, 6.4
Torsion, 3.9
Torsional resistance, 3.12
Torsional shearing strain, 3.10
Torsional shearing stress, 3.10
Transfer of axes, 2.7
Trusses, 4.14
    Influence lines for, 4.17

U

Ultimate strength, 1.6
Unit weight, 6.4
Unsymmetric bending, 3.37

V

Varignon's theorem, 4.8
Vector quantities, 4.2

W

Walls, design of, 6.25
Water, 6.7
Welding, 7.46
    Types of, 7.46
Welding code, 7.46
Welds, 7.46
    Complete penetration groove, 7.47
    Fillet, 7.46
    Nominal strength of, 7.47
    Types of, 7.46

Y

Yield point, 1.5

NOTES

NOTES

NOTES

NOTES

NOTES

NOTES